Lecture Notes in Computer Science   12959

More information about this subseries at http://www.springer.com/series/7412

Carole H. Sudre · Roxane Licandro ·
Christian Baumgartner ·
Andrew Melbourne · Adrian Dalca ·
Jana Hutter · Ryutaro Tanno ·
Esra Abaci Turk · Koen Van Leemput ·
Jordina Torrents Barrena ·
William M. Wells · Christopher Macgowan (Eds.)

Uncertainty for Safe Utilization of Machine Learning in Medical Imaging, and Perinatal Imaging, Placental and Preterm Image Analysis

3rd International Workshop, UNSURE 2021
and 6th International Workshop, PIPPI 2021
Held in Conjunction with MICCAI 2021
Strasbourg, France, October 1, 2021
Proceedings

 Springer

Editors
Carole H. Sudre (iD)
University College London/
King's College London
London, UK

Christian Baumgartner (iD)
University of Tübingen
Tübingen, Germany

Adrian Dalca (iD)
Massachusetts General Hospital
Harvard Medical School, MIT
Cambridge, MA, USA

Ryutaro Tanno
Microsoft Research/
University College London
London, UK

Koen Van Leemput (iD)
Technical University Denmark
Kongens Lyngby, Denmark

Harvard Medical School
Cambridge, MA, USA

William M. Wells
Harvard Medical School/Brigham
and Women's Hospital
Boston, MA, USA

Roxane Licandro (iD)
Medical University of Vienna and TU Wien
Vienna, Austria

Andrew Melbourne
King's College London
London, UK

Jana Hutter (iD)
King's College London
London, UK

Esra Abaci Turk
Boston Children's Hospital
Boston, MA, USA

Jordina Torrents Barrena
Hewlett Packard
Barcelona, Spain

Christopher Macgowan
The Hospital For Sick Children
University of Toronto
Toronto, ON, Canada

ISSN 0302-9743   ISSN 1611-3349 (electronic)
Lecture Notes in Computer Science
ISBN 978-3-030-87734-7   ISBN 978-3-030-87735-4 (eBook)
https://doi.org/10.1007/978-3-030-87735-4

LNCS Sublibrary: SL6 – Image Processing, Computer Vision, Pattern Recognition, and Graphics

This Springer imprint is published by the registered company Springer Nature Switzerland AG
The registered company address is: Gewerbestrasse 11, 6330 Cham, Switzerland

UNSURE 2021 Preface

The Third Workshop on Uncertainty for Safe Utilization of Machine Learning in Medical Imaging (UNSURE 2020), was prepared as a satellite event of the 24th International Conference on Medical Image Computing and Computer Assisted Intervention (MICCAI 2021).

With an ever-increasing diversity in machine learning techniques for medical imaging applications, the need to quantify and acknowledge the limitations of a given technique has been a growing topic of interest in the MICCAI community over the last few years. Since its inception, the purpose of this workshop has been to develop awareness and encourage research in the field of uncertainty modeling to enable safe implementation of machine learning tools in the clinical world.

The proceedings of UNSURE 2021 include 13 high-quality papers that were selected from 18 submissions following a double-blind review process. Each submission of 8 to 10 pages was reviewed by three members of the Program Committee, formed by 29 experts in the field of deep learning, Bayesian modeling, and Gaussian processes.

The accepted papers cover the fields of uncertainty quantification and modeling, as well as their application to clinical pipelines, notably focusing on uncertainty in out-of-distribution and domain shift problems as well as questions around annotation uncertainty. Two keynote presentations, from experts Ender Konukoglu, ETH Zurich, Switzerland, and Roland Wiest, University Hospital Bern, Switzerland, further contributed to placing this workshop at the interface between methodological advances and clinical applicability.

We hope this workshop highlighted both theoretical and practical challenges in communicating uncertainties, and further encourages research to (a) improve safety in the application of machine learning tools and (b) assist in the translation of such tools to clinical practice.

We would like to thank all the authors for submitting their manuscripts to UNSURE 2021 as well as the Program Committee members for the quality of their feedback and dedication to the review process.

August 2021

Carole H. Sudre
Christian F. Baumgartner
Adrian Dalca
Ryutaro Tanno
Koen Van Leemput
William M. Wells

UNSURE 2021 Organization

Program Committee Chairs

Christian Baumgartner	University of Tubingen
Adrian Dalca	Harvard Medical School and Massachusetts Institute of Technology, USA
Carole H. Sudre	University College London and King's College London, UK
Ryutaro Tanno	Microsoft Research and University College London, UK
Koen Van Leemput	Harvard Medical School, USA/Technical University of Denmark, Denmark
William M. Wells	Harvard Medical School, USA

Program Committee

Alejandro Granados	King's College London, UK
Alireza Mehrtash	Brigham and Women's Hospital, USA
Arunkumar Kannan	University of British Columbia, Canada
Azat Garifullin	Lappeeranta University of Technology, Finland
Danil Grzech	Imperial College London, UK
Daniel Coelho de Castro	Imperial College London, UK
Eleni Chiou	University College London, UK
Evan Yu	Cornell University, USA
Felix Bragman	King's College London, UK
Hongxiang Lin	University College London, UK/Zhejiang Lab, China
Ivor Simpson	University of Sussex, UK
Jinwei Zhang	Cornell University, USA
Jorge Cardoso	King's College London, UK
Leo Joskowicz	Hebrew University of Jerusalem, Israel
Liane Canas	King's College London, UK
Malte Hoffmann	Harvard Medical School, USA
Mark Graham	King's College London, UK
Max-Heinrich Laves	Leibniz Universitat Hannover, Germany
Pedro Borges	King's College London, UK
Pieter Van Molle	Ghent University, Belgium
Raghav Mehta	McGill University, Canada
Reuben Dorent	King's College London, UK
Robin Camarasa	Erasmus MC, The Netherlands
Roger Soberanis-Mukul	Technische Universitat Munchen, Germany
Tanya Nair	McGill University, Canada
Thomas Varsavsky	King's College London, UK

Tim Adler   DKFZ, Germany
Yukun Ding   University of Notre Dame, USA
Zhilu Zhang   Cornell University, USA

PIPPI 2021 Preface

The application of sophisticated analysis tools to fetal, infant, and paediatric imaging data is of interest to a substantial proportion of the MICCAI community. The main objective of this workshop is to bring together researchers in the MICCAI community to discuss the challenges of image analysis techniques as applied to the fetal and infant setting. Advanced medical image analysis allows the detailed scientific study of conditions such as prematurity and the study of both normal singleton and twin development in addition to less common conditions unique to childhood. This workshop brings together methods and experience from researchers and authors working on these younger cohorts and provides a forum for the open discussion of advanced image analysis approaches focused on the analysis of growth and development in the fetal, infant, and paediatric period.

The papers in this volume constitute the proceedings of the 6th International Workshop on Perinatal, Preterm and Paediatric Image Analysis (PIPPI 2021), held in conjunction with MICCAI 2021, the 24th International Conference on Medical Image Computing and Computer-Assisted Intervention. The conference was planned to take place in Strasbourg, France, but changed to an online event due to the COVID-19 pandemic. The 14 contributions from the PIPPI 2021 workshop were carefully reviewed and selected from 20 submissions. We would like to thank everyone involved in this year's workshop and we hope that we can meet again in person at the next PIPPI event.

August 2021

Roxane Licandro
Andrew Melbourne
Jana Hutter
Esra Abaci Turk
Jordina Torrents Barrena
Christopher Macgowan

PIPPI 2021 Organization

Organizing Committee

Roxane Licandro	TU Wien and Medical University of Vienna, Austria
Andrew Melbourne	Kings College London, UK
Jana Hutter	Kings College London, UK
Esra Abaci Turk	Boston Children's Hospital, USA
Jordina Torrents Barrena	Hewlett Packard, Spain
Christopher Macgowan	University of Toronto, Canada

Program Committee

Elisenda Eixarch	Barcelona Children's Hospital, Spain
Lucas Fidon	King's College London, UK
Miguel Angel González Ballester	Pompeu Fabra University, Spain
András Jakab	University Children's Hospital Zurich, Switzerland
Karim Lekadir	Universitat de Barcelona, Spain
Gemma Piella	Pompeu Fabra University, Spain
Ernst Schwartz	Medical University of Vienna, Austria
Paddy Slator	University College London, UK
Daniel Sobotka	Medical University of Vienna, Austria
Johannes Steinweg	King's College London, UK
Dimitra Flouri	King's College London, UK
Jeffrey Stout	Boston Children's Hospital, USA
Athena Taymourtash	Medical University of Vienna, Austria
Alena Uus	King's College London, UK
Veronika Zimmer	King's College London, UK
Daan Christiaens	KU Leuven, Belgium
Meritxell Bach Cuadra	Universtiy of Lausanne, Switzerland
Logan Williams	King's College London, UK
Kelly Payette	University Children's Hospital Zurich, Switzerland
Jorge Perez-Gonzalez	Universidad Nacional Autonoma de Mexico, Mexico
Hongwei Li	Technical University of Munich, Germany
Lilla Zöllei	Massachusetts General Hospital, Harvard University, USA

Contents

UNSURE 2021: Uncertainty Estimation and Modelling and Annotation Uncertainty

Modal Uncertainty Estimation for Medical Imaging Based Diagnosis

Di Qiu[✉] and Lok Ming Lui

The Chinese University of Hong Kong, Shatin, Hong Kong
lmlui@math.cuhk.edu.hk

Abstract. Medical image based diagnosis is constantly faced with uncertainties. In an ambiguous scenario, different experts will reach different conclusions from their initial assumptions. It is thus important for machine learning models to be capable of proposing different plausible predictions, along with *meaningful* uncertainty measures. In this work we propose such a novel learning-based framework, named *modal uncertainty estimation* (MUE), to learn such one-to-many relationship with faithful uncertainty estimation in the medical image understanding tasks. Technically, MUE is based on conditional generative models, but it crucially uses a set of discrete latent variables, each representing a latent mode hypothesis that explains one type of input-output relationship. We justify the use of discrete latent variables by the multi-modal posterior collapse problem in the common conditional generative models. Consequently, MUE can estimate the uncertainty effectively. MUE demonstrates significantly more accurate uncertainty estimation for one-to-many relationship than the current state-of-the-art, and is more informative for practical use. We validate these points on both real and synthetic tasks.

1 Introduction

Making medical diagnosis solely from medical imaging can be a difficult task. Taking for example the medical image segmentation task in LIDC-IDRI [2,3,7], it is often likely for radiologists to reach different conclusions for the same CT Lung scan. The difference can probably be attributed to the different hypotheses they have about the patient. In such an ambiguous scenario, it is of great interest to know which one(s) out of the many possible segmentations would be more reasonable than the others? Abstractly, this problem can be formulated as follows. Suppose the observed image is $\mathbf{x}$, we want to estimate the conditional distribution of the segmentation $p(\mathbf{y}|\mathbf{x})$, based on the training sample pairs $(\mathbf{x}, \mathbf{y})$.

Electronic supplementary material The online version of this chapter (https://doi.org/10.1007/978-3-030-87735-4_1) contains supplementary material, which is available to authorized users.

C. H. Sudre et al. (Eds.): UNSURE 2021/PIPPI 2021, LNCS 12959, pp. 3–13, 2021.
https://doi.org/10.1007/978-3-030-87735-4_1

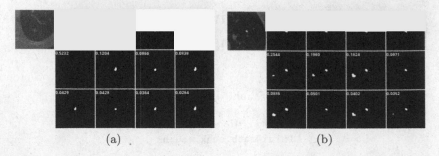

         (a)   (b)

Fig. 1. Visualization of our results on the highly ambiguous samples from LIDC-IDRI dataset. The first row shows the input samples and their segmentations, and the next two rows show the top-8 predictions from our method. The uncertainty estimation for each segmentation proposal is annotated on the upper-left corner. (a) three graders think the image doesn't contain lesion while one grader does. Our model successfully captures the ambiguity level. (b) our model not only captures the given four graders, but also proposed different solutions based on the training dataset. The segmentation proposals together with the uncertainty estimation will signal further examination and help better diagnosis of the patient.

Table 1. A quick comparison of the conceptual difference between cVAE and MUE.

Latent parameterization	cVAE: Gaussian	MUE: discrete
Latent probability reflects true uncertainty $p(\mathbf{y}|\mathbf{x})$?	No	Yes
How is $p(\mathbf{y}|\mathbf{x})$ approximated?	Conditional Gaussian prior and decoder transform	Mode classifier

It is extremely challenging to model the distribution $p(\mathbf{y}|\mathbf{x})$ explicitly and precisely. The space of images is very high dimensional with very complex structures, and it is difficult to adapt faithfully to every variation of the segmentation shape. However, for medical diagnosis it may be just as useful to faithfully capture the uncertainty associated to the possibly different *modalities* of the distribution $p(\mathbf{y}|\mathbf{x})$, i.e. the typical $\mathbf{y}$'s together with how much probability is concentrated around them. To this end, we introduce our *modal uncertainty estimation (MUE)* framework, which is able to both (1) predict the modal samples in the distribution, and (2) accurately evaluate their associated quantitative uncertainties.

Technically, MUE models the generation of $\mathbf{y}$ given $\mathbf{x}$ through an intermediate latent variable $\mathbf{c} \in \mathbb{R}^d$, d is a number of the dimension of the latent variable, as in conditional Variational Auto-Encoder (cVAE) [14,25]. But crucially, we require that the probability distribution of $\mathbf{c}$ faithfully reflects the uncertainty level of the generated $\mathbf{y}$. We identify the reason why the method based on Gaussian parametrization [15,25] of $\mathbf{c}$, the *de facto* choice for cVAE, fails for our purpose.

We henceforth propose to use discrete latent representation, which is the key to accurate uncertainty estimation. We summarize the key differences in Table 1.

There are several benefits to use a discrete latent space for **c** in our setting. First, the model can focus on generating the modal samples, rather than the complete set of samples. Second, there is no longer noise injection during training, so given the same **x**, the model is forced to use **c** to explain the variations in **y**, thus making it impossible for the model to ignore the latent variable unless the relationship between **x** and **y** is unambiguous. Third, the density value learned on the latent space is more interpretable, since the learned conditional prior distribution can better approximate the conditional posterior in the cVAE framework.

The main contributions of this work are: (1) We solve the MUE problem by using cVAE and justify the use of a discrete latent space from the perspective of multi-modal posterior collapse problem. (2) Our uncertainty estimation improves significantly over the existing state-of-art. (3) In contrast to models using noise inputs that require sampling at the testing stage, our model can directly produce results ordered by their latent mode hypothesis probabilities, and is thus more informative and convenient for practical use.

2 Related Work

Since the main objectives of many recent works on conditional generative models [9,11,28,29] are the visual quality and diversity of the outputs, they are not evaluated in terms of the quality of the approximation of the output distribution. In the field of medical imaging, the use of such models has not seen equal popularity. This could be attributed to a higher standard for faithful estimation of the distribution. In this direction, recently [15] has proposed Probabilistic U-Net, which has shown superior performance over various other methods [10,12,16,23,29]. However, as we will show, the Gaussian latent parameterization still bars Probabilistic U-Net from accurate conditional posterior approximation, due to the multi-modal posterior collapse problem [1,20]. This implies that the latent prior density will have no interpretation, and thus *the density value cannot be used to rank its prediction.*

The use of discrete latent variables [18] in neural network is a relatively new topic and has been less explored in medical imaging, while its use appears more natural in natural language processing. In [27], they designed the learned discrete representation for dialogue generation. There, the latent code is required to be "context free", where context is the input. This is in contrast to our assumption that the latent code should depend on the input.

3 Method

3.1 Architectures

Let $(\mathbf{x}, \mathbf{y})$ be the input-label pair. In medical imaging diagnosis tasks, the regression of the label from the input will be carried out by a U-Net [22], which consists

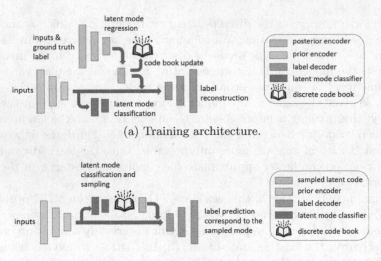

(a) Training architecture.

(b) Inference architecture.

Fig. 2. The architecture design of MUE, please refer to Sect. 3.1 for detail descriptions.

of an encoder E_θ and a decoder D_θ. We call E_θ the *prior encoder*. The novelty of our framework lies in the three additional ingredients: a latent code classifier that is contained inside the prior encoder E_θ, an auxiliary encoder E_ϕ called *posterior encoder*, and a discrete latent code book $\mathcal{C}$ as a matrix of size $|\mathcal{C}| \times d$, where $|\mathcal{C}|$ is the total number of codes and d is the dimension of the code. The code book will serve as the total set of additional information or *memory* to explain the one-to-many mapping. The architectures at the training time and inference time are slightly different. During training time we will sample a $(\mathbf{x}, \mathbf{y})$ to be the input of the posterior encoder, which will generate a latent code $\mathbf{e}$ of dimension d. We don't require that there must be multiple $\mathbf{y}$ corresponding to a single $\mathbf{x}$, as we can learn the uncertainty from different $\mathbf{y}$ for similar $\mathbf{x}$ in the dataset[4]. Before injecting $\mathbf{e}$ into the decoder, we will replace it by the closest neighbor in ℓ^2 distance in the code book $\mathcal{C}$. In the optimization step, the code $\mathbf{c}$ will be updated using exponential moving average of $\mathbf{e}$. The code is then replicated in spatial dimension and concatenated to the feature in one of the decoder layers.

To encourage the posterior encoder's outputs to approximate values in $\mathcal{C}$ as close as possible, we use an ℓ^2-penalization of the form $\beta \|\mathbf{e} - sg[\mathbf{c}]\|^2$ with parameter $\beta > 0$, and sg is the stop-gradient operation. The technique is the same with the VQ-VAE approach of [18; 21] for training neural networks with discrete latent space. At the same time, the prior encoder E_θ takes only $\mathbf{x}$ as the input, and its latent code classifier will classify which codes in the code book $\mathbf{C}$ are in correspondence with $\mathbf{x}$. For simplicity we used a soft-max classifier, though more complicated classification procedures can be incorporated [8]. Thus our loss function to be minimized for a single input pair $(\mathbf{x}, \mathbf{y})$ is

$$\mathcal{L}(\theta, \phi) = CE(E_\theta(\mathbf{x}), id_\mathbf{c}) + Recon(D_\theta(\mathbf{c}, \mathbf{x}), \mathbf{y}) + \beta\|E_\phi(\mathbf{x}, \mathbf{y}) - sg[\mathbf{c}]\|^2 \quad (1)$$

where CE denotes the cross entropy loss, $E_\theta(\mathbf{x})$ is the probability vector of dimension $|\mathcal{C}|$ and $id_\mathbf{c}$ is corresponding code index for the input-label pair $(\mathbf{x}, \mathbf{y})$. *Recon* denotes the label reconstruction loss between prediction $\hat{\mathbf{y}}$ and $\mathbf{y}$.

At inference time, we use the learned latent code classifier to output a conditional distribution given $\mathbf{x}$ on $\mathcal{C}$, where each of the code will give rise to a potentially different but sensible label prediction with the associated probability. The training and inference architectures are visualized in Fig. 2.

3.2 Variational Inference Interpretation

In the framework of cVAE, a latent variable $\mathbf{c}$ is generated from some prior distribution $p_\theta(\mathbf{c}|\mathbf{x})$ parametrized by a neural network. Then the label $\mathbf{y}$ is generated from some conditional distribution $p_\theta(\mathbf{y}|\mathbf{c}, \mathbf{x})$. The major distinction of our approach is that we assume $\mathbf{c}$ takes value in a finite set $\mathcal{C}$. Our goal is to learn the optimal parameters θ^* and the code book $\mathcal{C}$, so that possibly multiple latent codes corresponding to $\mathbf{x}$ can be identified, and label predictions $\hat{\mathbf{y}}$ can be faithfully generated. The latter means the marginal likelihood $p_\theta(\mathbf{y}|\mathbf{x})$ should be maximized.

The variational inference approach as in [14] starts by introducing a posterior encoding model $q_\phi(\mathbf{c}|\mathbf{x}, \mathbf{y})$ with parameters ϕ, which is used only during training. Since the label information is given, we will assume the posterior encoding model is *deterministic*, meaning there is no "modal uncertainty" for the posterior encoding model. So the posterior distribution will be a delta distribution for each input-label pair $(\mathbf{x}, \mathbf{y})$. The variational lower bound [6] states that

$$
\begin{aligned}
\log p_\theta(\mathbf{y}|\mathbf{x}) &\geq \mathbb{E}_{q_\phi(\mathbf{c}|\mathbf{x}, \mathbf{y})}\left[\log \frac{p_\theta(\mathbf{c}, \mathbf{y}|\mathbf{x})}{q_\phi(\mathbf{c}|\mathbf{x}, \mathbf{y})}\right] \\
&= -\mathbb{E}_{q_\phi(\mathbf{c}|\mathbf{x}, \mathbf{y})}\left[\log \frac{q_\phi(\mathbf{c}|\mathbf{y}, \mathbf{x})}{p_\theta(\mathbf{c}|\mathbf{x})}\right] + \mathbb{E}_{q_\phi(\mathbf{c}|\mathbf{x}, \mathbf{y})}[\log p_\theta(\mathbf{y}|\mathbf{c}, \mathbf{x})]
\end{aligned}
\quad (2)
$$

We further take a lower bound of Eq. (2) by observing that the entropy term is positive and is constant if $q_\phi(\mathbf{c}|\mathbf{x}, \mathbf{y})$ is deterministic. This yields a sum of a negative cross entropy and a conditional likelihood

$$\log p_\theta(\mathbf{y}|\mathbf{x}) \geq \mathbb{E}_{q_\phi(\mathbf{c}|\mathbf{x}, \mathbf{y})}[\log p_\theta(\mathbf{c}|\mathbf{x})] + \mathbb{E}_{q_\phi(\mathbf{c}|\mathbf{x}, \mathbf{y})}[\log p_\theta(\mathbf{y}|\mathbf{c}, \mathbf{x})] \quad (3)$$

Now we should maximize the lower bound (3). Since $\mathbf{c}$ takes value in the finite code book $\mathcal{C}$, the probability distribution $p_\theta(\mathbf{c}|\mathbf{x})$ can be estimated using multi-class classification, and the cross entropy term can be estimated efficiently using stochastic approximation.

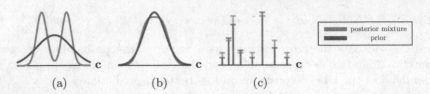

Fig. 3. Comparison between Gaussian latent representations and discrete latent representations in a multi-modal situation. Gaussian latents are structurally limited in such a setting. (a) The ideal situation when there is no posterior collapse as multiple modes appear, but the prior distribution is a poor approximation of the posterior. (b) Posterior collapse happens, and no multi-modal information is conveyed from the learned prior. (c) Discrete latent representation can ameliorate the posterior collapse problem while the prior can approximate the posterior more accurately when both are restricted to be discrete.

3.3   Why Discrete Latent Space

In the following we will show that Gaussian parametrization put a dilemma between model training and sample generation for VAE, as a form of what is known as the *posterior collapse* problem in the literature [1,20]. This issue is particularly easy to understand in our setting, where we assume there are multiple $\mathbf{y}$'s for a given $\mathbf{x}$.

Let us recall that one key ingredient of the VAE framework is to minimize the KL-divergence between the latent prior distribution $p(\mathbf{c}|\mathbf{x})$ and the latent variational approximation $p_\phi(\mathbf{c}|\mathbf{x}, \mathbf{y})$ of the posterior. Here ϕ denotes the model parameters of the "recognition model" in VAE. It does not matter if the prior is fixed $p(\mathbf{c}|\mathbf{x}) = p(\mathbf{c})$ [14] or learned $p(\mathbf{c}|\mathbf{x}) = p_\theta(\mathbf{c}|\mathbf{x})$ [25], as long as both prior and variational posterior are parameterized by Gaussians. Now suppose for a particular $\mathbf{x}$, there are two modes $\mathbf{y}_1, \mathbf{y}_2$ for the corresponding predictions. Since the minimization is performed on the entire training set, $p(\mathbf{c}|\mathbf{x})$ is forced to approximate a *posterior mixture* $p(\mathbf{c}|\mathbf{x}, \mathbf{y}_{(.)})$ of two Gaussians from mode $\mathbf{y}_1$ and $\mathbf{y}_2$. In the situation when the minimization is successful, meaning the KL divergence is small, the mixture of the variational posteriors must be close to a Gaussian, *i.e.* posterior collapsed as in Fig. 3(b), and hence the multi-modal information is lost. Putting it in contrapositive, if multi-modal information is to be conveyed by the variational posterior, then the minimization will not be successful, meaning higher KL divergence. The situation is schematically illustrated in Fig. 3 in one dimension. Note that *in both cases the density values of the prior cannot reflect the uncertainty level of the outputs.* We will quantitative demonstrate this phenomenon in Sect. 4.2.

4   Experiments

In both experiments below we compare with the state-of-the-art method Probabilistic U-Net [15] with the same model complexity. For all experiments, we fix

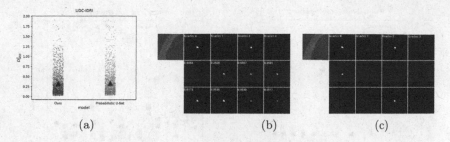

(a)   (b)   (c)

Fig. 4. Evaluation results on LIDC segmentation task. (a) The small dots represent test data instances' D_{GED}^2 values and the triangles mark the mean values. Our performance is competitive with the state-of-the-art in this empirical metric. (b, c) Comparison with [15], which cannot directly output uncertainty level.

the ℓ^2 penalization weight $\beta = 0.25$, number of initial candidate codes $|\mathcal{C}| = 512$, with dimension 128, and use the Adam optimizer [13] with its default setting. We found a small β as we chose here leads to more codes being used initially, which may be beneficial for the model to explore the latent space before convergence. The learned code $\mathbf{c}$ is replicated in the spatial dimension and concatenated to the last layer of the decoder. We release our Pytorch 1.4 [19] implementation to promote future research[1].

4.1   Results on LIDC-IDRI Benchmark

We use the LIDC-IDRI dataset provided by [2,3,7], which contains 1018 lung CT scans from 1010 patients. Each scan has lesion segmentations by 4 (out of a total of 12) expert graders. The identities of the graders for each scan are unknown from the dataset. The graders are often in disagreement about whether the scan contains lesion tissue. We use the same train/test split as in [15].

Some testing results predicted by our model to have *high uncertainty* are illustrated in Fig. 1. The first row is the input and its four grader's segmentations, and the last two rows are our top-8 predictions, where the probability associated to each latent code is annotated on the upper-left corner. We can see that MUE can capture the uncertainty that is contained in the segmentation labels with notable probability scores, as well as other type of segmentations that seem plausible without further information.

We follow the practice of [15] to adopt the *generalized energy distance* empirical metric D_{GED}^2 [5,24,26], treating the four graders' segmentations as the unbiased samples from the ground truth distribution. Note that this assumption may not be correct, but will nevertheless gives a comparison between our model and Probabilistic U-Net. The lower the value of D_{GED}^2, the closer the predicted samples and the ground truth samples. We report the results on the entire testing dataset in Fig. 4(a). For our model, the mean D_{GED}^2 of all testing data is 0.3354, the standard deviation is 0.2947. Our performance is thus competitive

---

[1] https://github.com/sylqiu/modal_uncertainty.

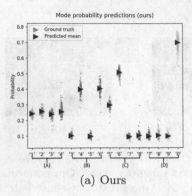 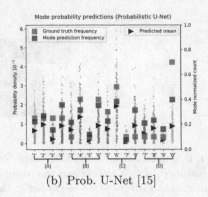

(a) Ours   (b) Prob. U-Net [15]

Fig. 5. Quantitative comparison on the MNIST guessing task. The small dots represent the predictions for 1000 testing samples. MUE in (a) successfully produces accurate uncertainty estimate for each mode. While Probabilistic U-Net's sample density provides no useful information about the uncertainty level, as shown in the left axis. We also count the frequencies for each category and plot it on the right axis. However, the approximation is far less accurate than ours.

with that of Probabilistic U-Net, whose mean is 0.3470 and the standard deviation is 0.3139. Moreover, our model can give faithful uncertainty estimation directly for each input scan, unlike Probabilistic U-Net, as shown in Fig. 4(b) and 4(c).

4.2 Quantitative Analysis on Synthetic Task

MNIST guess game. To test the ability for multi-modal prediction quantitatively, we design a simple guessing game using the MNIST dataset [17] as follows. We are shown a collection of images and only one of them is held by the opponent. The image being held is not fixed and follows certain probability distribution.

The task is to develop a generative model to understand the mechanism of which image is being held based on the previously seen examples. In details, the input $\mathbf{x}$ will be an image that consists of four random digits, and belongs to one of the four categories: (A) $(1, 2, 3, 4)$; (B) $(3, 4, 5, 6)$; (C) $(5, 6, 7, 8)$; (D) $(7, 8, 9, 0)$. The number represents the label of the image sampled. The output $\mathbf{y}$ will be an image of the same size but only one of the input digit is present. Specifically, for (A) the probability distribution is $(0.25, 0.25, 0.25, 0.25)$; for (B) $(0.1, 0.4, 0.1, 0.4)$; for (C) $(0.3, 0.5, 0.1, 0.1)$; for (D) $(0.1, 0.1, 0.1, 0.7)$. Note that the distribution of the output, conditioned on each category's input, consists of four modes, and is designed to be different for each category. We require the model to be trained *solely* based on the observed random training pairs $(\mathbf{x}, \mathbf{y})$, and thus no other information like digit categories should be used. The model would therefore need to learn to discriminate each category and assign the correct outputs with corresponding probabilities.

Thus for instance, an input image of Category (A) will be the combination of four random samples from Digit 1 to 4 in that order, and the output can be the same digit 1 in the input with probability 0.25, or it can be the same digit 2 with probability 0.25, and so forth. Training images from MNIST are used to form the training set, and testing images from MNIST are used to form the testing set.

MUE performs much better quantitatively, as shown in Fig. 5 with the results on 1000 random testing samples. We classify both models' outputs into the ground truth modes and aggregate the corresponding probabilities. We can see in Fig. 5(a) that MUE successfully discovered the distributional properties of the one-to-many mappings, and provides accurate uncertainty estimate. In contrast, due to the Gaussian latent parametrization, neither the individual density of each input nor their averages can provide useful information, as shown by the left axis of Fig. 5(b). By the right axis of Fig. 5(b) we also count the mode frequencies for each category for Probabilistic U-Net. However, even calculated on the entire testing dataset, the distribution approximation is still far from accurate compared to ours. Note that MUE can directly output the uncertainty estimate for *each input* accurately. This clearly demonstrates MUE's superiority and practical value.

5   Discussion and Conclusion

We have proposed MUE, a novel framework for quantitatively identifying the ambiguities for medical image understanding. Crucially we have used a set of learned discrete latent variables to explain the one-to-many input-output relationship, with faithful probability measures that reflects the ambiguity. We have extensively validated our method's performance and usefulness on both real and synthetic tasks, and demonstrate superior performance over the state-of-the-art methods.

References

1. Alemi, A., Poole, B., Fischer, I., Dillon, J., Saurous, R.A., Murphy, K.: Fixing a broken ELBO. In: International Conference on Machine Learning, pp. 159–168 (2018)
2. Armato III, S.G., et al.: The lung image database consortium (LIDC) and image database resource initiative (IDRI): a completed reference database of lung nodules on CT scans. Med. Phys. **38**(2), 915–931 (2011)
3. Armato III, S.G., et al.: Data from LIDC-IDRI (2015)
4. Baumgartner, C.F., et al.: PHiSeg: capturing uncertainty in medical image segmentation. In: Shen, D., et al. (eds.) MICCAI 2019. LNCS, vol. 11765, pp. 119–127. Springer, Cham (2019). https://doi.org/10.1007/978-3-030-32245-8_14
5. Bellemare, M.G., et al.: The cramer distance as a solution to biased Wasserstein gradients. arXiv preprint arXiv:1705.10743 (2017)
6. Bishop, C.M.: Pattern Recognition and Machine Learning. Springer, Heidelberg (2006)
7. Clark, K., et al.: The cancer imaging archive (TCIA): maintaining and operating a public information repository. J. Digit. Imaging **26**(6), 1045–1057 (2013)

8. Gal, Y., Ghahramani, Z.: Dropout as a Bayesian approximation: representing model uncertainty in deep learning. In: international Conference on Machine Learning, pp. 1050–1059 (2016)
9. Huang, X., Liu, M.Y., Belongie, S., Kautz, J.: Multimodal unsupervised image-to-image translation. In: Proceedings of the European Conference on Computer Vision (ECCV), pp. 172–189 (2018)
10. Ilg, E., et al.: Uncertainty estimates and multi-hypotheses networks for optical flow. In: Proceedings of the European Conference on Computer Vision (ECCV), pp. 652–667 (2018)
11. Isola, P., Zhu, J.Y., Zhou, T., Efros, A.A.: Image-to-image translation with conditional adversarial networks. In: CVPR (2017)
12. Kendall, A., Badrinarayanan, V., Cipolla, R.: Bayesian SegNet: model uncertainty in deep convolutional encoder-decoder architectures for scene understanding. arXiv preprint arXiv:1511.02680 (2015)
13. Kingma, D.P., Ba, J.: Adam: a method for stochastic optimization. In: Bengio, Y., LeCun, Y. (eds.) 3rd International Conference on Learning Representations, ICLR 2015, San Diego, CA, USA, 7–9 May 2015, Conference Track Proceedings (2015). http://arxiv.org/abs/1412.6980
14. Kingma, D.P., Welling, M.: Auto-encoding variational bayes. In: Bengio, Y., LeCun, Y. (eds.) 2nd International Conference on Learning Representations, ICLR 2014, Banff, AB, Canada, 14–16 April 2014, Conference Track Proceedings (2014). http://arxiv.org/abs/1312.6114
15. Kohl, S., et al.: A probabilistic u-net for segmentation of ambiguous images. In: Advances in Neural Information Processing Systems, pp. 6965–6975 (2018)
16. Lakshminarayanan, B., Pritzel, A., Blundell, C.: Simple and scalable predictive uncertainty estimation using deep ensembles. In: Advances in Neural Information Processing Systems, pp. 6402–6413 (2017)
17. LeCun, Y., Cortes, C.: MNIST handwritten digit database (2010). http://yann.lecun.com/exdb/mnist/
18. van den Oord, A., Vinyals, O., et al.: Neural discrete representation learning. In: Advances in Neural Information Processing Systems, pp. 6306–6315 (2017)
19. Paszke, A., et al.: Automatic differentiation in PyTorch (2017)
20. Razavi, A., van den Oord, A., Poole, B., Vinyals, O.: Preventing posterior collapse with delta-VAEs. In: International Conference on Learning Representations (2018)
21. Razavi, A., van den Oord, A., Vinyals, O.: Generating diverse high-fidelity images with VQ-VAE-2. In: Advances in Neural Information Processing Systems, pp. 14837–14847 (2019)
22. Ronneberger, O., Fischer, P., Brox, T.: U-net: convolutional networks for biomedical image segmentation. In: Navab, N., Hornegger, J., Wells, W.M., Frangi, A.F. (eds.) MICCAI 2015. LNCS, vol. 9351, pp. 234–241. Springer, Cham (2015). https://doi.org/10.1007/978-3-319-24574-4_28
23. Rupprecht, C., et al.: Learning in an uncertain world: representing ambiguity through multiple hypotheses. In: Proceedings of the IEEE International Conference on Computer Vision, pp. 3591–3600 (2017)
24. Salimans, T., Zhang, H., Radford, A., Metaxas, D.: Improving GANs using optimal transport. In: International Conference on Learning Representations (2018)
25. Sohn, K., Lee, H., Yan, X.: Learning structured output representation using deep conditional generative models. In: Advances in Neural Information Processing Systems, pp. 3483–3491 (2015)
26. Székely, G.J., Rizzo, M.L.: Energy statistics: a class of statistics based on distances. J. Stat. Plann. Inference **143**(8), 1249–1272 (2013)

27. Zhao, T., Lee, K., Eskenazi, M.: Unsupervised discrete sentence representation learning for interpretable neural dialog generation. arXiv preprint arXiv:1804.08069 (2018)
28. Zheng, C., Cham, T.J., Cai, J.: Pluralistic image completion. In: Proceedings of the IEEE Conference on Computer Vision and Pattern Recognition, pp. 1438–1447 (2019)
29. Zhu, J.Y., et al.: Toward multimodal image-to-image translation. In: Advances in Neural Information Processing Systems, pp. 465–476 (2017)

Accurate Simulation of Operating System Updates in Neuroimaging Using Monte-Carlo Arithmetic

Ali Salari[1][(✉)], Yohan Chatelain[1], Gregory Kiar[2], and Tristan Glatard[1]

[1] Department of Computer Science and Software Engineering, Concordia University, Montréal, QC, Canada
m_alari@encs.concordia.ca
[2] Center for the Developing Brain, Child Mind Institute, New York, NY, USA

Abstract. Operating system (OS) updates introduce numerical perturbations that impact the reproducibility of computational pipelines. In neuroimaging, this has important practical implications on the validity of computational results, particularly when obtained in systems such as high-performance computing clusters where the experimenter does not control software updates. We present a framework to reproduce the variability induced by OS updates in controlled conditions. We hypothesize that OS updates impact computational pipelines mainly through numerical perturbations originating in mathematical libraries, which we simulate using Monte-Carlo arithmetic in a framework called "fuzzy libmath" (FL). We applied this methodology to pre-processing pipelines of the Human Connectome Project, a flagship open-data project in neuroimaging. We found that FL-perturbed pipelines accurately reproduce the variability induced by OS updates and that this similarity is only mildly dependent on simulation parameters. Importantly, we also found between-subject differences were preserved in both cases, though the between-run variability was of comparable magnitude for both FL and OS perturbations. We found the numerical precision in the HCP pre-processed images to be relatively low, with less than 8 significant bits among the 24 available, which motivates further investigation of the numerical stability of components in the tested pipeline. Overall, our results establish that FL accurately simulates results variability due to OS updates, and is a practical framework to quantify numerical uncertainty in neuroimaging.

Keywords: Computational reproducibility · Neuroimaging pipelines · Monte-Carlo arithmetic

1 Introduction

Numerical round-off and cancellation errors are ubiquitous in floating-point computations. In neuroimaging, they contribute to results uncertainty along with other sources of variability, including population selection, scanning devices,

C. H. Sudre et al. (Eds.): UNSURE 2021/PIPPI 2021, LNCS 12959, pp. 14–23, 2021.
https://doi.org/10.1007/978-3-030-87735-4_2

sequence parameters, acquisition noise, and methodological flexibility [2,3]. Numerical errors manifest particularly through variations in elementary mathematical libraries resulting from operating system (OS) updates. Indeed, due to implementation differences, mathematical functions available in different OS versions provide slightly different results. The impact of such epsilonesque differences on image analysis depends on the conditioning of the problem and the pipeline's numerical implementation. In neuroimaging, established image processing pipelines have been shown to be substantially impacted: for instance, differences in cortical thicknesses measured by the same Freesurfer version in different execution platforms were shown to reach statistical significance in some brain regions [9], and Dice coefficients as low as 0.6 were observed between FSL or Freesurfer segmentations obtained in different platforms [8,18]. Such observations threaten the validity of neuroimaging results by revealing systematic instabilities.

Despite its possible implications on results validity, the effect of OS updates remains seldom studied due to (1) the lack of closed-form expressions of condition numbers for complex pipelines and non-differentiable non-linear analyses, (2) the technical challenge associated with experimental studies involving multiple OS distributions and versions, (3) the uncontrolled nature of OS updates. As a result, the effect of OS updates on neuroimaging analyses is generally neglected or handled through the use of software containers (Docker or Singularity), static executable builds, or similar approaches. While such techniques improve experiment portability, they only mask numerical instabilities and do not tackle them. Numerical perturbations are bound to reappear due to security updates [14], obsoleting software [17], or parallelization. Therefore, the mechanisms through which numerical instabilities propagate need to be investigated and eventually addressed.

This paper presents "fuzzy libmath" (FL), a framework to simulate OS updates in controlled conditions, allowing software developers to evaluate the robustness of their tools with respect to likely-to-occur numerical perturbations. As we hypothesize that numerical perturbations resulting from OS updates primarily come from implementation differences in elementary mathematical libraries, we leverage Monte-Carlo arithmetic (MCA) [16] to introduce controlled amounts of noise in these libraries. FL enables MCA in mathematical functions used by existing pipelines without the need to modify or recompile them. To demonstrate the approach, we study the effect of common OS updates on the numerical precision of structural MRI pre-processing pipelines of the Human Connectome Project [19], a major neuroimaging initiative.

2   Simulating OS Updates with Monte-Carlo Arithmetic

MCA models floating-point roundoff and cancellations errors through random perturbations, allowing for the estimation of error distributions from independent random result samples. MCA simulates computations at a given virtual precision using the following perturbation:

$$inexact(x) = x + 2^{e_x - t}\xi \tag{1}$$

where e_x is the exponent in the floating-point representation of x, t is the virtual precision and ξ is a random uniform variable of $(-\frac{1}{2}, \frac{1}{2})$.

MCA allows for three perturbation modes: Random Rounding (RR) introduces the perturbation in function outputs, simulating roundoff errors; Precision Bounding (PB) introduces the perturbation in function operands, allowing for the detection of catastrophic cancellations; and, Full MCA combines RR and PB, resulting in the following perturbation:

$$mca\_mode(x \circ y) = inexact_{RR}(inexact_{PB}(x) \circ inexact_{PB}(y)) \tag{2}$$

To simulate OS updates, we introduce random perturbations in the GNU mathematical library, the main mathematical library in GNU/Linux systems. Instrumenting mathematical libraries with MCA raises a number of issues as many functions assume deterministic arithmetic. For instance, applying random perturbations around a discontinuity or within piecewise approximations results in large variations and a total loss of significance that are not relevant in our context. Therefore, we have applied MCA to proxy mathematical functions wrapping those in the original library, such that only the outputs of the original functions were perturbed but not their inputs or the implementations themselves. This technique allows us to control the magnitude of the perturbation as perceived by the application.

We instrumented the GNU mathematical library with MCA using Verificarlo [5], a tool that (1) uses the Clang compiler to generate an LLVM (http://llvm.org) Intermediate Representation (IR) of the source code, (2) replaces floating-point operations in the IR by a call to the Verificarlo API, and (3) compiles the modified IR to an executable using LLVM. The perturbation applied by the Verificarlo API can be configured at runtime, for instance to change the virtual precision applied to single- and double-precision floating-point values.

The resulting MCA-instrumented mathematical library, "fuzzy libmath" (FL), is loaded in the pipeline using LD_PRELOAD, a Linux mechanism to force-load a shared library into an executable. As a result, functions defined in fuzzy libmath transparently overload the original ones without the need to modify or recompile the analysis pipeline. Fuzzy libmath functions call the original functions through dlsym, a function that returns the memory address of a symbol. To trigger MCA instrumentation, a floating-point zero is added to the output of the original function and the result of this sum is perturbed and returned.

Finally, we measure results precision as the number of significant bits among result samples, as defined in [16]:

$$s = -\log_2 \left| \frac{\sigma}{\mu} \right| \tag{3}$$

where σ and μ are the observed cross-sample standard deviation and average.

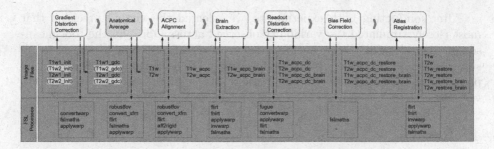

Fig. 1. PreFreeSurfer pipeline steps.

3   HCP Pipelines and Dataset

We apply the methodology described above to the minimal structural pre-processing pipeline associated with the Human Connectome Project (HCP) dataset [7], entitled "PreFreeSurfer". This pipeline consists of many independent components, including: spatial distortion correction, brain extraction, cross-modal registration, and alignment to standard space. Each high-level component of this pipeline (Fig. 1) consists of several function calls using FSL, the FMRIB Software Library [12]. The pipeline requires T1w and T2w images for each subject. A full description of the pipeline is available at [7].

It should be noted that the PreFreeSurfer pipeline uses both single and double precision functions from the GNU mathematical library. Among the preprocessing steps in the pipeline, it has been shown that linear and non-linear registrations implemented in FSL FLIRT [11,13] and FNIRT [1] are the most sensitive to numerical instabilities [18].

We selected 20 unprocessed subjects from the HCP data release S500 available in the ConnectomDB repository. We selected these subjects from different subject types to cover execution paths sufficiently. For each, the available data consisted of 1 or 2 T1w and T2w images each, with spatial dimensions of $256 \times 320 \times 320$ and voxel resolution of 0.7 mm. Acquisition protocols and parameters are detailed in [19]. Two distinct experimental configurations were tested:

Operating Systems (OS): subjects were processed on three different Linux operating systems inside Docker images: CentOS7 (glibc v.2.17), CentOS8 (glibc v.2.28), and Ubuntu20 (glibc v.2.31).

Fuzzy libmath (FL): the dataset was processed on an Ubuntu20 system using fuzzy libmath. The virtual precision (t) for the perturbations was swept from 53 bits (the full mantissa for double-precision data) down to 1 bit by steps of 2. For $t \geq 24$ bits, only double-precision was altered and single-precision was set to 24 bits, and for $t < 24$ bits, both double- and single-precision simultaneously were changed. Three FL-perturbed samples were generated for each subject and virtual precision, to match the number of OS samples.

After conducting both experiments, we selected the virtual precision that most closely simulated the variability observed across OSes via the root-mean-square error (RMSE) between the number of significant bits per voxel in all subjects and conditions. This precision is referred to as the global nearest virtual precision and was used to compare results obtained in both the FL and OS versions.

4   Results

The fuzzy libmath source code, Docker image specifications, and analysis code to reproduce the results are available at https://github.com/big-data-lab-team/MCA-libmath-paper. All experiments were conducted on the Béluga HPC computing cluster made available by Compute Canada through Calcul Québec. Béluga is a general-purpose cluster with 872 available nodes. All nodes contain 2× Intel Gold 6148 Skylake @ 2.4 GHz (40 cores/node) CPU, and node memory can range between 92 to 752 GB. The average processing time of the pipeline without FL instrumentation was 69 min (average of 3 executions). The FL perturbation increased it to 93 min.

We ensured that the pipeline does not use pseudo-random numbers by processing each subject twice on the same operating system. To validate that FL was correctly instrumented with Verificarlo, we used Veritracer [4], a tool for tracing the numerical quality of variables over time. For one subject, the traces showed that the number of significant bits in the function outputs varied over time, confirming the instrumentation with MCA. Throughout the pipeline execution, Veritracer reported approximately 4 billion calls to FL, with the following ratio of calls: 47.12% `log`, 40.96% `exp`, 6.92% `expf`, 3.39% `logf`, 1.55% `sincosf`, and 0.06% of cumulated calls to `atan2f`, `pow`, `sqrt`, `exp2f`, `powf`, `log10f`, `log10`, `cos`, and `asin`. We also checked that long double types were not used.

4.1   Fuzzy Libmath Accurately Simulates the Effect of OS Updates

Fuzzy libmath accurately reproduced the effect of OS updates, both globally (Fig. 2a) and locally (Fig. 2b). The distributions of significant bits in the atlas

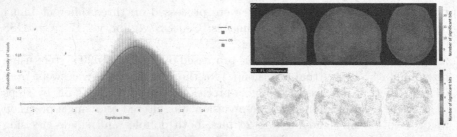

(a) Distribution of significant bits   (b) Significance map (subject average)

Fig. 2. Comparison of OS and FL effects on the precision of PreFreeSurfer results for n = 20 subjects. FL samples were obtained at the global nearest virtual precision of t = 37 bits.

registered T1w images were nearly identical ($p > 0.05$, KS test) on the average and individual subject distributions for 15/20 subjects, after correcting for multiple comparisons. Locally, the spatial distribution of significant digits also appeared to be preserved. Losses in significance were observed mainly at the brain-skull interface and between brain lobes, indicating spatial dependency of numerical properties.

The average number of significant bits in either the FL or OS conditions were 7.76 out of 24 available, which corresponds to 2.32 significant (base 10) digits. This relatively low precision motivates future investigations of the stability of pipeline components, in particular for image registration.

4.2   Fuzzy Libmath Preserves Between-Subjects Image Similarity

Numerically-perturbed samples remained primarily clustered by individual subjects (Fig. 3), indicating that neither FL nor OS perturbations were impactful enough to blur the differences between subjects. Notably, the similarity between subjects was also preserved by the numerical perturbation, leading to the same subject ordering in the dendrograms. However, the average RMSE within samples of a given subject was approximately 13× lower than the average RMSE between different subjects. The fact that between-subject variabilities were nearly on the same order of magnitude as OS and FL variability demonstrates the potential severity of these instabilities.

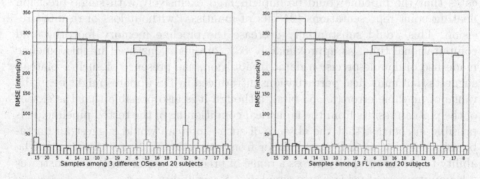

Fig. 3. RMSE-based hierarchical clustering of OS (left) and FL (right) samples. Colors identify different subjects, showing that similarities between subjects are preserved by the numerical perturbations. Horizontal gray lines represent average RMSEs between (top line) and within (bottom line) subject clusters. (Color figure online)

4.3   Results Are Stable Across Virtual Precision

The FL results presented previously were obtained at the global nearest virtual precision of t = 37 bits, determined as the precision which minimized the RMSE between FL and OS average maps of significant bits. We varied the virtual precision in steps of 2 between t = 1 and t = 53 bits (Fig. 4). On average, no

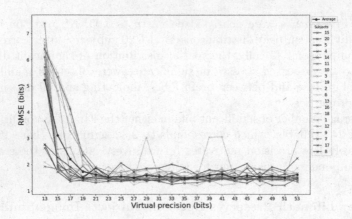

Fig. 4. Comparison of RMSE values computed between OS and FL results for different virtual precisions.

noticeable RMSE change was observed between the FL and OS variability for precisions ranging from $t = 21$ to $t = 53$ bits, which shows that FL can robustly approximate OS updates.

The observed plateau suggests the existence of an "intrinsic precision" for the pipeline, above which no improvement in results precision is expected. For the tested pipeline, this intrinsic precision was observed at $t = 21$ bits, which indicates that the pipeline could be implemented exclusively with single-precision floating-point representations (24 bits of mantissa) without loss of results precision. This would substantially decrease the pipeline memory footprint and computational time, as approximately 88% of operations used in this pipeline made use of double-precision data. In addition, the presence of such a plateau suggests that numerical perturbations introduced by OS updates might be in the range of machine error ($t = 53$ bits), although it is also possible that the extent of the plateau results from the numerical conditioning of the tested pipeline. It is possible in contrast that the absence of such a plateau would suggest an unstable pipeline that would benefit either from correction or larger datatypes. The ability to capture stability across a range of precisions importantly demonstrates a key advantage of using FL to simulate OS variability.

The relationship between RMSE of individual subjects was generally consistent with the average line, with the notable exception of subject 18. The observed discrepancies between this subject and potential others might be leveraged for quality control checks and, as a result, inform tool development.

The pipeline failed to complete for at least one subject below the virtual precision of $t = 13$ bits, also referred to as the tolerance of the pipeline. Specifically, 51% of pipeline executions crashed among all subjects for precisions ranging from 1–11 bits, and there was no relationship between tolerance-level and precision. The error raised was in the Readout Distortion Correction portion of the pipeline, and appears to stem from the FSL FAST tissue segmentation. The specific source of the error within this component is presently unknown, but is an open question for further exploration.

5 Conclusion and Discussion

We demonstrated fuzzy libmath as an accurate method to simulate variability in neuroimaging results due to OS updates. Alongside this evaluation, fuzzy libmath can be used by pipeline developers or consumers to evaluate the numerical uncertainty of tools and results. Such evaluations may also help decrease pipeline memory usage and computational time through the controlled use of reduced numerical precision. Fuzzy libmath does not require any modification of the pipeline as it operates on the level of shared libraries. The accuracy of the simulations were shown to be robust across a wide range of virtual precisions, which reinforces the applicability of the method.

The proposed technique is directly applicable to MATLAB code executed with GNU Octave, to Python programs executed on Linux, and to C programs that depend on GNU libmath. Numerical noise can be introduced in other libraries, such as OpenBLAS or NumPy, using our https://github.com/verificarlo/fuzzy environment.

A commonly used approach to address instabilities resulting from OS version updates in practice is to sweep the issue under the rug of software containers or static linking. While such solutions are undoubtedly helpful to improve code portability or strict re-executability, a more honest position is to consider computational results as realizations of random variables depending on numerical error. The presented technique enables estimating result distributions, a first step toward making analyses reproducible across heterogeneous execution environments. While this work did not investigate the precise cause of numerical instabilities by tracing the system function calls, this is a topic for future work.

The tested OS versions span a timeframe of 7 years (2012–2020) and focused on GNU/Linux, a widely-used platform in neuroimaging [10]. Given that our experiments focused on numerical perturbations applied to mathematical functions, which are implemented similarly across OSes, our findings are likely to generalize to OS/X or MS Windows, although future work would be needed to confirm that. The tested pipeline is the official solution of the HCP project to pre-process data, and is considered the state-of-the-art. This pipeline assembles software components from the FSL toolbox consistent with common practice in neuroimaging, such as in fMRIPrep [6] or the FSL feat workflow [12], to which fuzzy libmath can be directly applied. Efforts are on-going to use fuzzy libmath in fMRIPrep software tests, to guarantee that bug fixes do not perturb results beyond numerical uncertainty.

The fact that the induced numerical variability preserves image similarity between subjects is reassuring and, in fact, exciting. OS updates provide a convenient, practical target to define a virtual precision leading to a detectable but still reasonable numerical perturbation. However, it is also of importance that OS- and FL-induced variability were on a similar order of magnitude as subject-level effects. This suggests that the preservation of relative between-subject differences may not hold in all pipelines, and such a comparison could be used to evaluate the robustness of a pipeline to OS instabilities. The fact that the results observed across OS versions and FL perturbations arise from equally-valid numerical operations also suggests that the observed variability may contain meaningful signal. In particular, signal measured from these perturbations might be leveraged to enhance biomarkers, as suggested in [15] where augmenting a diffusion MRI dataset with numerically-perturbed samples was shown to improve age classification.

References

1. Andersson, J.L., Jenkinson, M., Smith, S., et al.: Non-linear registration, aka spatial normalisation FMRIB. Technical report TR07JA2, FMRIB Analysis Group of the University of Oxford (2007)
2. Botvinik-Nezer, R., et al.: Variability in the analysis of a single neuroimaging dataset by many teams. Nature **582**(7810), 84–88 (2020)
3. Bowring, A., Maumet, C., Nichols, T.E.: Exploring the impact of analysis software on task fMRI results. Hum. Brain Mapp. **40**, 1–23 (2019)
4. Chatelain, Y., de Oliveira Castro, P., Petit, E., Defour, D., Bieder, J., Torrent, M.: VeriTracer: context-enriched tracer for floating-point arithmetic analysis. In: 2018 IEEE 25th Symposium on Computer Arithmetic (ARITH), pp. 61–68. IEEE (2018)
5. Denis, C., de Oliveira Castro, P., Petit, E.: Verificarlo: checking floating point accuracy through Monte Carlo arithmetic. In: 2016 IEEE 23nd Symposium on Computer Arithmetic (ARITH), pp. 55–62 (2016)
6. Esteban, O., et al.: fMRIPrep: a robust preprocessing pipeline for functional MRI. Nat. Methods **16**(1), 111–116 (2019)
7. Glasser, M.F., et al.: The minimal preprocessing pipelines for the Human Connectome Project. NeuroImage **80**, 105–124 (2013)
8. Glatard, T., et al.: Reproducibility of neuroimaging analyses across operating systems. Front. Neuroinform. **9**, 12 (2015)
9. Gronenschild, E.H.B.M., et al.: The effects of FreeSurfer version, workstation type, and Macintosh operating system version on anatomical volume and cortical thickness measurements. PloS ONE **7**(6), e38234 (2012)
10. Hanke, M., Halchenko, Y.O.: Neuroscience runs on GNU/Linux. Front. Neuroinform. **5**, 8 (2011)
11. Jenkinson, M., Bannister, P., Brady, M., Smith, S.: Improved optimization for the robust and accurate linear registration and motion correction of brain images. NeuroImage **17**(2), 825–841 (2002)
12. Jenkinson, M., Beckmann, C.F., Behrens, T.E., Woolrich, M.W., Smith, S.M.: FSL. NeuroImage **62**(2), 782–790 (2012)
13. Jenkinson, M., Smith, S.: A global optimisation method for robust affine registration of brain images. Med. Image Anal. **5**(2), 143–156 (2001)
14. Kaur, B., Dugré, M., Hanna, A., Glatard, T.: An analysis of security vulnerabilities in container images for scientific data analysis. GigaScience **10**(6), giab025 (2021)
15. Kiar, G., Chatelain, Y., Salari, A., Evans, A.C., Glatard, T.: Data augmentation through Monte Carlo arithmetic leads to more generalizable classification in connectomics. bioRxiv (2020)

16. Parker, D.S.: Monte Carlo arithmetic: exploiting randomness in floating-point arithmetic. Computer Science Department, University of California, Los Angeles (1997)
17. Perkel, J.M.: Challenge to scientists: does your ten-year-old code still run? Nature **584**(7822), 656–658 (2020)
18. Salari, A., Kiar, G., Lewis, L., Evans, A.C., Glatard, T.: File-based localization of numerical perturbations in data analysis pipelines. GigaScience **9**(12), giaa106 (2020)
19. Van Essen, D.C., et al.: The WU-Minn human connectome project: an overview. NeuroImage **80**, 62–79 (2013)

Leveraging Uncertainty Estimates to Improve Segmentation Performance in Cardiac MR

Tewodros Weldebirhan Arega[✉], Stéphanie Bricq, and Fabrice Meriaudeau

ImViA Laboratory, Université Bourgogne Franche-Comté, Dijon, France

Abstract. In medical image segmentation, several studies have used Bayesian neural networks to segment and quantify the uncertainty of the images. These studies show that there might be an increased epistemic uncertainty in areas where there are semantically and visually challenging pixels. The uncertain areas of the image can be of a great interest as they can possibly indicate the regions of incorrect segmentation. To leverage the uncertainty information, we propose a segmentation model that incorporates the uncertainty into its learning process. Firstly, we generate the uncertainty estimate (sample variance) using Monte-Carlo dropout during training. Then we incorporate it into the loss function to improve the segmentation accuracy and probability calibration. The proposed method is validated on the publicly available EMIDEC MICCAI 2020 dataset that mainly focuses on segmentation of healthy and infarcted myocardium. Our method achieves the state of the art results outperforming the top ranked methods of the challenge. The experimental results show that adding the uncertainty information to the loss function improves the segmentation results by enhancing the geometrical and clinical segmentation metrics of both the scar and myocardium. These improvements are particularly significant at the visually challenging and difficult images which have higher epistemic uncertainty. The proposed system also produces more calibrated probabilities.

Keywords: Cardiac MRI Segmentation · Myocardial scar · Uncertainty · Bayesian deep learning

1 Introduction

Cardiac magnetic resonance (CMR) is a set of magnetic resonance imaging (MRI) used to provide anatomical and functional information of the heart. Late Gadolinium Enhancement (LGE), sometimes called delayed-enhancement MRI, is one type of CMR which is gold standard for the quantification of myocardial

Electronic supplementary material The online version of this chapter (https://doi.org/10.1007/978-3-030-87735-4_3) contains supplementary material, which is available to authorized users.

C. H. Sudre et al. (Eds.): UNSURE 2021/PIPPI 2021, LNCS 12959, pp. 24–33, 2021.
https://doi.org/10.1007/978-3-030-87735-4_3

infarction. Myocardial infarction, also called heart attack, is the interruption of coronary blood supply to certain myocardial area which leads to irreversible death of myocardial tissue [11]. No-reflow phenomenon is an incident that usually appears in a proportion of patients with acute myocardial infarction following re-perfusion therapy of an occluded coronary artery [1].

Recently, deep learning based semi-automatic and fully-automatic methods have been proposed to segment myocardial scar (infarction) from LGE images. Zabihollahy *et al.* [23] used manual segmentation for myocardium and 2D Fully Convolutional Network to segment scar from the myocardium. Zhang [24], Ma [15] and Girum *et al.* [8] used a two stage cascaded segmentation framework to automatically segment myocardial scar and tested their method on EMIDEC dataset. In the first stage, Zhang [24] used a 2D nnUNet [9] to get a coarse segmentation. In the second stage, a 3D nnUNet is utilized to further refine the segmentation result. Ma [15] used a 2D nnUNet[9] to first segment the whole heart as region of interest (ROI) and then utilized a second 2D nnUNet to segment the myocardial infarction from the ROI. Arega *et al.* [2] also used a cascaded framework of three networks to automatically segment scar from multi-sequence CMR. The main problem with these cascaded methods is that they can be time consuming and computationally expensive.

Bayesian deep learning have been used in segmentation task to provide a prediction as well as quantify the uncertainty associated with each prediction. Recently, several studies have employed Monte Carlo Dropout to estimate uncertainty for medical image segmentation [16–18,20,21]. Monte Carlo (MC) dropout is an uncertainty estimation method proposed by Gal and Ghahramani [7]. It is done by training a network with dropout and taking the Monte Carlo samples of the prediction using dropout at test time. Nair *et al.* [17] explored MC dropout based uncertainty estimates for multiple sclerosis lesion detection and segmentation. They improved the segmentation results by filtering and excluding the most uncertain voxels. Similarly, Sander *et al.* [21] applied MC Dropout based method for cardiac MRI segmentation and showed that the uncertainty maps are close to the reported segmentation errors and they improved the segmentation results by correcting the uncertain pixels. These previous studies [10,16,17,20,21] mostly focused on the correlations between predictive uncertainty and the segmentation accuracy and how the uncertainty metrics can be used to improve the segmentation by filtering the most uncertain predictions. However, these methods did not leverage the uncertainty information during training to enhance the segmentation result. In the area of computer vision, Kendall *et al.* [13] utilized homoscedastic aleatoric uncertainty to weight the losses of multi-task problems.

In this paper, we proposed a segmentation model that generates uncertainty estimates during training using MC-dropout. Then it leverages these uncertainty estimates to improve the segmentation results by incorporating them to the loss function. Uncertainty information can possibly indicate the regions of incorrect segmentation [21,22]. We hypothesized that by incorporating this information as part of the learning process, it can help the network to improve the segmentation results by correcting the segmentation errors that have high epistemic uncertainty. The proposed method was evaluated on the publicly available EMIDEC MICCAI 2020 dataset [14]. It achieved the state of the art results outperforming the top ranked methods of the challenge. The experimental results showed that the uncertainty information was indeed beneficial in enhancing the segmentation performance. We also observed that the improvements were more significant at the semantically and visually challenging images which have higher epistemic uncertainty. Assessing the probability calibration, we showed that the proposed method produced more calibrated probabilities than the baseline method.

2   Materials and Methods

2.1   Dataset

The Automatic Evaluation of Myocardial Infarction from Delayed-Enhancement Cardiac MRI challenge (EMIDEC)[1] is a MICCAI 2020 challenge that focuses on cardiac MRI segmentation. More specifically, it involves the segmentation of healthy myocardium, infarction (scar) and no-reflow regions. The dataset consists of LGE images of 100 patients for training. From these cases, 67 are pathological cases and the remaining 33 are normal cases. The testing set includes 50 patients in which 33 are pathological and 17 are normal cases. Each case has 5 to 10 short-axis slices covering the left ventricle from base to apex with the following characteristics: slice thickness of 8 mm, distance between slices of 10 mm and spatial resolution ranging from 1.25×1.25 mm^2 to 2×2 mm^2 [14]. As a pre-processing step, we normalized the intensity of every patient image to have zero-mean and unit-variance and we resampled all the volumes to have a voxel spacing of 1.458 mm $\times$ 1.458 mm $\times$ 10.0 mm.

2.2   Methods

Various Bayesian deep learning methods are used to estimate uncertainties in images. Among the most widely used Bayesian deep learning methods in medical images is Monte-Carlo dropout (MC-dropout). In MC-dropout, a network with dropout is trained, then during testing the network is sampled N times in order to get N segmentation samples. From these N segmentation samples, the uncertainty measure (sample variance) is computed. In our method, we used MC-dropout during training in order to get the uncertainty estimates. During training, the model is sampled N times and the mean of these samples is used as the final segmentation as can be seen from Fig. 1. The uncertainty metric is computed from the N Monte-Carlo dropout samples. It can be calculated per pixel or per structure [18]. In this work, we used the pixel-wise uncertainty and image-level uncertainty. Pixel-wise uncertainty is computed per pixel. Sample variance is one of the pixel-wise uncertainty measures. It is calculated as the variance of the N Monte-Carlo prediction samples of a pixel. Each pixel i has

---

[1] http://emidec.com/.

N sigmoid predictions $(y_{i,1}...y_{i,N})$. From these predictions, the mean μ_i is computed (Eq. 1). In Eq. 2, σ_i^2 is the sample variance of each pixel i of the image [17]. In order to compute the image-level uncertainty, the per-pixel uncertainty is averaged over all pixels of the image as shown in Eq. 4. In this equation, I is the total number of pixels of the image.

$$\mu_i = \frac{1}{N} \sum_n (y_{i,n}) \tag{1}$$

$$\sigma_i^2 = \frac{1}{N} \sum_n (y_{i,n} - \mu_i)^2 \tag{2}$$

As stated by [21] and [22], uncertainty information indicates potential mis-segmentations and the most uncertain part of the segmentation results cover regions of incorrect segmentations. In order to leverage this uncertainty information, we proposed to include it as part of the loss function so that the network will learn to correct the possible mis-segmentations. Hence, the total loss is computed as a sum of the segmentation loss and uncertainty loss as can be seen from Fig. 1. The segmentation loss is the weighted average of cross-entropy (CE) loss and Dice loss (Eq. 3). For the uncertainty loss, we first computed the image level uncertainty (Eq. 4). Then, it is added to the segmentation loss with a hyper-parameter value alpha (α) that controls the contribution of the uncertainty loss to the total loss (Eq. 5).

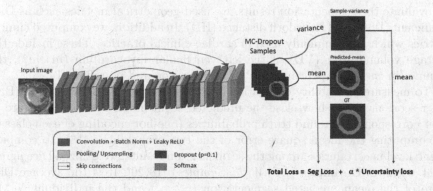

Fig. 1. The proposed method

$$L_{Seg} = \lambda_{Dice} L_{Dice} + \lambda_{CE} L_{CE} \tag{3}$$

$$L_{Uncertainty} = \frac{1}{I} \sum_i (\sigma_i^2) \tag{4}$$

$$L_{Total} = L_{Seg} + \alpha \times L_{Uncertainty} \tag{5}$$

For the segmentation network, we used a 3D UNet [9] architecture with dropout placed at the middle layers of the network (Fig. 1) as suggested by the literatures [4,6,12,18]. The dropout rate was set at 0.1. The UNet's encoder and decoder consists of 8 convolutional layers where each convolution is followed by batch normalization and Leaky ReLU (negative slope of 0.01) activation function.

2.3 Training

The weights of the segmentation network are optimized using Stochastic gradient descent (SGD) with nesterov momentum ($\mu = 0.99$) with an initial learning rate of 0.01. The mini-batch size was 5 and the model was trained for 1000 epochs on a five-fold cross validation scheme. For the segmentation loss, we set a weighting factor of 1.0 for Dice loss and 1.0 for CE loss as they provided the best results. In order to generate the segmentation uncertainty (sample variance), we used 5 Monte Carlo samples (the N value in Eq. 1). The weighting factor (α) for the uncertainty loss (in Eq. 5) is empirically selected to be 3.0 after experimenting with different weighting factors. The training was done on NVIDIA Tesla V100 GPUs using Pytorch deep learning framework based on nnU-Net implementation [9].

3 Results and Discussion

To evaluate the segmentation results, we used geometrical metrics such as Dice coefficient (DSC) and Hausdorff distance (HD). In addition, we computed clinical metrics which are commonly used in cardiac clinical practice. These include the average volume error (VD) of the left ventricular myocardium (in cm^3), the volume (in cm^3) and percentage (PD) of infarction and no-reflow [14].

To measure probability calibration of the models, we used Brier score (BS). Brier score measures how close the predicted segmentation probabilities are to their corresponding ground truth probabilities (one-hot encoding of each classes) by computing the mean square error of the two probabilities [18]. To compare image level uncertainties among the segmentation results, we utilized Dice agreement within MC samples ($DiceWithinSamples$) [18,20]. It is the average Dice score of the mean predicted segmentation (S_{mean}) and the individual N MC prediction samples as shown in Eq. 6. Note that $Dice_{WithinSamples}$ is inversely related to uncertainty.

$$Dice_{\text{WithinSamples}} = \frac{1}{N} \sum_n Dice(S_{mean}, S_n) \qquad (6)$$

3.1 Ablation Study

To evaluate the effect of adding uncertainty information to the segmentation loss, we compared the model that uses only segmentation loss which is called *baseline* with the model that uses combined loss of segmentation loss and uncertainty loss which is referred as *proposed*. Both networks have the same architecture and the comparison is done on the test dataset. For the ablation study, most of the comparisons are done on the main two classes that are healthy myocardium and infarction. The comparison on all the three classes can be found in the supplementary material.

As can be seen from the Table 1, the addition of uncertainty information into the segmentation loss enhanced the segmentation accuracy. It increased the DSC of scar (infarction) by 3% and that of myocardium by around 0.2%. It also improved the HD and the average volume error of both scar and myocardium. The segmentation enhancement is more significant on scar than on myocardium. This can be explained by the fact that scar has more irregular shape, smaller area and visually challenging pixels which may result in higher uncertainty compared to myocardium (Fig. 2 (b)). The relatively high standard deviation in the metrics of the infarction can be caused due to the availability of normal cases with no pathology (17 cases) in the test dataset. This can result in fluctuation of the metrics of infarction. For example, the Dice of infarction will be 0 if few false positive pixels are present during the segmentation of the normal cases.

The apical and basal slices of the left ventricle are more difficult to segment than mid-ventricular images even for human experts [3,19]. Particularly at the apical slices, the MRI resolution is very low that it is even difficult to resolve size of small structures (first row in Fig. 3). Assessing the segmentation performance and uncertainties at different slice positions of the left ventricle, it can be observed that the apical slices have the highest epistemic uncertainty (lowest *DiceWithinSamples*) among the slices (Fig. 2 (b)). Similarly, in the comparison of segmentation performance, most of the improvements due to the addition of uncertainty information (proposed method) are predominantly on the apical slices (Fig. 2 (a)). The DSC increased by 2% for scar and by almost 1% for myocardium in the apical slices. While the segmentation performance of the proposed method at the mid and basal slices are similar or slightly better than the baseline method. This tells us that the addition of uncertainty information to the loss function is more advantageous to the semantically and visually challenging images which generate higher epistemic uncertainty. This confirms our initial assumption about the proposed method.

Table 1. Comparison of myocardium and scar (infarction) segmentation performance of the baseline method and the proposed method in terms of geometrical and clinical metrics obtained on the test set (50 cases). The values mentioned are mean (standard deviation). The best results are in bold. VD is the volume error. For DSC, the higher the value the better whereas for HD, Brier score (BS) and VD the lower is the better.

Method	Myocardium			Infarction		
	DSC (%)	HD (mm)	BS (10^{-2})	DSC (%)	VD (cm^3)	BS (10^{-2})
Baseline	88.0 (2.63)	12.1(7.79)	4.03 (2.45)	65.0 (29.7)	3.04 (5.0)	1.19 (1.81)
Proposed	**88.2 (2.55)**	**11.8 (7.26)**	**3.86 (2.8)**	**67.6 (28.8)**	**2.99 (4.55)**	**1.18 (1.83)**

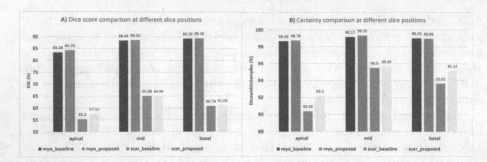

Fig. 2. Dice score (A) and certainty (B) comparison of the baseline and proposed method at different slice locations. Myo_baseline and Scar_baseline refer to myocardium and scar Dice score or certainty of the baseline method respectively. Similarly, Myo_proposed and Scar_proposed refer to myocardium and scar Dice score or certainty of the proposed method.

Figure 3 shows examples of the segmentation results of baseline and proposed method at apical, mid-ventricular and basal slices. At the apical slice, one can see that the segmentation result of the baseline method has a lot of errors. In the generated uncertainties (sample variance), the incorrectly segmented regions have higher uncertainty. The proposed method, which utilizes the sample variance as part of the loss, minimized the segmentation errors of the baseline. Similarly, our proposed method produced more robust segmentation results at the mid and basal slices. From the results, we can say that the uncertainty captures relevant information that can be leveraged to improve the segmentation result.

Regarding the probability calibration, the proposed method produced more calibrated probabilities than the baseline method on both myocardium and scar

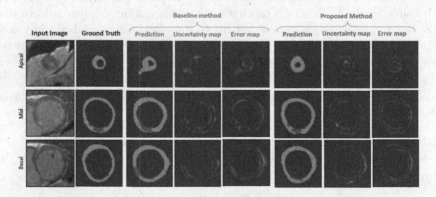

Fig. 3. Qualitative results comparison of the proposed method with the baseline on a typical cardiac MRI. The generated uncertainty is sample variance. The error map is computed from the pixel-wise CE. Scar (green) and myocardium (yellow). (Color figure online)

as it yielded lower Brier score. This suggests that using MC-dropout during training and the addition of uncertainty information to the loss can improve not only the segmentation accuracy but also the calibration of the probabilities.

3.2   Comparison with State of the Art

Table 2 shows the comparison of the proposed method with state of the art methods on EMIDEC challenge. One can observe that the proposed method outperformed the state of the art methods on most of the geometrical and clinical metrics. Our proposed method yielded much better results in all metrics than Feng *et al.* [5], which used a dilated 2D Unet. Zhang [24] and Ma [15] employed nnU-Net based segmentation pipeline which is similar to the proposed method's pipeline. However, the proposed method, which utilizes a novel loss function that took into account the uncertainty generated during training, outperformed these two top ranked methods. In the segmentation of infarction, the proposed method reduced the average volume error from 3.12 cm^3 to 2.99 cm^3 and the percentage from 2.38% to 2.29% compared to Zhang's [24] method. In terms of the Dice score of infarction, Zhang's [24] method achieved better results. This can be due to the usage of a cascaded 2D-3D framework where the 2D nnU-Net segments the heart region as ROI and the 3D nnU-Net segments the pathological area from the pre-segmented region. However, Zhang's method is more computationally expensive.

Table 2. Comparison of segmentation performance with state of the art methods on EMIDEC challenge's test set (50 cases). Bold results are the best.

Authors	Myocardium			Infarction			NoReflow		
	DSC (%)	VD (cm^3)	HD (mm)	DSC (%)	VD (cm^3)	PD(%)	DSC (%)	VD (cm^3)	PD (%)
Zhang	87.86	9.26	13.01	**71.24**	3.12	2.38	78.51	0.635	0.38
Ma	86.28	10.2	14.31	62.24	4.87	3.50	77.76	0.830	0.49
Feng et al.	83.56	15.2	33.77	54.68	3.97	2.89	72.22	0.883	0.53
Proposed	**88.22**	**9.23**	**11.78**	67.64	**2.99**	**2.29**	**81.00**	**0.601**	**0.37**

4   Conclusion

In this paper, we proposed a segmentation model that generates uncertainty estimates during training using MC-dropout method and utilizes the uncertainty information to enhance the segmentation results by incorporating it into the loss function. The proposed method was evaluated on the publicly available EMIDEC dataset. It achieved state of the art results outperforming the top ranked methods. Assessing the segmentation performance of the proposed method at different slice positions, we observed that the Dice scores of the more challenging apical slices increased much more than the other slice positions. Furthermore, the improvements in the more difficult scar segmentation was higher than that of myocardium segmentation. In the quantitative and qualitative results, we demonstrated that the uncertainty information was indeed advantageous in enhancing the segmentation performance and the improvements were more significant at the semantically and visually challenging images which have higher epistemic uncertainty. In addition, the proposed method produced more calibrated segmentation probabilities.

The main limitation of our method is that it takes more time to train than the baseline method as it uses MC-dropout during training to generate the uncertainty estimates. However, once it is trained, the inference time is exactly the same as the baseline method. Future work will focus on utilizing the uncertainty estimates generated by other Bayesian methods such as variational inference to improve the segmentation performance. We will also extend the evaluation onto other challenging public datasets.

Acknowledgments. This work was supported by the French National Research Agency (ANR), with reference ANR-19-CE45-0001-01-ACCECIT. Calculations were performed using HPC resources from DNUM CCUB (Centre de Calcul de l'Université de Bourgogne). We also thank the Mesocentre of Franche-Comté for the computing facilities.

References

1. Abbas, A., Matthews, G.H., Brown, I.W., Shambrook, J., Peebles, C., Harden, S.: Cardiac MR assessment of microvascular obstruction. Br. J. Radiol. **88**(1047), 20140470 (2015)
2. Arega, T.W., Bricq, S.: Automatic myocardial scar segmentation from multi-sequence cardiac MRI using fully convolutional densenet with inception and squeeze-excitation module. In: Zhuang, X., Li, L. (eds.) MyoPS 2020. LNCS, vol. 12554, pp. 102–117. Springer, Cham (2020). https://doi.org/10.1007/978-3-030-65651-5_10
3. Bernard, O., et al.: Deep learning techniques for automatic MRI cardiac multi-structures segmentation and diagnosis: is the problem solved? IEEE Trans. Med. Imaging **37**, 2514–2525 (2018)
4. Blundell, C., Cornebise, J., Kavukcuoglu, K., Wierstra, D.: Weight uncertainty in neural network. In: International Conference on Machine Learning, pp. 1613–1622. PMLR (2015)
5. Feng, X., Kramer, C.M., Salerno, M., Meyer, C.H.: Automatic scar segmentation from DE-MRI using 2D dilated UNet with rotation-based augmentation. In: Puyol Anton, E., et al. (eds.) STACOM 2020. LNCS, vol. 12592, pp. 400–405. Springer, Cham (2021). https://doi.org/10.1007/978-3-030-68107-4_42
6. Fortunato, M., Blundell, C., Vinyals, O.: Bayesian recurrent neural networks. arXiv preprint arXiv:1704.02798 (2017)
7. Gal, Y., Ghahramani, Z.: Dropout as a Bayesian approximation: representing model uncertainty in deep learning. In: International Conference on Machine Learning, pp. 1050–1059. PMLR (2016)
8. Girum, K.B., Skandarani, Y., Hussain, R., Grayeli, A.B., Créhange, G., Lalande, A.: Automatic myocardial infarction evaluation from delayed-enhancement cardiac MRI using deep convolutional networks. In: Puyol Anton, E., et al. (eds.) STACOM 2020. LNCS, vol. 12592, pp. 378–384. Springer, Cham (2021). https://doi.org/10.1007/978-3-030-68107-4_39

9. Isensee, F., Jaeger, P.F., Kohl, S.A., Petersen, J., Maier-Hein, K.H.: nnU-Net: a self-configuring method for deep learning-based biomedical image segmentation. Nat. Methods **18**(2), 203–211 (2021)
10. Jungo, A., Reyes, M.: Assessing reliability and challenges of uncertainty estimations for medical image segmentation. In: Shen, D., et al. (eds.) MICCAI 2019. LNCS, vol. 11765, pp. 48–56. Springer, Cham (2019). https://doi.org/10.1007/978-3-030-32245-8_6
11. Kate Meier, C., Oyama, M.A.: Chapter 41 - Myocardial infarction. In: Silverstein, D.C., Hopper, K. (eds.) Small Animal Critical Care Medicine, pp. 174–176. W.B. Saunders, Saint Louis (2009). https://doi.org/10.1016/B978-1-4160-2591-7.10041-4, https://www.sciencedirect.com/science/article/pii/B9781416025917100414
12. Kendall, A., Badrinarayanan, V., Cipolla, R.: Bayesian SegNet: model uncertainty in deep convolutional encoder-decoder architectures for scene understanding. arXiv preprint arXiv:1511.02680 (2015)
13. Kendall, A., Gal, Y., Cipolla, R.: Multi-task learning using uncertainty to weigh losses for scene geometry and semantics. In: 2018 IEEE/CVF Conference on Computer Vision and Pattern Recognition, pp. 7482–7491 (2018)
14. Lalande, A., et al.: Emidec: a database usable for the automatic evaluation of myocardial infarction from delayed-enhancement cardiac MRI. Data **5**(4), 89 (2020)
15. Ma, J.: Cascaded framework for automatic evaluation of myocardial infarction from delayed-enhancement cardiac MRI. arXiv preprint arXiv:2012.14556 (2020)
16. Mehrtash, A., Wells, W.M., Tempany, C.M., Abolmaesumi, P., Kapur, T.: Confidence calibration and predictive uncertainty estimation for deep medical image segmentation. IEEE Trans. Med. Imaging **39**(12), 3868–3878 (2020)
17. Nair, T., Precup, D., Arnold, D.L., Arbel, T.: Exploring uncertainty measures in deep networks for multiple sclerosis lesion detection and segmentation. Med. Image Anal. **59**, 101557 (2020)
18. Ng, M., et al.: Estimating uncertainty in neural networks for cardiac MRI segmentation: a benchmark study. arXiv preprint arXiv:2012.15772 (2020)
19. Petitjean, C., Dacher, J.N.: A review of segmentation methods in short axis cardiac MR images. Med. Image Anal. **15**(2), 169–184 (2011)
20. Roy, A.G., Conjeti, S., Navab, N., Wachinger, C.: Inherent brain segmentation quality control from fully ConvNet Monte Carlo sampling. In: Frangi, A.F., Schnabel, J.A., Davatzikos, C., Alberola-López, C., Fichtinger, G. (eds.) MICCAI 2018. LNCS, vol. 11070, pp. 664–672. Springer, Cham (2018). https://doi.org/10.1007/978-3-030-00928-1_75
21. Sander, J., de Vos, B.D., Wolterink, J.M., Išgum, I.: Towards increased trustworthiness of deep learning segmentation methods on cardiac MRI. In: Medical Imaging 2019: Image Processing, vol. 10949, p. 1094919. International Society for Optics and Photonics (2019)
22. Wang, G., Li, W., Ourselin, S., Vercauteren, T.: Automatic brain tumor segmentation based on cascaded convolutional neural networks with uncertainty estimation. Front. Comput. Neurosci. **13**, 56 (2019)
23. Zabihollahy, F., White, J.A., Ukwatta, E.: Myocardial scar segmentation from magnetic resonance images using convolutional neural network. In: Medical Imaging 2018: Computer-Aided Diagnosis, vol. 10575, p. 105752Z. International Society for Optics and Photonics (2018)
24. Zhang, Y.: Cascaded convolutional neural network for automatic myocardial infarction segmentation from delayed-enhancement cardiac MRI. arXiv preprint arXiv:2012.14128 (2020)

Improving the Reliability of Semantic Segmentation of Medical Images by Uncertainty Modeling with Bayesian Deep Networks and Curriculum Learning

Sora Iwamoto[1], Bisser Raytchev[1(✉)], Toru Tamaki[2], and Kazufumi Kaneda[1]

[1] Graduate School of Advanced Science and Engineering, Hiroshima University, Higashihiroshima, Japan
bisser@hiroshima-u.ac.jp
[2] Department of Computer Science, Nagoya Institute of Technology, Nagoya, Japan

Abstract. In this paper we propose a novel method which leverages the uncertainty measures provided by Bayesian deep networks through curriculum learning so that the uncertainty estimates are fed back to the system to resample the training data more densely in areas where uncertainty is high. We show in the concrete setting of a semantic segmentation task (iPS cell colony segmentation) that the proposed system is able to increase significantly the reliability of the model.

Keywords: Bayesian deep learning · Uncertainty · Curriculum learning

1 Introduction

Although in recent years deep neural networks have achieved state-of-the-art performance on many medical image analysis tasks, even surpassing human-level performance in certain cases [6,9], their extensive adoption in clinical settings has been hampered by their false over-confidence when confronted with out-of-distribution (OOD) test samples (samples that lie far away from the data which they have been trained with). This is due to the fact that the probability vector obtained from the softmax output is erroneously interpreted as model confidence [8]. It may be amusing if a deep net trained with cats and dogs images classifies a human as a dog with 98% probability, but similar mistake due to encountering test samples lying outside the data distribution of a cancer detection system can lead to life-threatening situations, thus the reluctance of some medical professionals to adopt such systems wholeheartedly.

In order to address this problem different risk-aware Bayesian networks have been proposed [8,13,14], which rather than point estimates, as in popular deep learning models, are able to output uncertainty estimates, which can provide information about the *reliability* of the trained models, thus allowing the users

© Springer Nature Switzerland AG 2021
C. H. Sudre et al. (Eds.): UNSURE 2021/PIPPI 2021, LNCS 12959, pp. 34–43, 2021.
https://doi.org/10.1007/978-3-030-87735-4_4

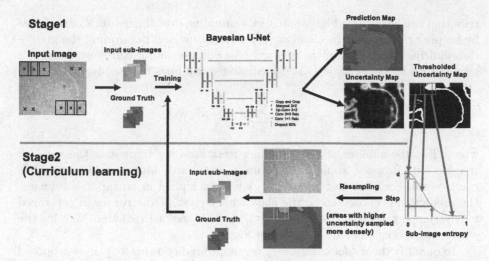

Fig. 1. Overview of the proposed system.

to take necessary actions to ensure safety when the model is under-confident or falsely over-confident.

In this paper, we endeavor to take this work one step further, i.e. not merely to provide uncertainty measures, but to leverage those through curriculum learning [2], so that the uncertainty estimates are fed back to the system to resample the training data more densely in areas where the uncertainty is high. We show, in the setting of a concrete semantic segmentation task, that the reliability of the model can be significantly increased without decreasing segmentation accuracy, which can potentially lead to wider acceptance of deep learning models in clinical settings where safety is first priority.

2   Methods

Figure 1 provides an overview of the proposed method which consists of two stages. During stage 1 a large image which needs to be segmented (the class of each pixel needs to be determined) is input into the system. We assume large bio-medical images with predominantly local texture information (one example would be colonies of cells), where local texture statistics are more important for the correct segmentation and distant areas might not be correlated at all. We extract sub-images of size $d \times d$ pixels from the large image, and these are sent to a Bayesian U-Net for learning the segmentation end-to-end, using ground truth segmentation map images provided by experts. In the lack of additional information the sub-images can be extracted from the large image in a sliding window manner, scanning the large image left-to-right and top-to-bottom using a predefined step s.

To obtain the model's uncertainty of its prediction, we apply Monte Carlo (MC) Dropout [5,8,13] to a U-Net [15]. An approximate predictive posterior distribution can be obtained by Monte Carlo sampling over the network parameters by keeping the dropout mechanism [16] at test time and performing the prediction multiple times during the forward pass (dropout approximately integrates over the model's weights [8]). The predictive mean for a test sample $\mathbf{x}^*$ is

$$\mu_{pred} \approx \frac{1}{T} \sum_{t=1}^{T} p\left(y^* \mid \mathbf{x}^*, \hat{\mathbf{w}}_t\right) \tag{1}$$

where T is the number of MC sampling iterations, $\hat{\mathbf{w}}_t$ represents the network weights with dropout applied to the units at the t-th MC iteration, and y^* is class vector. For each test sample $\mathbf{x}^*$, which is a pixel in an input sub-image, the prediction is chosen to be the class with largest predictive mean (averaged for overlapping areas across sub-images). In this way a Prediction Map for the whole image is calculated, as shown in Fig. 1.

To quantify the *model uncertainty* we adopt predictive entropy H as proposed in [7]:

$$H\left(\mathbf{y}^* \mid \mathbf{x}^*, \mathcal{D}\right) = -\sum_c p\left(y^* = c \mid \mathbf{x}^*, \mathcal{D}\right) \log p\left(y^* = c \mid \mathbf{x}^*, \mathcal{D}\right) \tag{2}$$

where c ranges over the classes. Since the range of the uncertainty values can vary across different datasets $\mathcal{D}$ or models, similarly to [13] we adopt the normalized entropy $H_{norm} \in [0, 1]$, computed as $H_{norm} = \frac{H - H_{min}}{H_{max} - H_{min}}$. In this way, an Uncertainty Map is calculated for the whole image, which can be thresholded using a threshold H_T to obtain the *Thresholded Uncertainty Map* (see Fig. 1) where each pixel's prediction is considered *certain* if the corresponding value in the map is larger than H_T and *uncertain* otherwise.

In stage 2 of the proposed method, we utilize Curriculum Learning [2] to leverage the information in the Uncertainty Map about the uncertainty of the model to improve its reliability (methods for evaluating the reliability of a model are explained in Sect. 3). The main idea is that *areas where segmentation results are uncertain need to be sampled more densely than areas where the model is certain*. The uncertainty $H(S)$ of sub-images S is calculated as the average entropy obtained from the uncertainty values in the Uncertainty Map *pmfs* $p^{(i)}$ for each pixel (indexed by i) corresponding to the sub-image. Using $H(S)$ as a measure of the current sub-image's uncertainty, the position of the next location where to resample a new sub-image is given by

$$f(H(S)) = d \exp\{-(H(S))^2 / 2\sigma^2\}. \tag{3}$$

where d is the size of the sub-image and σ the width of the Gaussian. This process is illustrated in Fig. 1, starting at the upper left corner of the input image, uncertainty for the current sub-image is calculated and the step size in pixels to move in the horizontal direction is calculated by Eq. 3. The whole image is resampled this way to re-train the model and this process (stage 2) can be repeated several times until no further improvement in reliability is obtained.

Additionally, we propose a second method (Method 2, single-staged), which does not use curriculum learning, i.e. consists of a single training stage. This method, rather than resampling the training set, directly uses the values in the Uncertainty Map to improve model reliability. This method initially trains the Bayesian U-Net for 5 epochs using cross-entropy loss (several other losses have also been tried as shown in the experiments), after which generates an Uncertainty Map similarly to the curriculum learning based method, and continues the training for 5 more epochs augmenting the training loss with a term which tries directly to minimize the uncertainty values of the Uncertainty Map - for this reason we call it *Uncertainty Loss*. Note that the term that minimizes the uncertainty is added to the cross-entropy term *after* the cross-entropy loss has been minimized for several epochs, so that this does not encourage overconfident false predictions.

3   Experiments

In this section we evaluate the performance of the proposed methods on a dataset which consists of 59 images showing colonies of undifferentiated and differentiated iPS cells obtained through phase-contrast microscopy. The task we have to solve is to segment the input images into three categories: Good (undifferentiated), Bad (differentiated) and Background (BGD, the culture medium). Several representative images together with ground-truth provided by experts can be seen in Fig. 2. All images in this dataset are of size 1600×1200 pixels.

Network Architecture and Hyperparameters: We used a Bayesian version of U-Net [12,13,15], the architecture of which can be seen in Fig. 1, with 50% dropout applied to all layers of both the encoder and decoder parts.

The learning rate was set to $1e-4$ for the Stage 1 learning, and to $1e-6$ for the curriculum learning (beyond stage 2). For the optimization procedure we used ADAM [10] ($\beta_1 = 0.9, \beta_2 = 0.999$), batch size was 12, training for 10 epochs per learning stage, while keeping the model weights corresponding to minimal loss on the validation sets. All models were implemented using Python 3.6.10, TensorFlow 2.3.0 [1] and Keras 2.3.1 [4]. Computation was performed using NVIDIA GeForce GTX1080 Ti. Code is available from https://github.com/sora763/Uncertainty. The size of the sub-images was fixed to $d = 160$ (i.e. 160×160 pixels). The width of the Gaussian in Eq. 3 was empirically set to $\sigma = 0.4$ for all experimental results. The values of the other system parameters used in Sect. 2 were set to $T = 10$, $s = 10$, $H_T = 0.5$ throughout the experiments.

Evaluation Procedure and Criteria: Evaluation was done through 5-fold cross validation by splitting the data into training, validation and test sets in proportions 3:3:1. Each method was evaluated by using the metrics described below, which are adopted from [12]. In a Bayesian setting there are four different possible cases for an inference: it can be (a) incorrect and uncertain (True Positive, TP); (b) correct and uncertain (False Positive, FP); (c) correct and certain (True Negative, TN); (d) incorrect and certain (False Negative, FN).

Table 1. Experimental results comparing reliability (NPV, TPR and UA) and segmentation accuracy (IoU) for baseline (first row) and curriculum learning based methods using different loss functions (second to fifth rows).

Stage	Loss	NPV	TPR	UA	IoU
1	CE	0.867 ± 0.026	0.313 ± 0.021	0.859 ± 0.027	0.783 ± 0.039
4	CE	0.917 ± 0.021	0.352 ± 0.009	0.899 ± 0.022	0.796 ± 0.039
2	Dice	**0.955 ±0.010**	**0.431 ±0.011**	**0.921 ± 0.015**	0.794 ± 0.043
2	SS	0.894 ± 0.020	0.326 ± 0.015	0.883 ± 0.021	**0.797 ± 0.037**
2	CE+Dice	0.904 ± 0.019	0.340 ± 0.010	0.890 ± 0.020	**0.797 ± 0.042**

Correctness can be obtained by comparing the Prediction Map with the Ground Truth, while certain/uncertain values can be obtained from the Thresholded Uncertainty Map described in Sect. 2.

1. Negative Predictive Value (NPV)
   The model should predict correctly if it is certain about its prediction. This can be evaluated by the following conditional probability and corresponds to the NPV measure in a binary test:

$$P(correct \mid certain) = \frac{P(correct, certain)}{P(certain)} = \frac{TN}{TN + FN} \qquad (4)$$

2. True Positive Rate (TPR)
   The model should be uncertain if the prediction is incorrect. This can be evaluated by the following conditional probability and corresponds to the TPR measure in a binary test:

$$P(uncertain \mid incorrect) = \frac{P(uncertain, incorrect)}{P(incorrect)} = \frac{TP}{TP + FN} \qquad (5)$$

3. Uncertainty Accuracy (UA)
   Finally, the overall accuracy of the uncertainty estimation can be measured as the ratio of the desired cases (TP and TN) over all possible cases:

$$UA = \frac{TP + TN}{TP + TN + FP + FN} \qquad (6)$$

For all metrics described above, higher values indicate a model that performs better. Additionally, overall segmentation performance was evaluated using the mean Intersection-over-Union (IoU, or Jaccard index), as is common for segmentation tasks. For each score the average and standard deviation obtained from 5-fold cross-validation are reported.

Experimental Results for the Curriculum Learning: Table 1 shows the results obtained by the proposed method based on resampling with curriculum learning, compared with the single-stage baseline (first row) using the same U-Net trained with cross-entropy (CE) loss without uncertainty modelling through

Table 2. Experimental results comparing reliability (NPV, TPR and UA) and segmentation accuracy (IoU) for three baseline single-stage learning methods using different loss functions (first 3 rows) and the proposed single-stage method (Uncertainty Loss).

Stage	Loss	NPV	TPR	UA	IoU
1	CE	0.867 ± 0.026	0.313 ± 0.021	0.859 ± 0.027	0.783 ± 0.039
1	SS	0.913 ± 0.023	0.356 ± 0.013	0.896 ± 0.025	0.791 ± 0.039
1	CE+Dice	0.913 ± 0.016	0.358 ± 0.021	0.895 ± 0.019	0.792 ± 0.042
1	Uncertainty Loss	**0.935 ±0.018**	**0.382 ± 0.025**	**0.910 ±0.018**	**0.798 ± 0.041**

curriculum learning. The second row in the table reports results obtained with curriculum learning using cross-entropy loss over 4 learning stages (first same as the baseline and next 3 stages using curriculum learning). Third row corresponds to results using Dice loss [11] for the second stage (curriculum learning), fourth row corresponds to results using the Sensitivity-Specificity (SS) loss [3] for the second stage (curriculum learning), and the last row corresponds to results using both cross-entropy and Dice loss for the second stage (curriculum learning). Regarding the system reliability evaluation, the results show that a big improvement is achieved when using the proposed curriculum learning method: in the case when Dice loss is used which achieves best performance, TPR improved by 12%, NPV by 9% and UA by 6%. It was found that when using Dice loss, SS loss or CE+Dice loss for the curriculum learning necessitates only a single stage of curriculum learning (no significant improvement observed after that), while if the original CE loss is used 3 stages of curriculum learning were needed for best results (note that using Dice loss instead of CE during the first stage resulted in much inferior results). Additionally, segmentation performance (as measured by IoU) also improved by about 1% on average for all curriculum learning method compared with the baseline.

Figure 2 shows segmentation results and uncertainty maps for several images from the iPS dataset. The first two columns show instances where both segmentation accuracy and reliability was improved significantly by curriculum learning in comparison with the baseline method, while the last column shows a case without much improvement.

Experimental Results for the Single-Stage Learning: Table 2 shows the results obtained by the second proposed method (shown in the last row of the table), compared with three different single-stage baseline methods (first 3 rows) using the same U-Net trained with cross-entropy (CE) loss without uncertainty modelling (row 1 in the table), Sensitivity-Specificity (SS) loss without uncertainty modelling (row 2), and using both cross-entropy and Dice loss without uncertainty modelling (row 3 in the table). The results indicate that the proposed method outperforms all three baseline methods both in terms of reliability and segmentation accuracy. However, regarding reliability performance, this method did not perform as well as the curriculum learning based method.

Figure 3 shows segmentation results and uncertainty maps for the same images shown in Fig. 2, this time comparing the proposed single-stage Method 2

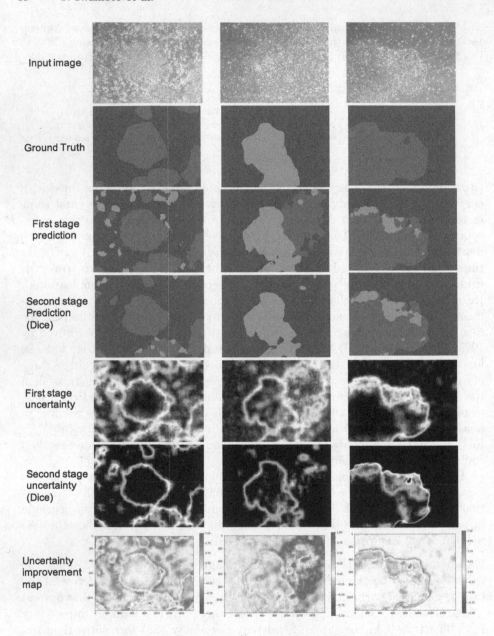

Fig. 2. Segmentation results and uncertainty maps for several images from the iPS dataset. In the segmentation results (2nd to 4th rows from the top) red corresponds to Good colonies, green to Bad colonies and blue to the culture medium. In the uncertainty maps (5th and 6th rows) high uncertainty level is represented by high intensity values. First two columns show instances where both segmentation accuracy and reliability was improved significantly by curriculum learning (4th and 6th rows) in comparison with the baseline method (3rd and 5th rows), while the last column shows a case without much improvement. The last row is a heat map where reduction in uncertainty between first and second stage of learning is shown in red and increase in blue. (Best viewed in color)

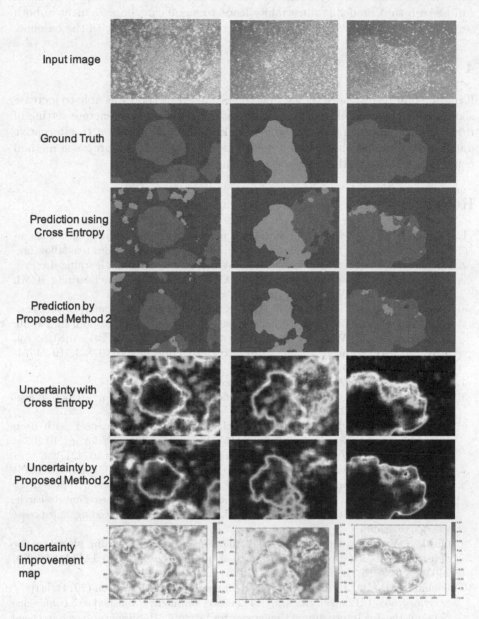

Fig. 3. Comparison of segmentation results and uncertainty maps for a baseline method using cross-entropy without uncertainty modelling (3rd row and 5th row) and proposed single-stage Method 2 (4th row and 6th row). The last row is a heat map where reduction in uncertainty between baseline method and Method 2 is shown in red and increase in blue. (Best viewed in color).

with the baseline using cross-entropy without uncertainty modeling. Here again can be seen that modeling uncertainty leads to significant improvement in both segmentation accuracy and decrease in uncertainty compared with the baseline.

4   Conclusion

Experimental results have shown that the proposed method was able to increase significantly the reliability of the segmentation model in the concrete setting of iPS cell colony segmentation. Further work includes application to alternative datasets and evaluation whether a hybrid model between both proposed method could lead to even further increase in reliability.

References

1. Abadi, M., et al.: TensorFlow: large-scale machine learning on heterogeneous systems (2015). https://www.tensorflow.org/. software available from tensorflow.org
2. Bengio, Y., Louradour, J., Collobert, R., Weston, J.: Curriculum learning. In: Proceedings of the 26th Annual International Conference on Machine Learning, ICML 2009, pp. 41–48. ACM, New York (2009)
3. Brosch, T., Yoo, Y., Tang, L.Y.W., Li, D.K.B., Traboulsee, A., Tam, R.: Deep convolutional encoder networks for multiple sclerosis lesion segmentation. In: Navab, N., Hornegger, J., Wells, W.M., Frangi, A.F. (eds.) MICCAI 2015. LNCS, vol. 9351, pp. 3–11. Springer, Cham (2015). https://doi.org/10.1007/978-3-319-24574-4_1
4. Chollet, F.: Keras (2015). https://github.com/fchollet/keras
5. DeVries, T., Taylor, G.W.: Leveraging uncertainty estimates for predicting segmentation quality (2018)
6. Esteva, A., et al.: Dermatologist-level classification of skin cancer with deep neural networks. Nature **542**(7639), 115–118 (2017). https://doi.org/10.1038/nature21056. https://app.dimensions.ai/details/publication/pub.1074217286
7. Gal, Y.: Uncertainty in deep learning. Ph.D. thesis, University of Cambridge (2016)
8. Gal, Y., Ghahramani, Z.: Dropout as a Bayesian approximation: representing model uncertainty in deep learning. In: Proceedings of Machine Learning Research, vol. 48, pp. 1050–1059. PMLR, New York (2016). http://proceedings.mlr.press/v48/gal16.html
9. Gulshan, V., et al.: Development and validation of a deep learning algorithm for detection of diabetic retinopathy in retinal fundus photographs. JAMA **316**(22), 2402–2410 (2016). https://doi.org/10.1001/jama.2016.17216
10. Kingma, D.P., Ba, J.: Adam: A method for stochastic optimization (2014). http://arxiv.org/abs/1412.6980. cite arxiv:1412.6980 Comment: Published as a conference paper at the 3rd International Conference for Learning Representations, San Diego (2015)
11. Milletari, F., Navab, N., Ahmadi, S.: V-net: fully convolutional neural networks for volumetric medical image segmentation. CoRR abs/1606.04797 (2016). http://arxiv.org/abs/1606.04797
12. Mobiny, A., Nguyen, H.V., Moulik, S., Garg, N., Wu, C.C.: Dropconnect is effective in modeling uncertainty of Bayesian deep networks. CoRR abs/1906.04569 (2019). http://arxiv.org/abs/1906.04569

13. Mobiny, A., Singh, A., Van Nguyen, H.: Risk-aware machine learning classifier for skin lesion diagnosis. J. Clin. Med. **8**(8), 1241 (2019)
14. Neal, R.M.: Bayesian Learning for Neural Networks. Springer, Heidelberg (1996). https://doi.org/10.1007/978-1-4612-0745-0
15. Ronneberger, O., Fischer, P., Brox, T.: U-Net: convolutional networks for biomedical image segmentation. In: Navab, N., Hornegger, J., Wells, W.M., Frangi, A.F. (eds.) MICCAI 2015. LNCS, vol. 9351, pp. 234–241. Springer, Cham (2015). https://doi.org/10.1007/978-3-319-24574-4_28 http://lmb.informatik.uni-freiburg.de/Publications/2015/RFB15a
16. Srivastava, N., Hinton, G., Krizhevsky, A., Sutskever, I., Salakhutdinov, R.: Dropout: a simple way to prevent neural networks from overfitting. J. Mach. Learn. Res. **15**(1), 1929–1958 (2014)

Unpaired MR Image Homogenisation by Disentangled Representations and Its Uncertainty

Hongwei Li[1,2], Sunita Gopal[2], Anjany Sekuboyina[1,2], Jianguo Zhang[3(✉)],
Chen Niu[4], Carolin Pirkl[2], Jan Kirschke[5], Benedikt Wiestler[5],
and Bjoern Menze[1,2]

[1] Department of Quantitative Biomedicine, University of Zurich, Zürich, Switzerland
[2] Department of Informatics, Technical University of Munich, Munich, Germany
{hongwei.li,bjoern.menze}@tum.de
[3] Department of Computer Science and Engineering, Southern University of Science
and Technology, Shenzhen, China
zhangjg@sustech.edu.cn
[4] Department of Medical Imaging, First Affiliated Hospital of Xi'an Jiaotong
University, Xi'an, China
[5] Klinikum rechts der Isar, Technical University of Munich, Munich, Germany

Abstract. Inter-scanner and inter-protocol differences in MRI datasets
are known to induce significant quantification variability. Hence data
homogenisation is crucial for a reliable combination of data or observations from different sources. Existing homogenisation methods rely on
pairs of images to learn a mapping from a source domain to a reference
domain. In real-world, we only have access to unpaired data from the
source and reference domains. In this paper, we successfully address this
scenario by proposing an unsupervised image-to-image translation framework which models the complex mapping by disentangling the image
space into a common content space and a scanner-specific one. We perform image quality enhancement among two MR scanners, enriching
the structural information and reducing noise level. We evaluate our
method on both healthy controls and multiple sclerosis (MS) cohorts
and have seen both visual and quantitative improvement over state-of-
the-art GAN-based methods while retaining regions of diagnostic importance such as lesions. In addition, for the first time, we quantify the
uncertainty in the unsupervised homogenisation pipeline to enhance the
interpretability. Codes are available: https://github.com/hongweilibran/
Multi-modal-medical-image-synthesis.

1 Introduction

Currently Magnetic Resonance Imaging (MRI) is ubiquitous in neuro-radiology
with the use of high-field scanners from different manufacturers. However, inter-
scanner and inter-protocol differences in MRI datasets are known to impede

---

B. Wiestler and B. Menze—Equal contributions to this work.

© Springer Nature Switzerland AG 2021
C. H. Sudre et al. (Eds.): UNSURE 2021/PIPPI 2021, LNCS 12959, pp. 44–53, 2021.
https://doi.org/10.1007/978-3-030-87735-4_5

successful generalisation of image algorithms across scanners and centers [4]. Scanner-related variations, as well as differences in acquisition protocol parameters introduce sources of variability when quantifying pathology in brains [10] and tissue measurements (e.g. diffusion MRI) [16]. In the current era of big data, reliable combination of data acquired on multiple sources (e.g. now and in the past) will increase the statistical power and sensitivity of multi-centre studies. Unfortunately, contrasts, resolutions, artefacts, and noise level of historical datasets differ from ones acquired in modern diagnostic protocols, limiting their use in machine learning algorithms for analysing information from state-of-the-art acquisition protocols. Data homogenisation enables the transfer of rich information content from state-of-the-art acquisitions to acquisitions with reduced quality of data. In the context of MR imaging, homogenisation aims to bring low-resolution images to a reference image quality which is achieved by high performance scanners. This typically involves two tasks: i) resolution alignment (i.e. super-resolution) and ii) image style transfer. However, in most of the cases, images from different scanners or centers are not paired; thus solving the two tasks above remains challenging. Furthermore, the reliability of the homogenisation process is critical in clinical use and is to be explored.

Related Work. In recent years, deep learning methods have been introduced as a mapping function from one image domain to another to solve various image quality transfer tasks [11,12,15] and achieve superior results over traditional methods [1,2,13] in super resolution tasks. Existing deep-learning based homogenisation methods rely on pairs of images with different qualities to train a network in a fully supervised manner. In order to obtain paired data, downsampling and data simulation [8] are popularly employed to generate synthetic low-quality images. However, those strategies often fails to introduce real-world characteristics from the source domain and therefore struggle to generalise when given real-world input. As a consequence, the models trained on simulated data become less effective when applied to practical scenarios. Especially in homogenisation tasks, the image contrast, noise and voxel size are difficult to be simulated using traditional strategies. Generative adversarial network (GANs) and their extensions [3,5,20] have been proposed to learn the data distribution from either paired or unpaired datasets scenarios. In particular, unpaired image-to-image translation has recently been explored in medical imaging [7,19]. Further disentangled representation as a learning strategy to model the factors of variation in data, is explored in natural image translation tasks [6,18] by learning a many-to-many mapping. It shows a promising venue to tackle homogenisation tasks in which voxel size, contrast and image noise should be considered together.

Contributions. Our goal is to homogenise unpaired scans from different MR scanners. 1) We identify that using disentangled representations for data homogenisation with GANs allow us to tackle complex underlying distributions among two domains. 2) We provide both qualitative and quantitative results from the homogenisation task, showing the superiority of our method over state-of-the-art unsupervised GAN-based methods. 3) We draw uncertainty measures from the translation that makes the proposed method reliable for clinical use.

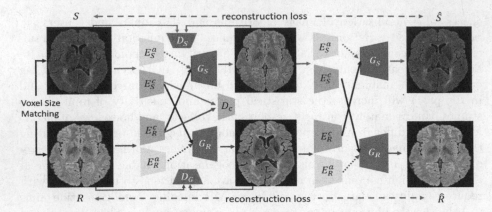

Fig. 1. Method overview. Images from individual scanners are encoded into a shared-content feature space that captures anatomical information and a scanner-specific feature space. To disentangle the features, the encoders and generator are trained in a cross-cycle reconstruction fashion. Multiple encoders and generators are optimised in an end-to-end fashion.

2 Methodology

The homogenisation process includes two tasks: i) unpaired super-resolution and ii) unpaired image style transfer. Without loss of generality, we simplify this task to tackle unpaired images from two MR scanners.

2.1 Problem Definition

The general goal is to map one image domain to a reference domain. Let $\mathcal{S}$ denote a source image space and $\mathcal{R}$ a referenced image space. Considering the complexity of real-world datasets, we assume a many-to-many mapping function exists from the source space to referenced space:: $\Phi : S \rightarrow R$. We designate an image from an old scanner to be from $\mathcal{S}$ and that from a new scanner to be from $\mathcal{R}$. Note that these images exhibit different visual appearances but with shared structure, e.g. the anatomical structure of brain. We hypothesise that both image domains share a content space which preserves anatomical structure and each individual domain contains an attribute space related to scanner characteristics. Inspired by disentangled representations [14], we decompose the image space into a shared content space $\mathcal{C}$ that is domain-invariant and a scanner-specific attribute space $\mathcal{A}$. Therefore, a shared domain-invariant space that preserves the anatomical information and a scanner-specific attribute space can be found to recover the underlying mapping between different scanners. Furthermore, based on the disentangling representation, for the first time, we explore the uncertainty quantification of the unsupervised domain translation by sampling the attribute space, which is difficult in previous pipelines such as CycleGAN and UNIT [18, 20].

2.2 Data Homogenisation Framework

Our data homogenisation framework via disentangled representations consists of two modules: Disentangled Representation Learning Module (DR-Module) and Uncertainty Quantification Module (IUQ-Module).

DR-Module. The first module embeds input images from all domains onto a shared content space $\mathcal{C}$, and scanner-specific attribute spaces, $\mathcal{S}_S$ and $\mathcal{S}_R$. As illustrated in Fig. 1, our framework consists of content encoders $\{E_S^c, E_R^c\}$, attribute encoders $\{E_S^a, E_R^a\}$, generators $\{G_S, G_R\}$, and domain discriminators $\{D_S, D_R\}$ for both domains, and a content discriminators $\{D_c\}$. A content discriminator D_c is trained to distinguish the extracted content representations between two domains. Similar to Lee's work [6], we train the encoder and generator for self reconstruction, where the reconstruction loss is minimised to encourage the encoders and generators to be invertible. The GAN component for cross-domain translation is trained to encourage the disentanglement of the latent space, decomposing it into content and attributes subspaces. Specifically, the generators are trying to fool the discriminators by successful cross-domain generation with swapped attribute space. The training loss are as follows:

Content Adversarial Loss: To obtain a shared content space $\mathcal{C}$, we enforce the content representation to be mapped onto the same space by introducing a content discriminator D_c which aims to distinguish the domain labels of the encoded content features.

$$\mathcal{L}_{adv}^c(E_S^c, E_R^c, D_c) = \mathbb{E}_{x_i \sim S}[log\ D_c(E_S^c(x))] + \mathbb{E}_{y_i \sim R}[log\ (1 - D_c(E_R^c(y)))] \quad (1)$$

Cross-Domain Reconstruction Loss: To enhance the disentangled representation learning process, we used a cross-cycle consistency proposed in [6].

$$\mathcal{L}_{c-rec}(G_S, G_R, E_S^c, E_S^a, E_R^c, E_R^a) = \mathbb{E}_{x,y}[||x - G_S(E_R^c(v), E_S^a(u))||_1] + \\ \mathbb{E}_{x_i \sim R}[x - ||G_R(E_R^c(x), E_R^a(x))||_1] \quad (2)$$

where $u = G_S(E_R^c(y), E_S^a(x))$, $v = G_R(E_S^c(y), E_S^a(x))$

Self-reconstruction Loss: We encourage a cycle-consistency to reconstruct the images from individual domains.

$$\mathcal{L}_{s-rec} = \mathbb{E}_{x \sim S}[||x - G_S(E_S^c(x), E_S^a(x))||_1] + \mathbb{E}_{y \sim R}[||y - G_R(E_R^c(y), E_R^a(y))||_1] \quad (3)$$

Domain Adversarial Loss: We impose domain adversarial loss $\mathcal{L}_{adv}^d$ where two discriminators D_S and D_R attempt to discriminate between real images and generated images in each domain, while S_S and G_R attempt to generate realistic images.

The full objective function of our framework is:

$$\mathcal{L}_{total} = \min_{G,E^C,E^a} \max_{D^S,D^G,D^C} \lambda_{adv}(\mathcal{L}^c_{adv} + \mathcal{L}^d_{adv}) + \lambda_{rec}(\mathcal{L}_{c-rec} + \mathcal{L}_{s-rec}) \quad (4)$$

where λ_{adv} and λ_{rec} control the importance of the adversarial training and reconstruction.

IUQ-Module. Based on the disentangled content and attributes, during the inference stage, a source image is mapped to shared content space by using encoder E_S and one image from the reference domain is used to compute the attribute vector by encoder E_R. Finally we use G_R to generate the image by using the above shared space feature and scanner-specific attribute vector. Attribute vector is image-specific, i.e., different images in the reference space result in slightly different attribute representations. This induces a distribution over the translation of one content representation from the source space. Therefore, we randomly sample the attribute space from $\mathcal{R}$ and generate various images per one image from $\mathcal{S}$, and compute the mean as the final translation and the variance as the translation uncertainty.

Implementations. The voxel size matching between two scanners is performed by bicubic interpolation. Common-space encoders consist of convolutional layers and residual layers followed by batch normalisation, while attribute encoders consist of convolutional layers, a global average pooling layer, and a fully-connected layer. Generators take the attribute vector (of length 8) and common-space feature (feature map of $256 \times 68 \times 68$) as inputs. For training, we use the Adam optimizer with a batch size of 5 and a learning rate of 0.0001. In all experiments, we set the hyper-parameters as follows: $\lambda_{adv} = 1$, $\lambda_{rec} = 10$. We train and save the best-performing model based on the total loss. Discriminators are convolutional neural networks for binary classification. Experiments are run on two Nvidia Titan Xp GPUs. The training time is eight hours for the DR-Module. The inference stage takes 5 mins for 200 samples per source image.

2.3   Visual Rating and Evaluation Protocol

Quantitative analysis of images generated from unsupervised image to image translation cannot be done metrics such as PSNR and MAE [17] due to lack of paired images. Additionally in MR images, these algorithms will not be able to tell us useful information such as the preservation of clinically relevant structures. In order to get an qualitative assessment of the images generated from our approach, we design a visual rating system which compares our images with those generated with state of the art methods such as CycleGAN and UNIT. In each trial, neuro-radiologists were presented with three sets of images, where each set consists of an original FLAIR image from an old scanner on the left and the synthetic image generated by CycleGAN, UNIT or our approach. The rater does not know which method is used and the order of the sets is shuffled to prevent any bias. The rater is asked to rate on the following criterion on both healthy subjects and MS patients: a) good resolution and low noise level for healthy subjects, b) good pathological information for MS patients, and c) good structural information for both groups. The experts were asked to rate on a five star scale where 'one' represents 'poor' and 'five' represents 'excellent'. Five neuroradiologists with a mean of 7+ years of experience rated the image quality.

Table 1. Data characteristics of the MRI datasets including two scanners and two groups of subjects.

Scanner Name	Voxel Size (m3)	Volume Size	Healthy Controls	MS Patients
3T Achiva	0.98 × 0.98 × 1.00	132 × 256 × 83	40	30
3T Ingenia	0.71 × 0.71 × 0.71	132 × 256 × 83	40	40

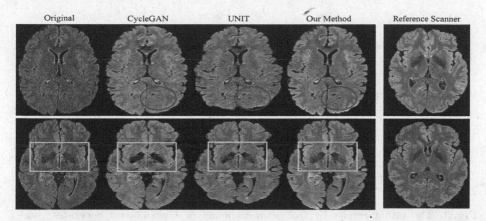

Original   CycleGAN   UNIT   Our Method   Reference Scanner

Fig. 2. Qualitative results on healthy-control group. Comparing with other unsupervised methods, ours visually enhances the brain structure and reduces noise. Note that the reference scan only indicates the image style and not its content.

3   Experiments

Datasets and Evaluation Metric. We evaluate the proposed method on two MR scanners as shown in Table 1. Specifically we homogenise the scans from the 'old' scanner Philips Achieva to the new scanner Philips Ingenia. All the scans are acquired with a multiple sclerosis (MS) patients study protocol and separated into healthy controls group and MS patients group. For pre-processing, scans are skull-stripped and subjected to bias-field correction. For each group, 80% scans are used for training, 20% for testing. The axial slices from Philips Achieva are upsampled using bicubic interpolation consider the voxel size difference and cropped to 270 × 270.

Intuitively, after data homogenisation, the brain structure from two domains are expected to be similar. To quantitatively compare our method with state-of-the-art on unpaired data, we introduce Maximum Structure Similarity (m-SSIM) as an extension of the traditional structure similarity (SSIM) metric to handle unpaired similarity measurement. Given two sets of unpaired data $\mathcal{X}$ and $\mathcal{Y}$, we take the maximum SSIM for all possible combinations of the elements:

$$m-SSIM = max\{SSIM(x,y)|x \in \mathcal{X}, y \in \mathcal{Y}\} \tag{5}$$

Table 2. Comparison with state-of-the-art unsupervised methods using *m-SSIM*.

Methods	UNIT [9]	CycleGAN [20]	Ours
Healthy control	0.761	0.784	**0.854**
MS patients	0.755	0.768	**0.823**

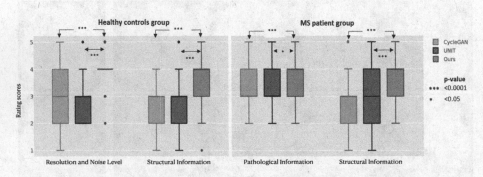

Fig. 3. The rating results by five neuro-radiologists. Our method generates good-quality images and significantly outperforms other unsupervised methods.

Qualitative and Quantitative Results. We firstly evaluate our method on healthy-control group and compare with state-of-the-art methods CycleGAN and UNIT. As shown in Fig. 2, visually our method enhances the brain structure and reduce image noise comparing with other unsupervised approaches. We also observe that CycleGAN and UNIT struggle to remove the image noise and UNIT changes the brain structure. Table 2 shows the quantitative comparison of three methods using m-SSIM. For each homogenised slice, we calculate the m-SSIM by comparing it to the whole test set. The results of all testing slices are averaged. The proposed method outperforms CycleGAN and UNIT by a large margin.

Visual Evaluations by Neuroradiologists. The results on test set for the different metrics (i.e. resolution, noise level and structure preservation) for both healthy subjects and MS patients are presented in the boxplots in Fig. 3. For healthy controls group, we observed that CycleGAN produced a wider range in image quality compared with UNIT. We conduct Wilcoxon rank-sum tests on the paired rating scores of our method and other methods from 5 raters on 15

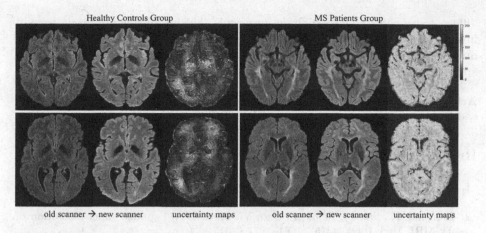

Fig. 4. Results on uncertainty quantification by sampling the attribute spaces. Headmaps represent the normalised standard deviation of the 200 predictions of each pixel. We found that in overall the healthy controls group show less uncertainty comparing to MS patients group and uncertain regions (e.g. boundary) are non-crucial for healthy controls.

observations (scores of 5 raters are averaged). Results show that the pair of rating scores on our approach with other two methods are all significantly different on both of the two metrics. This demonstrates that our method significantly outperforms CycleGAN and UNIT on healthy controls and MS patient group.

Uncertainty Quantification Results. By randomly sampling N images from the reference space, we obtain a distribution of the attribute space. In practice, we set $N = 200$ considering the computation complexity. The heatmaps in Fig. 4 depict the uncertainty obtained for both healthy-control and MS patients groups. Overall, we observed less uncertainty in the healthy-control group compared to the MS patients group. On the healthy-control group, we observed that the uncertain regions are non-crucial, e.g. the boundary. On the MS patients group, we found that the pathological regions (MS lesions) present higher uncertainty than other structures. This is in accordance with our expectations, as MS lesions represent outliers from the intensity profiles of the learned spatial anatomical representations and show a larger variability than anatomical variants. In a clinical setting, the uncertainty map informs the doctor accordingly, thus making our translation interpretable.

4 Conclusion and Discussion

This work presents a novel GAN-based approach for data homogenisation on FLAIR MRI sequence, with multi-rater experiments, and statistical tests. For the first time, we quantify the uncertainty in the unsupervised domain translation task. This approach potentially allows us to homogenise datasets from different scanners and with different protocols. As a promising approach, we will investigate its effect on improving the statistical power and sensitivity studies in multi-center research. The uncertainty maps allows us to interpret the outcome and helps in identifying outliers scans and MS lesions in scans.

Acknowledgement. This work was supported by Helmut Horten Foundation. B. W. and B. M. were supported through the DFG, SFB-824, sub-project B12.

References

1. Alexander, D.C., et al.: Image quality transfer and applications in diffusion MRI. Neuroimage **152**, 283–298 (2017)
2. Bahrami, K., Shi, F., Rekik, I., Gao, Y., Shen, D.: 7T-guided super-resolution of 3T MRI. Med. Phys. **44**(5), 1661–1677 (2017)
3. Choi, Y., et al.: StarGAN: unified generative adversarial networks for multi-domain image-to-image translation. In: CVPR, pp. 8789–8797 (2018)
4. Glocker, B., Robinson, R., Castro, D.C., Dou, Q., Konukoglu, E.: Machine learning with multi-site imaging data: an empirical study on the impact of scanner effects. arXiv preprint arXiv:1910.04597 (2019)
5. Goodfellow, I.: Nips 2016 tutorial: generative adversarial networks. arXiv preprint arXiv:1701.00160 (2016)
6. Lee, H.-Y., Tseng, H.-Y., Huang, J.-B., Singh, M., Yang, M.-H.: Diverse image-to-image translation via disentangled representations. In: Ferrari, V., Hebert, M., Sminchisescu, C., Weiss, Y. (eds.) ECCV 2018. LNCS, vol. 11205, pp. 36–52. Springer, Cham (2018). https://doi.org/10.1007/978-3-030-01246-5_3
7. Li, H., et al.: DiamondGAN: unified multi-modal generative adversarial networks for MRI sequences synthesis. In: Shen, D., et al. (eds.) MICCAI 2019. LNCS, vol. 11767, pp. 795–803. Springer, Cham (2019). https://doi.org/10.1007/978-3-030-32251-9_87
8. Lin, H., et al.: Deep learning for low-field to high-field MR: image quality transfer with probabilistic decimation simulator. In: Knoll, F., Maier, A., Rueckert, D., Ye, J.C. (eds.) MLMIR 2019. LNCS, vol. 11905, pp. 58–70. Springer, Cham (2019). https://doi.org/10.1007/978-3-030-33843-5_6
9. Liu, M.Y., Breuel, T., Kautz, J.: Unsupervised image-to-image translation networks. In: Advances in Neural Information Processing Systems, pp. 700–708 (2017)
10. Lysandropoulos, A.P., et al.: Quantifying brain volumes for multiple sclerosis patients follow-up in clinical practice-comparison of 1.5 and 3 tesla magnetic resonance imaging. Brain Behav. **6**(2), e00422 (2016)
11. Oktay, O., et al.: Multi-input cardiac image super-resolution using convolutional neural networks. In: Ourselin, S., Joskowicz, L., Sabuncu, M.R., Unal, G., Wells, W. (eds.) MICCAI 2016. LNCS, vol. 9902, pp. 246–254. Springer, Cham (2016). https://doi.org/10.1007/978-3-319-46726-9_29
12. Park, J., Hwang, D., Kim, K.Y., Kang, S.K., Kim, Y.K., Lee, J.S.: Computed tomography super-resolution using deep convolutional neural network. Phys. Med. Biol. **63**(14), 145011 (2018)
13. Shi, W., et al.: Cardiac image super-resolution with global correspondence using multi-atlas PatchMatch. In: Mori, K., Sakuma, I., Sato, Y., Barillot, C., Navab, N. (eds.) MICCAI 2013. LNCS, vol. 8151, pp. 9–16. Springer, Heidelberg (2013). https://doi.org/10.1007/978-3-642-40760-4_2

14. Siddharth, N., et al.: Learning disentangled representations with semi-supervised deep generative models. In: Advances in Neural Information Processing Systems, pp. 5925–5935 (2017)

15. Tanno, R., et al.: Bayesian image quality transfer with CNNs: exploring uncertainty in dMRI super-resolution. In: Descoteaux, M., Maier-Hein, L., Franz, A., Jannin, P., Collins, D.L., Duchesne, S. (eds.) MICCAI 2017. LNCS, vol. 10433, pp. 611–619. Springer, Cham (2017). https://doi.org/10.1007/978-3-319-66182-7_70

16. Tax, C.M., et al.: Cross-scanner and cross-protocol diffusion MRI data harmonisation: a benchmark database and evaluation of algorithms. Neuroimage **195**, 285–299 (2019)

17. Welander, P., Karlsson, S., Eklund, A.: Generative adversarial networks for image-to-image translation on multi-contrast MR images-a comparison of CycleGAN and unit. arXiv preprint arXiv:1806.07777 (2018)

18. Yang, J., Dvornek, N.C., Zhang, F., Chapiro, J., Lin, M.D., Duncan, J.S.: Unsupervised domain adaptation via disentangled representations: application to cross-modality liver segmentation. In: Shen, D., et al. (eds.) MICCAI 2019. LNCS, vol. 11765, pp. 255–263. Springer, Cham (2019). https://doi.org/10.1007/978-3-030-32245-8_29

19. Yurt, M., Dar, S.U.H., Erdem, A., Erdem, E., Çukur, T.: mustGAN: multistream generative adversarial networks for MR image synthesis. arXiv preprint arXiv:1909.11504 (2019)

20. Zhu, J.Y., et al.: Unpaired image-to-image translation using cycle-consistent adversarial networks. In: CVPR, pp. 2223–2232 (2017)

Uncertainty-Aware Deep Learning Based Deformable Registration

Irina Grigorescu[1,2(✉)], Alena Uus[1,2], Daan Christiaens[1,3],
Lucilio Cordero-Grande[1,2,5], Jana Hutter[1], Dafnis Batalle[1,4],
A. David Edwards[1], Joseph V. Hajnal[1,2], Marc Modat[2], and Maria Deprez[1,2]

[1] Centre for the Developing Brain, School of Biomedical Engineering and Imaging
Sciences, King's College London, London, UK
irina.grigorescu@kcl.ac.uk
[2] Biomedical Engineering Department, School of Biomedical Engineering
and Imaging Sciences, King's College London, London, UK
[3] Departments of Electrical Engineering, ESAT/PSI, KU Leuven, Leuven, Belgium
[4] Department of Forensic and Neurodevelopmental Science, Institute of Psychiatry,
Psychology and Neuroscience, King's College London, London, UK
[5] Biomedical Image Technologies, ETSI Telecomunicación, Universidad Politécnica
de Madrid and CIBER-BNN, Madrid, Spain

Abstract. We introduce an uncertainty-aware deep learning deformable
image registration solution for magnetic resonance imaging multi-channel
data. In our proposed framework, the contributions of structural and
microstructural data to the displacement field are weighted with spatially varying certainty maps. We produce certainty maps by employing
a conditional variational autoencoder image registration network, which
enables us to generate uncertainty maps in the deformation field itself.
Our approach is quantitatively evaluated on pairwise registrations of 36
neonates to a standard structural and/or microstructural template, and
compared with models trained on either single modality, or both modalities together. Our results show that by incorporating uncertainty while
fusing the two modalities, we achieve superior alignment in cortical gray
matter and white matter regions, while also achieving a good alignment
of the white matter tracts. In addition, for each of our trained models, we
show examples of average uncertainty maps calculated for 10 neonates
scanned at 40 weeks post-menstrual age.

Keywords: Multi-channel registration · Uncertainty · Certainty maps

1 Introduction

Tracking changes in the developing brain depends on precise inter-subject image
registration. However, most applications in this field rely on a single modality [3,
15], such as structural or diffusion data, to learn spatial correspondences between
images, without taking into account the complementary information provided by
using both. In general, T_2-weighted (T_2w) magnetic resonance imaging (MRI)

© Springer Nature Switzerland AG 2021
C. H. Sudre et al. (Eds.): UNSURE 2021/PIPPI 2021, LNCS 12959, pp. 54–63, 2021.
https://doi.org/10.1007/978-3-030-87735-4_6

scans have high contrast between different brain tissues and can delineate the cortical gray matter (cGM) region well, while diffusion weighted imaging (DWI) data primarily provides information on white matter (WM) structures.

Multi-channel registration which includes both structural and diffusion data has been shown to improve alignment [1, 7] of images. However, one of the main challenges of this approach is the low contrast or homogeneous intensity which characterises different anatomical regions on both structural (*e.g.,* deep gray matter) and microstructural MRI (*e.g.,* cortex). Classic approaches for fusing these channels are based on simple averaging [1], or on calculating certainty maps based on normalised gradients correlated to structural content [7].

In order to establish accurate correspondences between MR images acquired during the neonatal period, we propose an uncertainty-aware deep learning image registration framework that allows local certainty-based fusion of T_2w neonatal scans with DWI-derived fractional anisotropy (FA) maps. More specifically, we employ a conditional variational autoencoder (CVAE) image registration network [11] and use it to calculate uncertainty maps in the generated dense displacement fields. We predict the displacement fields and their uncertainty for each individual modality, empirically calculate certainty maps and use them in a combined model which we call T_2w+FA+uncert.

Throughout this work we use 2-D MRI mid-brain axial slices acquired as part of the developing Human Connectome Project[1] (dHCP), and T_2w and FA 36 weeks gestational age templates [17] for the fixed slices. We showcase the capabilities of our proposed framework on images of infants born and scanned at different gestational ages, and we compare the results against three different models, trained on T_2w-only, on FA-only and on T_2w+FA scans. Our results show that in terms of Dice scores and average Hausdorff distances, our proposed model performs better in WM and cGM regions when compared to the T_2w+FA model, and better than the T_2w-only model in terms of aligning white matter structures (*i.e.,* internal capsule).

2   Method

Data Acquisition. The structural (T_2w) imaging data used in this study was acquired with a Philips Achieva 3T scanner and a 32-channels neonatal head coil [8], using a turbo spin echo (TSE) sequence ($T_R = 12$ s, echo time $T_E = 156$ ms, and SENSE factors of 2.11 for the axial plane and 2.58 for the sagittal plane). Images were acquired with an in-plane resolution of 0.8×0.8 mm, slice thickness of 1.6 mm and overlap of 0.8 mm. All data was motion corrected [5] and super-resolution reconstructed to a 0.5 mm isotropic resolution [12]. The DWI scans were acquired using a monopolar spin echo echo-planar imaging (SE-EPI) Stejksal-Tanner sequence [9]. A multiband factor of 4 and a total of 64 interleaved overlapping slices (1.5 mm in-plane resolution, 3 mm thickness, 1.5 mm overlap) were used to acquire a single volume, with parameters $T_R = 3800$ ms, $T_E = 90$ ms. This data underwent denoising, outlier removal, motion correction, and it

---

[1] http://www.developingconnectome.org/.

was subsequently super-resolved to a 1.5 mm isotropic voxel resolution [4]. All resulting images were checked for abnormalities by a paediatric neuroradiologist, and infants with major congenital malformations were excluded.

Image Selection. For this study, we use a total of 363 T_2w and FA maps of neonates born between 23–42 weeks gestational age (GA) and scanned at term-equivalent age (37–45 weeks GA). The age distribution in our dataset is found in Fig. 1.

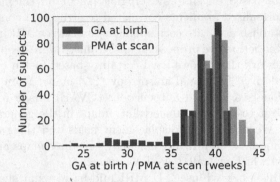

Fig. 1. Distribution of gestational ages at birth (GA) and post-menstrual ages at scan (PMA) in our imaging dataset.

Image Preprocessing. In order to use both the T_2w and FA axial slices in our registration networks, we first resampled both types of modalities into a template space of 1 mm isotropic resolution. We then affinely pre-registered both to a common 36 weeks gestational age atlas space [17] using the MIRTK software toolbox [14] and obtained the FA maps using the MRtrix3 toolbox [16]. Finally, we performed skull-stripping using the available dHCP brain masks [4], and we cropped the resulting images to a 128×128 size. Out of the 363 subjects in our dataset, we used 290 for training, 37 for validation and 36 for test, divided such that the GA at birth and post-menstrual age (PMA) at scan were kept as similar to the original distributions as possible. The validation set was used to inform us about our models' performance during training. All of our results are reported on the test set.

Network Architecture. In this study, we employ a CVAE [10] to model the registration probabilistically as proposed by [11]. Figure 2 shows the architecture at both train and inference time. In short, a pair of 2D MRI axial slices (**M** and **F**) are passed through the network to learn a velocity field v, while the *exponentiation layers* (with 4 *scaling-and-squaring* [2] steps) transform it into a topology-preserving deformation field ϕ. A *Spatial Transformer* layer [6] is then used to warp (linearly resample) the moving images **M** and obtain the moved image $\mathbf{M}(\phi)$.

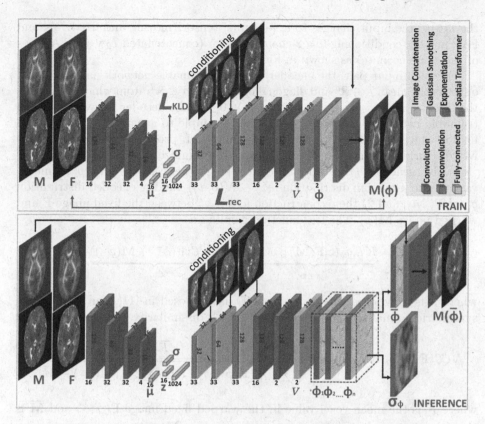

Fig. 2. We use a convolutional neural network based on the architecture proposed by [11]. During inference, we use the trained network to generate n dense displacement fields ϕ_i and create a mean displacement field $\overline{\phi}$, and its associated uncertainty map σ_ϕ.

The encoder branch is made up of four 2D convolutional layers of 16, 32, 32, and 4 filters, respectively, with a kernel size of 3^2, followed by *Leaky ReLU* ($\alpha = 0.2$) activations [18]. The bottleneck (μ, σ, z) is fully-connected, and we kept the latent code size (16) the same as in the original paper [11]. The decoder branch is composed of three 2D deconvolutional layers of 32 filters and a kernel size of 3^2 each, followed by *Leaky ReLU* ($\alpha = 0.2$) activations. The deconvolutional layers' feature maps are concatenated with the original-sized or downsampled versions of the moving input image. Two more convolutional layers (with 16 and 2 filters, respectively) are added, followed by a Gaussian smoothing layer (kernel size 21) which outputs the velocity field v.

Training. For this study, we train three separate models on different combinations of input data. The first model is trained on pairs of structural data (T_2w-only), the second model on microstructural data (FA-only), while the third model uses both modalities as input to the network (T_2w+FA). While training the latter, the input changes to a 4-channel tensor (moving and fixed T_2w and FA), and the conditioning to a 2-channel tensor (concatenated T_2w and FA slices of the same neonate) as shown in Fig. 2.

For each input pair, the encoder q_ω (with trainable network parameters ω) outputs the mean $\mu \in \mathbb{R}^d$ and diagonal covariance $\sigma \in \mathbb{R}^d$, from which we sample the latent vector $z = \mu + \epsilon \cdot \sigma$, with $\epsilon \sim \mathcal{N}(0, I)$. The decoder network p_γ (with trainable network parameters γ) uses the z-sample to generate a displacement field ϕ which, together with the moving image $\mathbf{M}$, produces the warped image $\mathbf{M}(\phi)$. During training, the optimizer aims to minimize: 1) the Kullback-Leibler (KL) divergence $\mathcal{L}_{KLD}$ in order to reduce the gap between the prior $p(z)$ (multivariate unit Gaussian distribution $p(z) \sim \mathcal{N}(0, I)$) and the encoded distribution $q_\omega(z | \mathbf{F}, \mathbf{M})$, and 2) the reconstruction loss $\mathcal{L}_{rec}$ between the fixed image $\mathbf{F}$ and warped image $\mathbf{M}(\phi)$. The loss function results in:

$$\mathcal{L} = \underbrace{KL[q_\omega(z|\mathbf{F},\mathbf{M}) \,||\, p(z)]}_{\mathcal{L}_{KLD}} + \lambda \underbrace{\mathcal{NCC}(\mathbf{F}(\phi^{-\frac{1}{2}}), \mathbf{M}(\phi^{\frac{1}{2}}))}_{\mathcal{L}_{rec}} \tag{1}$$

where λ is a hyperparameter set to 5000 as proposed in [11], and $\mathcal{NCC}$ is the symmetric normalised cross correlation (NCC) dissimilarity measure defined as:

$$\mathcal{NCC}(\mathbf{F}(\phi^{-\frac{1}{2}}), \mathbf{M}(\phi^{\frac{1}{2}})) = -\frac{\sum_{\mathbf{x}\in\Omega}(\mathbf{F}(\phi^{-\frac{1}{2}}) - \overline{F}) \cdot (\mathbf{M}(\phi^{\frac{1}{2}}) - \overline{M})}{\sqrt{\sum_{\mathbf{x}\in\Omega}(\mathbf{F}(\phi^{-\frac{1}{2}}) - \overline{F})^2 \cdot \sum_{\mathbf{x}\in\Omega}(\mathbf{M}(\phi^{\frac{1}{2}}) - \overline{M})^2}}$$

where $\overline{F}$ is the mean voxel value in the warped fixed image $\mathbf{F}(\phi^{-\frac{1}{2}})$ and $\overline{M}$ is the mean voxel value in the warped moving image $\mathbf{M}(\phi^{\frac{1}{2}})$.

We train our models for 2500 epochs each, using the Adam optimizer with its default parameters ($\beta_1 = .9$ and $\beta_2 = .999$), a constant learning rate of $5 \cdot 10^{-4}$, and a L_2 weight decay factor of 10^{-5}. All networks were implemented in PyTorch.

Uncertainty-Aware Image Registration. To investigate uncertainty-aware image registration, we use our trained models to generate uncertainty maps. We achieve this at inference time by using the trained decoders to generate multiple displacement fields ϕ_i, as shown in Figs. 2 and 3. More specifically, for each subject in our test dataset, we first use the trained encoders to yield the μ and σ outputs. Then, we generate n latent vector $z = \mu + \epsilon \cdot \sigma$ samples and pass them through the trained decoder networks to generate n dense displacement fields ϕ_i. Throughout this work we set $n = 50$. From these, we obtain a mean displacement field $\overline{\phi}$ and a standard deviation displacement field σ_ϕ for each model.

For the uncertainty-aware image registration task, we combine the T_2w-only and the FA-only models into a single model, which we call T_2w+FA+uncert. This is achieved in a three-step process. First, we use the trained T_2w-only and FA-only models to generate n dense displacement fields ϕ_i, and create the modality-specific mean displacement fields $\overline{\phi}_{T2w}$ and $\overline{\phi}_{FA}$, and uncertainty maps $\sigma_{\phi T2w}$ and $\sigma_{\phi FA}$. Second, we calculate certainty maps $(\alpha_{\phi T2w}, \alpha_{\phi FA})$ using the following equations:

$$\alpha_{\phi_{T2w}} = \frac{1/\sigma_{\phi_{T2w}}}{1/\sigma_{\phi_{T2w}} + 1/\sigma_{\phi_{FA}}} \quad ; \quad \alpha_{\phi_{FA}} = \frac{1/\sigma_{\phi_{FA}}}{1/\sigma_{\phi_{T2w}} + 1/\sigma_{\phi_{FA}}} \quad (2)$$

Finally, the T_2w+FA+uncert model's displacement field is constructed by locally weighting the contributions from each modality with the 2D certainty maps: $\phi_{T2w+FA+uncert} = \alpha_{\phi_{T2w}} \odot \overline{\phi}_{T2w} + \alpha_{\phi_{FA}} \odot \overline{\phi}_{FA}$, where $\odot$ represents element-wise multiplication. A visual explanation of these steps can be seen in Fig. 3.

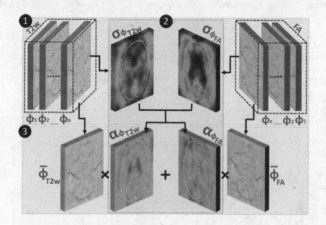

Fig. 3. The construction of certainty maps: 1) Create modality-specific mean displacement fields $\overline{\phi}_{T2w}$ and $\overline{\phi}_{FA}$, and uncertainty maps $\sigma_{\phi_{T2w}}$ and $\sigma_{\phi_{FA}}$. 2) Create modality specific certainty maps $\alpha_{\phi_{T2w}}$ and $\alpha_{\phi_{FA}}$ using Eq. 2. 3) Create the $\phi_{T2w+FA+uncert}$ displacement field by locally weighting the contributions from each modality with the 2D certainty maps.

3   Results

To validate whether our proposed model (T_2w+FA+uncert) can outperform the other three models (T_2w-only, FA-only, T_2w+FA), we carry out a quantitative evaluation on our test dataset. Each subject and template had the following tissue label segmentations obtained using the Draw-EM pipeline [13]: cerebrospinal fluid (CSF), cGM, WM, ventricles, deep gray matter (dGM), and a WM structure called the internal capsule (IC). These labels were propagated from each subject into the template space using the predicted deformation fields. The resulting Dice scores and average Hausdorff distances calculated between the warped labels and the template labels are summarised in Fig. 4, where the initial pre-alignment is shown in pink, the T_2w-only results are shown in green, the FA-only results in light blue, the T_2w+FA model in magenta, and our proposed model's results in purple (T_2w+FA+uncert). The yellow diamond points to the best performing model for each tissue type and metric.

In terms of Dice scores, our proposed model performs similarly well to the T_2w-only model in the cGM, WM, ventricles and the dGM structures, and outperforms the T_2w+FA model in the cGM, and WM labels. At the same time, the T_2w+FA+uncert obtains the lowest average Hausdorff distances in the cGM and WM regions, again outperforming the T_2w+FA model. The IC is best aligned using the FA-only model, where the T_2w-only performs worse, while the models which incorporate FA maps perform comparably well.

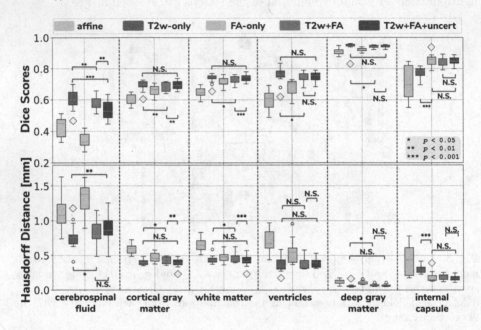

Fig. 4. The results on our test dataset for all four methods, together with the initial affine alignment. The yellow diamond highlights the model which performed best for its respective tissue type and metric. (***N.S.*** *means "not-significant"*) (Color figure online)

Next, we used data from 10 neonatal subjects scanned around 40 weeks PMA to produce average uncertainty maps for our three trained models. For each subject $j \in [1, 10]$ and model $m \in \{T_2$w-only, FA-only, T_2w+FA$\}$, we obtained an uncertainty map $\sigma^j_{\phi_m}$, and averaged them across the subjects. Figure 5 shows these average uncertainty maps (in the template space), overlaid on top of the fixed images which were used for training. By combining the uncertainty maps from the trained T_2w-only and FA-only models, we can obtain modality-dependent certainty maps (see Eq. 2) which are shown on the last row.

The T_2w-only model (first row in Fig. 5) yields high uncertainty in dGM regions (cyan arrow) where there is little contrast, as well as in difficult brain areas, such as the cGM folds. Similarly, the FA-only model (first row in Fig. 5) shows high uncertainty in low contrast cortical areas (yellow arrow) and low uncertainty in the high contrast WM structures such as the IC region (cyan

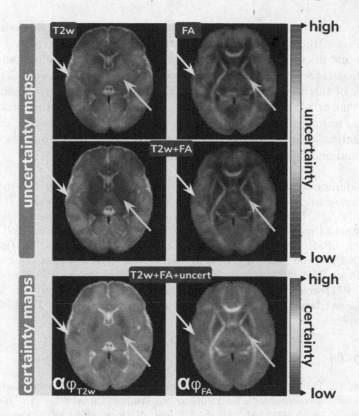

Fig. 5. Average uncertainty maps for the T_2w- and FA-only (on the first row) and the T_2w+FA models (on the second row). The cyan arrows point to the same region of dGM, while the yellow arrows point to the same region of cGM for all models and modalities, respectively. The last row shows the modality-dependent certainty maps obtained from the T_2w-only and FA-only models. (Color figure online)

arrow). When using both modalities (second row in Fig. 5), the uncertainty becomes low in the dGM regions, as the model is being helped by the extra FA channel, but becomes higher in the cGM regions (yellow arrows) when compared to the T_2w-only model. This could be caused by the high uncertainty in the low contrast cortical regions of the FA channel. The average certainty maps are shown on the last row, where we can see that in our proposed model the combined displacement field will depend more on the FA-only model in the WM tracts regions as seen in the $\alpha_{\phi_{FA}}$ certainty map (cyan arrow), and on the T_2w-only model for the cortical regions (yellow arrows).

4   Discussion and Future Work

This paper presents a novel solution for multi-channel registration, which combines FA and T_2w data driven displacement fields based on their respective uncertainty maps. The quantitative evaluation performed on 36 neonatal subjects from the dHCP project showed that the proposed certainty based fusion of structural and microstructural channels improves overall alignment when compared to models trained on either single-channel or multi-channel data. The main limitations of this work are: the use of a single latent code size and smoothing kernel, the use of 2-D axial slices only, and no comparison with other probabilistic registration frameworks (*e.g.,* [6]). Future work will focus on further improving the registration accuracy in the cortical regions, on adapting our work to 3-D datasets, and on exploring the aforementioned limitations.

Acknowledgments. This work was supported by the Academy of Medical Sciences Springboard Award [SBF004\1040], Medical Research Council (Grant no. [MR/K006355/1]), European Research Council under the European Union's Seventh Framework Programme [FP7/20072013]/ERC grant agreement no. 319456 dHCP project, the EPSRC Research Council as part of the EPSRC DTP (grant Ref: [EP/R513064/1]), the Wellcome/EPSRC Centre for Medical Engineering at King's College London [WT 203148/Z/16/Z], the NIHR Clinical Research Facility (CRF) at Guy's and St Thomas', and by the National Institute for Health Research Biomedical Research Centre based at Guy's and St Thomas' NHS Foundation Trust and King's College London.

References

1. Avants, B., Duda, J.T., Zhang, H., Gee, J.C.: Multivariate normalization with symmetric diffeomorphisms for multivariate studies. In: Ayache, N., Ourselin, S., Maeder, A. (eds.) MICCAI 2007. LNCS, vol. 4791, pp. 359–366. Springer, Heidelberg (2007). https://doi.org/10.1007/978-3-540-75757-3_44
2. Balakrishnan, G., Zhao, A., Sabuncu, M.R., Guttag, J., Dalca, A.V.: VoxelMorph: a learning framework for deformable medical image registration. IEEE Trans. Med. Imaging **38**(8), 1788–1800 (2019)
3. Barnett, M.L., et al.: Exploring the multiple-hit hypothesis of preterm white matter damage using diffusion MRI. NeuroImage: Clin. **17**, 596–606 (2018)
4. Christiaens, D., et al.: Scattered slice SHARD reconstruction for motion correction in multi-shell diffusion MRI. NeuroImage **225**, 117437 (2021). https://doi.org/10.1016/j.neuroimage.2020.117437
5. Cordero-Grande, L., Hughes, E.J., Hutter, J., Price, A.N., Hajnal, J.V.: Three-dimensional motion corrected sensitivity encoding reconstruction for multi-shot multi-slice MRI: Application to neonatal brain imaging. Magn. Reson. Med. **79**(3), 1365–1376 (2018)
6. Dalca, A.V., Balakrishnan, G., Guttag, J., Sabuncu, M.R.: Unsupervised learning for fast probabilistic diffeomorphic registration. In: Frangi, A.F., Schnabel, J.A., Davatzikos, C., Alberola-López, C., Fichtinger, G. (eds.) MICCAI 2018. LNCS, vol. 11070, pp. 729–738. Springer, Cham (2018). https://doi.org/10.1007/978-3-030-00928-1_82
7. Forsberg, D., Rathi, Y., Bouix, S., Wassermann, D., Knutsson, H., Westin, C.-F.: Improving registration using multi-channel diffeomorphic demons combined with certainty maps. In: Liu, T., Shen, D., Ibanez, L., Tao, X. (eds.) MBIA 2011. LNCS, vol. 7012, pp. 19–26. Springer, Heidelberg (2011). https://doi.org/10.1007/978-3-642-24446-9_3

8. Hughes, E.J., et al.: A dedicated neonatal brain imaging system. Magn. Reson. Med. **78**(2), 794–804 (2017)
9. Hutter, J., et al.: Time-efficient and flexible design of optimized multishell HARDI diffusion. Magn. Reson. Med. **79**(3), 1276–1292 (2018)
10. Kingma, D.P., Rezende, D.J., Mohamed, S., Welling, M.: Semi-supervised learning with deep generative models (2014). https://arxiv.org/abs/1406.5298
11. Krebs, J., Mansi, T., Mailhé, B., Ayache, N., Delingette, H.: Unsupervised probabilistic deformation modeling for robust diffeomorphic registration. In: Stoyanov, D., et al. (eds.) DLMIA/ML-CDS -2018. LNCS, vol. 11045, pp. 101–109. Springer, Cham (2018). https://doi.org/10.1007/978-3-030-00889-5_12
12. Kuklisova-Murgasova, M., Quaghebeur, G., Rutherford, M.A., Hajnal, J.V., Schnabel, J.A.: Reconstruction of fetal brain MRI with intensity matching and complete outlier removal. Med. Image Anal. **16**, 1550–1564 (2012)
13. Makropoulos, A., Gousias, I.S., Ledig, C., Aljabar, P., Serag, A., Hajnal, J.V., Edwards, A.D., Counsell, S.J., Rueckert, D.: Automatic whole brain MRI segmentation of the developing neonatal brain. IEEE Trans. Med. Imaging **33**(9), 1818–1831 (2014)
14. Rueckert, D., Sonoda, L.I., Hayes, C., Hill, D.L.G., Leach, M.O., Hawkes, D.J.: Nonrigid registration using free-form deformations: application to breast MR images. IEEE Trans. Med. Imaging **18**(8), 712–721 (1999)
15. Schuh, A., et al.: Unbiased construction of a temporally consistent morphological atlas of neonatal brain development. bioRxiv (2018)
16. Tournier, J.D., et al.: MRtrix3: a fast, flexible and open software framework for medical image processing and visualisation. NeuroImage **202**, 116137 (2019)
17. Uus, A., et al.: Multi-channel 4D parametrized atlas of macro- and microstructural neonatal brain development. Front. Neurosci. **15**, 721 (2021). https://doi.org/10.3389/fnins.2021.661704
18. Xu, B., Wang, N., Chen, T., Li, M.: Empirical evaluation of rectified activations in convolutional network (2015). https://arxiv.org/abs/1505.00853

Monte Carlo Concrete DropPath for Epistemic Uncertainty Estimation in Brain Tumor Segmentation

Natalia Khanzhina[(✉)] [ID], Maxim Kashirin, and Andrey Filchenkov [ID]

Machine Learning Lab, ITMO University,
49 Kronverksky Pr, 197101 St. Petersburg, Russia
{nehanzhina,mikashirin,afilchenkov}@itmo.ru

Abstract. Well-calibrated uncertainty is crucial for medical imaging tasks. However, Monte Carlo (MC) Dropout - one of the most common methods for epistemic uncertainty estimation in deep neural networks (DNN), has been found ineffective for multi-path DNN, such as NASNet, and has been recently bypassed by DropPath and ScheduledDropPath.

In this work, we propose two novel model calibration frameworks for uncertainty estimation: *MC ScheduledDropPath* and *MC Concrete DropPath*. Particularly, MC ScheduledDropPath drops out paths in DNN cells during test-time, which has proven to improve the model calibration. At the same time, the MC Concrete DropPath method applies concrete relaxation for DropPath probability optimization, which was found to even better regularize and calibrate DNNs at scale. We further investigate both methods on the problem of brain tumour segmentation and demonstrate a significant Dice score improvement and better calibration ability as compared to state-of-the-art baselines.

Keywords: Brain tumor segmentation · Medical imaging · Image segmentation · Deep learning · Bayesian deep learning · Uncertainty estimation · Epistemic uncertainty · Model calibration

1 Introduction

A brain tumor, or intracranial tumor, which is an abnormal mass of tissue with uncontrollably growing and multiplying cells, is one of the widespread death causes. Survival rates vary and depend on several factors, including the type of brain tumor; the classification includes more than 150 types. Over the past 20 years, the number of diagnoses related to brain tumors has increased significantly [1,15,27]. The type, size and location of a brain tumor determines the treatment.

The most common method for diagnosing and analyzing brain tumors is Magnetic Resonance Imaging (MRI). Radiologists make diagnosis by laborious viewing numerous brain scans for each patient. This visual method is fraught with

© Springer Nature Switzerland AG 2021
C. H. Sudre et al. (Eds.): UNSURE 2021/PIPPI 2021, LNCS 12959, pp. 64–74, 2021.
https://doi.org/10.1007/978-3-030-87735-4_7

possible omissions of tumors nidi. This leads to erroneous diagnoses and, consequently, incorrect treatment, which can cause irreparable damage to the patient's health. Thus, routine work on the analysis of brain MRI requires automation, both at the stage of tumor localization and classification [3,22]. The classification problem raises due to the different grades of tumors. However, despite the importance of the classification task, the primary and higher priority is the task of localizing brain tumors in MRI, often solved by segmentation. The segmentation here is more convenient than detection (which is the object bounding box prediction) as brain tumors have an arbitrary shape and relatively large volume compared with lung nodules.

Currently, the task of brain tumor segmentation in MRI is most effectively solved with deep neural networks (NN), namely, convolutional NNs (CNN) [36,37,37]. However, training NNs requires a huge amount of data [25]. At the same time, labeling images with tumors for the segmentation is extremely time-consuming, and there are relatively few datasets for certain diseases [24]. Moreover, NNs tend to overfit, especially with a small amount of data [30]. Overfitting leads to high models confidence on new data and can result in undesirable errors in the diagnosis of brain tumors.

To overcome this problem, methods to assess model confidence or uncertainty in its predictions are needed. The uncertainty assessment can free radiologists from the laborious work of detecting tumors in the cases where the model is confident, and allow to focus only on those areas of the image where the uncertainty is high.

Researchers distinguish several types of uncertainty. Uncertainty caused by a small amount of training data or its unrepresentativeness is called **epistemic**, or model uncertainty [16]. Epistemic uncertainty estimation applied to the task of brain tumor segmentation in MRI has been studied by the research community [6,14,20,28]. Recent works in this area are grounded on Monte Carlo-based approximate Variational Inference, referred to as Monte-Carlo Dropout (MC Dropout, MCDO) [7].

Dropout was originally proposed for the regularization of fully-connected NNs, however, it is known to be poorly applicable to CNNs. For CNNs other techniques such as DropBlock [9], DropLayer [13], and DropFilter [34] are developed and they surpass Dropout in terms of accuracy and regularization power. These techniques are also called Structured Dropouts [40]. Following the same direction, Larsson et al. proposed the DropPath technique, which drops out the whole paths in convolutional multi-path cells (sometimes it also referred to as DropLayer, although they are not the same) [17]. The technique was advanced by Zoph et al. to train the state-of-the-art NASNet model [41]. They suggested disabling paths in cells with a probability that increases linearly during training. This technique, which was called ScheduledDropPath, showed its efficiency compared to Dropout and DropPath for the ImageNet classification task.

Although the DropBlock, DropLayer, and DropFilter techniques have been adapted to estimate epistemic uncertainty based on Monte Carlo (MC) sampling [40], no similar work has been done for DropPath and ScheduledDropPath.

To close this gap, we propose two new methods for estimating epistemic uncertainty. They incorporate the DropPath and ScheduledDropPath to the approximate Variational Inference based on the test-time MC sampling. Moreover, we put forward the DropPath technique by optimizing its probability as follows. For each network cell, the probability of dropping out the path is calculated based on a continuous relaxation. This is provided by replacing the Bernoulli distribution over the dropping out probability with the Concrete distribution, which was introduced by Gal et al. [8] for the case of Dropout.

The paper research questions are the following:

RQ1. Can *MC DropPath*-based methods with an irregular probability p effectively calibrate the model and improve the accuracy of tumor segmentation?

RQ2. Can the continuous relaxation be effectively applied to the dropping out probability optimization in DropPath?

RQ3. Is DropPath probability continuous relaxation better than the traditional DropPath scheduling in NASNet applied to brain tumor segmentation?

Thus, the contributions of this work are the following:

1. A new technique Concrete DropPath for NNs regularization;
2. A new method Monte Carlo Concrete DropPath, *MC Concrete DropPath*, for estimating epistemic uncertainty;
3. A new method Monte Carlo ScheduledDropPath, *MC ScheduledDropPath*, for estimating epistemic uncertainty;
4. A comparative study of new methods performance on the brain tumors MRI segmentation task.

In this paper, we investigate the research questions on the task of segmenting brain tumors based on BraTS dataset. Besides novel uncertainty estimation methods, we also introduce several uncommon for this task techniques: ImageNet pre-training, double model head for the whole tumor and its border, and the NASNet-based segmentation model. We experimentally studied various segmentation U-Net-like baselines, based on several backbones (DenseNet, Xception, NASNet) and compared our new methods with other uncertainty estimation methods on the NASNet [41] as the best backbone. The proposed methods showed promising performance on our problem both in terms of accuracy and calibration metrics. Moreover, they can be applied to other computer vision tasks, as well as to other CNN architectures with multi-path cells. The source code is available at https://github.com/Vole1/MC-CDP-BraTS2018.

2   Background

The MC sampling based on several Dropout techniques was already used for epistemic uncertainty estimation in deep NNs. Thus, in this paper we examine our proposed uncertainty estimation MC DropPath-based methods compared to other MC Dropout techniques, namely MC DropFilter [34] and regular MC Dropout [10]. Following [40], we call them Structured Dropouts. We do not study MC DropBlock [9] as it is known to be less effective for the model calibration [40].

2.1 Monte Carlo Dropout

We assume we are given a dataset $\mathcal{D}_{train} = (\boldsymbol{X}, \boldsymbol{Y}) = \{(X_i, Y_i)\}_{i=1}^{M}$.

The approximating distribution $q_\theta(\omega)$ for NN $f(\omega, X)$ with Dropout applied before every layer l equals to $\prod_l q_{Ml}(W_l)$ with M_l the mean weights matrix. The approximating distribution for the layer l is $q_{Ml}(W) = M_l^K \cdot z^K$ with $z^K \sim Bern(1-p)$ the dropping out probability vector, where p is the Dropout rate. This technique is called MC Dropout. Monte Carlo sampling from $q_\theta(\omega)$ for the model $f(\omega, X)$ aims to approximate the predictive distribution $p(Y|X)$:

$$p(Y = C|X, \mathcal{D}_{train}) = \int p(Y = C|X, \theta)p(\theta|\mathcal{D}_{train})d\theta \approx \frac{1}{N} \sum_{k=1}^{K} p(Y|X, \theta^{(n)}).$$

Therefore, using Dropout for the model sampling with different dropout masks during inference is a way to obtain a Bayesian NN from any DNN.

2.2 ScheduledDropPath

ScheduledDropPath is originally proposed by NASNet authors [41]. NASNet is a very deep NN, which is state-of-the-art for the task of image classification. NAS-Net contains two types of cells: normal and reduction. Each cell is a multi-path module, which structure was found using the neural architecture search (NAS). The authors claim that the Dropout regularization decreased the model performance. Thus, they adopted FractalNet regularization technique called Drop-Path [17]. DropPath can be considered as an extension of DropFilter idea to the level of paths in multi-path NNs. However, the original DropPath did not work well. Therefore, they modified DropPath to drop paths out with linear schedule during training and called the new technique ScheduledDropPath.

2.3 Concrete Dropout

The continuous relaxation was successfully applied for Dropout [8]. Dropout probability p is usually fixed (often equal to 0.5) as a model hyperparameter and does not change during training as its grid-searching is computationally expensive. Gal et al. [8] proposed to assume Dropout probability p as a parameter to optimize. To make it differentiable, they changed the discrete Bernoulli distribution of p probability to the continuous relaxation using the Concrete distribution [19]. As here the Bernoulli distribution is one-dimensional, the Concrete distribution is reduced to the sigmoid distribution:

$$\tilde{z} = \text{sigmoid}\left(\frac{1}{t} \cdot (\log p - \log(1-p) + \log u - \log(1-u))\right) \qquad (1)$$

with $\tilde{z}$ a random variable relaxation, $u \sim \mathcal{U}(0,1)$, t a temperature.

Sigmoid pushes random variable $\tilde{z}$ to the boundaries 0 and 1. The resulting function 1 is differentiable with respect to p, thus, allowing to backpropagate its value.

3   Monte Carlo Concrete DropPath

In this study, we propose the way to turn an arbitrary multi-path NN into a Bayesian NN to estimate epistemic uncertainty. This can be achieved using test-time MC sampling of NN with ScheduledDropPath. However, our experiments revealed that simple *MC ScheduledDropPath* does not perform significantly better than other calibration techniques. Thus, we also introduce a new DropPath technique based on the Concrete distribution [19]. This new regularization technique is called *Concrete DropPath* and the corresponding uncertainty estimation method is *MC Concrete DropPath*.

We tested both of our methods on NASNet, although they can be trivially extended to any multi-path NN, such as FractalNet [17], Inception-ResNet-v2 [33], ResNeXt [38] etc.

The Bayesian multi-path NN objective is the following:

$$\hat{\mathcal{L}}_{MC}(\theta) = -\frac{1}{M} \sum_{i \in S} \log p\left(y_i | f^\omega(x_i)\right) + \frac{1}{N} KL(q_\theta(\omega) || p(\omega)), \tag{2}$$

where $\log p(y_i | f^\omega(x_i))$ is the model likelihood, S is a random sample of M data points and $KL(q_\theta(\omega) || p(\omega))$ is the Kullback-Leibler (KL) divergence between prior distribution $p(\omega)$ and approximate posterior distribution $q_\theta(\omega)$ with parameters represented by Concrete DropPath or ScheduledDropPath.

While the model likelihood is a usual loss function, which is combined binary cross-entropy and Dice losses in our case, KL divergence is the sum of KL divergences for every NN cell parameters distribution. For a cell c KL is defined as:

$$KL(q_{M_c}(W) || p(W)) \propto \frac{(1 - p_c)}{2s^2} ||M_c||^2 - K\mathcal{H}(p_c), \tag{3}$$

where $\mathcal{H}(p_c)$ is entropy of probability p_c, M_c is the mean cell weights matrix, $q_{M_c}(W) = M_c^K \cdot D^K$ with $D^K \sim Bern(1 - p_c)$ the path dropping out probability vector, K is the number of paths in cell c, s^2 is the variance of the prior distribution.

In MC ScheduledDropPath, the second term $K\mathcal{H}(p)$ is omitted because p is fixed. But for MC Concrete DropPath, we optimize p, replacing the Bernoulli distribution with the Concrete distribution for every cell c as in Sect. 2.3.

Finally, to estimate epistemic uncertainty, we sample from the approximating distribution $q_\theta(\omega)$ N times during testing in a MC manner forming the prediction vector $\mathcal{Y} = \{\hat{y}_1, \dots, \hat{y}_N\}$. Then $\mathcal{Y}$ can be used for the predictive mean evaluation and epistemic uncertainty estimation based on Mutual Information, for example.

4   Experiments

4.1   Dataset

We evaluated the models on BraTS 2018 dataset [4, 22]. It contains 44,175 multimodal scans of patients with brain tumors. Each scan consists of 4 modes, namely

T1-weighted (T1), post-contrast T1-weighted (T1c), T2-weighted (T2), T2 Fluid Attenuated Inversion Recovery (FLAIR). Labels include the whole tumor (WT), active tumor, and tumor core tissue. Here, we only consider the WT segmentation task as the most common.

4.2 Models

The segmentation model used in this study is U-Net-like NN with two heads: one for the tumor prediction and the auxiliary head for the tumor border prediction. Seferbekov et al. [29] showed that such an approach penalizes the model more for errors on tumor border and leads to a more accurate segmentation result. Our experiments demonstrated that accounting the auxiliary head in metrics evaluation does not improve the segmentation WT score. Thus, the auxiliary head is only used for model training, which required the tumor label preprocessing to make two targets per each sample: with WT and WT border.

To test our proposed method, we have chosen the baseline backbone for U-Net over the following: Xception [5]; DenseNet [11]; NASNet [41]. The first two baselines recently showed high performance compared to popular architectures for the brain tumor segmentation task [39]. We tested them compared to NASNet to ensure that NASNet is effective for the segmentation task.

As NASNet was the best baseline (see Table 1), we examined the following uncertainty estimation methods based on it: NASNet MC Dropout (NASNet MCDO), NASNet MC Concrete Dropout (NASNet MCCDO), NASNet MC DropFilter (NASNet MCDF), NASNet MC ScheduledDropPath (NASNet MCSDP, with the proposed method), NASNet MC Concrete DropPath (NASNet MCCDP, with the proposed method), deep ensemble of four NASNets trained with ScheduledDropPath. For MC Dropout, MC DropFilter, MC Scheduled-DropPath models we used dropping out probability $p=0.3$ as it provides the best results according to our experiments.

ImageNet Pre-training. We pre-trained all the models on ImageNet dataset, which is a common practice in computer vision problems. ImageNet is the 3-channel images dataset. However, the dataset used in our experiments includes 4 MRI modes. Therefore, the forth channel for the models was formed using the transfer learning, which required modifications of the NN on the framework level to train it. To our knowledge, there is no standard interface for such modifications. We performed the forth channel initialization with the first channel weights. As ImageNet pre-training is uncommon for medical domain, in this study we tested some of the models (NASNet MCDO and NASNet MCCDP) with and without pre-training to prove its effectiveness.

4.3 Experimental Setup

We implemented all the models in the `Tensorflow 2` framework. For the experiments, we used single GeForce GTX 1080 Ti GPU per each model's training and evaluation. Training of the single model took 3 days.

For segmentation accuracy evaluation Dice coefficient is traditionally used [42]. Based on it, the training was performed using Dice loss, which is calculated as $(1 - Dice)$ and combined with binary cross-entropy. For the model calibration we used three metrics, namely Negative Log Likelihood (NLL) [18], in the form of binary cross-entropy (BCE), Brier score (BS) [26], Expected Calibration Error (ECE) [2,23,40], and Mean Entropy score (MES) [2]. Also we compared the two best models by the area under curve (AUC) of Dice score, Filtered True Positive (FTP) AUC, and Filtered True Negative (FTN) AUC metrics calculated with different uncertainty thresholds according to the BraTS 2019 uncertainty task evaluation protocol and [21].

The initial Concrete DropPath probability p value was 0.0001, which was optimized during training. We trained all the models with Adam optimizer with learning rate (lr) equal to 0.0001, lr decay and weight decay equal to 0.0001. The train/test/validation split was 75/15/10, respectively. All the reported results were obtained on the validation set. For MC techniques, the number of samples per image N was equal to 20, according to the experimental results in [35]. For the uncertainty estimation visualization we used Mutual Information (MI).

5   Results and Discussion

Table 1 presents the evaluation results of different baselines and the proposed MC Concrete DropPath and MC ScheduledDropPath methods compared to other MC models and the deep ensemble. The Table shows that the best segmentation backbone is NASNet, as it achieved the highest Dice score. Thus, the uncertainty estimation methods were applied to NASNet-based U-Net, their results are presented in the second and third parts of the Table.

The *MC Concrete DropPath* (MCCDP) model outperformed all other models by *Dice coefficient*. The proposed method (with and without pre-training) along with MCCDO had the best calibration by *MES* , which is significantly lower (by 2–3 times) than *MES* of other models. Moreover, our MCCDP method provided one of the best *ECE* scores, reducing its value almost twice compared to the deep ensemble. The *BS* calibration metric is on the same level with the deep ensemble. To summarize, *MC Concrete DropPath* superiors other methods by aggregate accuracy and calibration metrics. This can be considered as the positive answer to the **RQ1**. The *MC ScheduledDropPath* (MCSDP) with regular cell-wise p value achieved relatively good calibration metrics and improved the Dice score compared to the baseline NASNet. This confirms the answer to the **RQ1**. At the same time, *MC Concrete DropPath* provided much better calibration and segmentation scores than *MC ScheduledDropPath*, answering **RQ3**. This proves the effectiveness of the Concrete relaxation for DropPath, answering **RQ2**.

Despite the deep ensemble achieved close segmentation accuracy and slightly lower Brier score, our method *MC Concrete DropPath* is less memory and time consuming and twice better calibrates the model according to ECE and MES. Moreover, *MC Concrete DropPath* achieved better Dice AUC, FTP AUC and FTN AUC, which confirms its effectiveness (see Fig. [31]).

Table 1. Results of the proposed methods and other uncertainty estimation methods on different baselines on BraTS 2018. WT DC is Whole tumor Dice coefficient. The first part presents baselines comparison, the second part presents existing uncertainty estimation methods, the third part presents the results of the proposed methods. All the results obtained using 5-fold cross-validation.

Model	WT DC, %	NLL	BS	ECE	MES
Xception	84.71	0.0292	–	–	–
DenseNet	85.06	0.0269	–	–	–
NASNet	**85.42**	0.0248	–	–	–
NASNet MCDO (no pre-train)	78.13	0.0246	0.0035	0.0096	0.0007
NASNet MCDO	84.44	0.0156	0.0023	0.0101	0.0005
NASNet MCCDO	86.06	0.0198	0.0020	**0.0067**	**0.0002**
NASNet MCDF	86.32	**0.0115**	0.0020	0.0121	0.0006
Deep ensemble of NASNets	**86.48**	0.0148	**0.0018**	0.0131	0.0004
NASNet MCSDP (our)	86.28	0.0155	0.0020	0.0097	0.0005
NASNet MCCDP (no pre-train)	81.32	0.0273	0.0030	**0.0069**	**0.0002**
NASNet MCCDP (our)	**86.74**	0.0173	**0.0019**	0.0071	0.0002

The Concrete DropPath probabilities p obtained for different NASNet cells are approximately equal to 0.1 (some p_c are higher, some – lower), which means that although the model is quite deep, the dataset is rather large enough and does not require stronger model regularization.

Finally, the results confirm that ImageNet pre-training improves the brain tumor segmentation task solving significanly, by 5–6% of Dice coefficient. Thus, ImageNet pre-training is feasible for computer vision medical domain tasks.

The visualized uncertainties can be found in [32]. The Figure shows that MCCDP technique provides better model calibration, than the deep ensemble: the tumor borders, presented by MCCDP epistemic uncertainty, are more precise and less bright (MI is lower, which is better). Such uncertainty visualization can help doctors to recognize cases when an additional tumor investigation is required.

6 Conclusion

In the paper, we have studied the brain MRI tumor segmentation task and presented two novel methods: MC Concrete DropPath and MC ScheduledDropPath addressing the problem of estimating epistemic uncertainty. We compared our methods with other MC Structured Dropouts and the deep ensemble.

As the baselines we implemented three U-Net models with different backbones, namely Xception, DenseNet, NASNet and evaluated them on BraTS 2018 dataset. Then, we tested our methods and other MC Structured Dropouts based on U-Net with NASNet, as it was the best backbone.

The results show that MC Concrete DropPath majorizes other methods by both accuracy and calibration metrics. Moreover, our method can be useful for a wide variety of applications, not limited to medical imaging tasks.

In the future, we plan to study the proposed methods on other multi-path NN architectures, such as Inception-ResNet-v2 [33] and ResNeXt [38]. Concrete DropPath can be effective not only for uncertainty estimation, but also as the regularization technique. Also we plan to incorporate our methods to the state-of-the-art BraTS approach nnU-net [12] and to extend them to a multi-class segmentation problems.

Acknowledgments. This work is financially supported by National Center for Cognitive Research of ITMO University and Russian Science Foundation, Grant 19-19-00696.

The authors thank Ilya Osmakov and Pavel Ulyanov for useful ideas, Alex Farseev, Inna Anokhina, and Tatyana Polevaya for useful comments.

References

1. Al-Shamahy, H.: Prevalence of CNS tumors and Histo-logical recognition in the operated patients: 10 years experience. Ann. Clin. Med. Case Rep. **6**(12), 1–8 (2021)
2. Ashukha, A., Lyzhov, A., Molchanov, D., Vetrov, D.: Pitfalls of in-domain uncertainty estimation and ensembling in deep learning (2019)
3. Bakas, S., et al.: Advancing the cancer genome atlas glioma MRI collections with expert segmentation labels and radiomic features. Sci. Data **4**(1), 1–13 (2017)
4. Bakas, S., et al.: Identifying the best machine learning algorithms for brain tumor segmentation, progression assessment, and overall survival prediction in the brats challenge. arXiv preprint arXiv:1811.02629 (2018)
5. Chollet, F.: Xception: deep learning with depthwise separable convolutions. In: Proceedings of the IEEE Conference on Computer Vision and Pattern Recognition, pp. 1251–1258 (2017)
6. Eaton-Rosen, Z., Bragman, F., Bisdas, S., Ourselin, S., Cardoso, M.J.: Towards safe deep learning: accurately quantifying biomarker uncertainty in neural network predictions. In: Frangi, A.F., Schnabel, J.A., Davatzikos, C., Alberola-López, C., Fichtinger, G. (eds.) MICCAI 2018. LNCS, vol. 11070, pp. 691–699. Springer, Cham (2018). https://doi.org/10.1007/978-3-030-00928-1_78
7. Gal, Y., Ghahramani, Z.: Dropout as a Bayesian approximation. arXiv preprint arXiv:1506.02157 (2015)
8. Gal, Y., Hron, J., Kendall, A.: Concrete dropout. In: Advances in Neural Information Processing Systems, vol. 30 (2017)
9. Ghiasi, G., Lin, T.Y., Le, Q.V.: DropBlock: a regularization method for convolutional networks. arXiv preprint arXiv:1810.12890 (2018)
10. Hinton, G.E., Srivastava, N., Krizhevsky, A., Sutskever, I., Salakhutdinov, R.: Improving neural networks by preventing co-adaptation of feature detectors. CoRR abs/1207.0580 (2012). http://arxiv.org/abs/1207.0580
11. Huang, G., Liu, Z., Van Der Maaten, L., Weinberger, K.Q.: Densely connected convolutional networks. In: Proceedings of the IEEE Conference on Computer Vision and Pattern Recognition, pp. 4700–4708 (2017)

12. Isensee, F., Jäger, P.F., Full, P.M., Vollmuth, P., Maier-Hein, K.H.: nnU-Net for brain tumor segmentation. In: Crimi, A., Bakas, S. (eds.) BrainLes 2020. LNCS, vol. 12659, pp. 118–132. Springer, Cham (2021). https://doi.org/10.1007/978-3-030-72087-2_11

13. Izmailov, P., Podoprikhin, D., Garipov, T., Vetrov, D., Wilson, A.G.: Averaging weights leads to wider optima and better generalization. arXiv preprint arXiv:1803.05407 (2018)

14. Jungo, A., Meier, R., Ermis, E., Herrmann, E., Reyes, M.: Uncertainty-driven sanity check: application to postoperative brain tumor cavity segmentation. arXiv preprint arXiv:1806.03106 (2018)

15. Kaneko, S., Nomura, K., Yoshimura, T., et al.: Trend of brain tumor incidence by histological subtypes in Japan: estimation from the brain tumor registry of t. J. Neurooncol. **60**, 61–69 (2002). https://doi.org/10.1023/A:1020239720852

16. Kendall, A., Gal, Y.: What uncertainties do we need in Bayesian deep learning for computer vision? In: Advances in Neural Information Processing Systems, pp. 5574–5584 (2017)

17. Larsson, G., Maire, M., Shakhnarovich, G.: FractalNet: ultra-deep neural networks without residuals. arXiv preprint arXiv:1605.07648 (2016)

18. Loquercio, A., Segu, M., Scaramuzza, D.: A general framework for uncertainty estimation in deep learning. IEEE Robot. Autom. Lett. **5**(2), 3153–3160 (2020)

19. Maddison, C.J., Mnih, A., Teh, Y.W.: The concrete distribution: a continuous relaxation of discrete random variables. arXiv preprint arXiv:1611.00712 (2016)

20. Mehta, R., Arbel, T.: RS-Net: regression-segmentation 3D CNN for synthesis of full resolution missing brain MRI in the presence of tumours. In: Gooya, A., Goksel, O., Oguz, I., Burgos, N. (eds.) SASHIMI 2018. LNCS, vol. 11037, pp. 119–129. Springer, Cham (2018). https://doi.org/10.1007/978-3-030-00536-8_13

21. Mehta, R., Filos, A., Gal, Y., Arbel, T.: Uncertainty evaluation metric for brain tumour segmentation. arXiv preprint arXiv:2005.14262 (2020)

22. Menze, B.H., et al.: The multimodal brain tumor image segmentation benchmark (BRATS). IEEE Trans. Med. Imaging **34**(10), 1993–2024 (2014)

23. Naeini, M.P., Cooper, G., Hauskrecht, M.: Obtaining well calibrated probabilities using Bayesian binning. In: Proceedings of the AAAI Conference on Artificial Intelligence, vol. 29 (2015)

24. Nalepa, J., Marcinkiewicz, M., Kawulok, M.: Data augmentation for brain-tumor segmentation: a review. Front. Comput. Neurosci. **13**, 83 (2019)

25. Northcutt, C.G., Athalye, A., Mueller, J.: Pervasive label errors in test sets destabilize machine learning benchmarks. arXiv preprint arXiv:2103.14749 (2021)

26. Ovadia, Y., et al.: Can you trust your model's uncertainty? Evaluating predictive uncertainty under dataset shift. arXiv preprint arXiv:1906.02530 (2019)

27. Porter, K.R., McCarthy, B.J., Freels, S., Kim, Y., Davis, F.G.: Prevalence estimates for primary brain tumors in the united states by age, gender, behavior, and histology. Neuro Oncol. **12**(6), 520–527 (2010)

28. Rousseau, A.J., Becker, T., Bertels, J., Blaschko, M.B., Valkenborg, D.: Post training uncertainty calibration of deep networks for medical image segmentation. In: 2021 IEEE 18th International Symposium on Biomedical Imaging (ISBI), pp. 1052–1056. IEEE (2021)

29. Seferbekov, S.: DSB 2018 [ods.ai] topcoders 1st place solution (2018). https://github.com/selimsef/dsb2018_topcoders

30. Srivastava, N., Hinton, G., Krizhevsky, A., Sutskever, I., Salakhutdinov, R.: Dropout: a simple way to prevent neural networks from overfitting. J. Mach. Learn. Res. **15**(1), 1929–1958 (2014)

31. Supplementary Material: The dice AUC, FTP AUC, and FTN AUC for deep ensemble and mc concrete droppath (2021). https://genome.ifmo.ru/files/papers_files/MICCAI2021/UNSURE/AUC.png

32. Supplementary Material: The prediction and uncertainty visualization (2021). https://genome.ifmo.ru/files/papers_files/MICCAI2021/UNSURE/MRI_uncertainty.png

33. Szegedy, C., Ioffe, S., Vanhoucke, V., Alemi, A.A.: Inception-v4, inception-resnet and the impact of residual connections on learning. In: Thirty-First AAAI Conference on Artificial Intelligence (2017)

34. Tompson, J., Goroshin, R., Jain, A., LeCun, Y., Bregler, C.: Efficient object localization using convolutional networks. In: Proceedings of the IEEE Conference on Computer Vision and Pattern Recognition, pp. 648–656 (2015)

35. Wang, G., Li, W., Aertsen, M., Deprest, J., Ourselin, S., Vercauteren, T.: Aleatoric uncertainty estimation with test-time augmentation for medical image segmentation with convolutional neural networks. Neurocomputing **338**, 34–45 (2019)

36. Wang, G., Li, W., Ourselin, S., Vercauteren, T.: Automatic brain tumor segmentation using cascaded anisotropic convolutional neural networks. In: Crimi, A., Bakas, S., Kuijf, H., Menze, B., Reyes, M. (eds.) BrainLes 2017. LNCS, vol. 10670, pp. 178–190. Springer, Cham (2018). https://doi.org/10.1007/978-3-319-75238-9_16

37. Wang, G., Li, W., Ourselin, S., Vercauteren, T.: Automatic brain tumor segmentation using convolutional neural networks with test-time augmentation. In: Crimi, A., Bakas, S., Kuijf, H., Keyvan, F., Reyes, M., van Walsum, T. (eds.) BrainLes 2018. LNCS, vol. 11384, pp. 61–72. Springer, Cham (2019). https://doi.org/10.1007/978-3-030-11726-9_6

38. Xie, S., Girshick, R., Dollár, P., Tu, Z., He, K.: Aggregated residual transformations for deep neural networks. In: Proceedings of the IEEE Conference on Computer Vision and Pattern Recognition, pp. 1492–1500 (2017)

39. Zeineldin, R.A., Karar, M.E., Coburger, J., Wirtz, C.R., Burgert, O.: DeepSeg: deep neural network framework for automatic brain tumor segmentation using magnetic resonance flair images. Int. J. Comput. Assist. Radiol. Surg. **15**(6), 909–920 (2020). https://doi.org/10.1007/s11548-020-02186-z

40. Zhang, Z., Dalca, A.V., Sabuncu, M.R.: Confidence calibration for convolutional neural networks using structured dropout. arXiv preprint arXiv:1906,09551 (2019)

41. Zoph, B., Vasudevan, V., Shlens, J., Le, Q.V.: Learning transferable architectures for scalable image recognition. In: Proceedings of the IEEE Conference on Computer Vision and Pattern Recognition, pp. 8697–8710 (2018)

42. Zou, K.H., et al.: Statistical validation of image segmentation quality based on a spatial overlap index1: scientific reports. Acad. Radiol. **11**(2), 178–189 (2004)

Improving Aleatoric Uncertainty Quantification in Multi-annotated Medical Image Segmentation with Normalizing Flows

M. M. Amaan Valiuddin[✉], Christiaan G. A. Viviers, Ruud J. G. van Sloun,
Peter H. N. de With, and Fons van der Sommen

Eindhoven University of Technology, 5612 AZ Eindhoven, The Netherlands
m.m.a.valiuddin@student.tue.nl

Abstract. Quantifying uncertainty in medical image segmentation applications is essential, as it is often connected to vital decision-making. Compelling attempts have been made in quantifying the uncertainty in image segmentation architectures, e.g. to learn a density segmentation model conditioned on the input image. Typical work in this field restricts these learnt densities to be strictly Gaussian. In this paper, we propose to use a more flexible approach by introducing Normalizing Flows (NFs), which enables the learnt densities to be more complex and facilitate more accurate modeling for uncertainty. We prove this hypothesis by adopting the Probabilistic U-Net and augmenting the posterior density with an NF, allowing it to be more expressive. Our qualitative as well as quantitative (GED and IoU) evaluations on the multi-annotated and single-annotated LIDC-IDRI and Kvasir-SEG segmentation datasets, respectively, show a clear improvement. This is mostly apparent in the quantification of aleatoric uncertainty and the increased predictive performance of up to 14%. This result strongly indicates that a more flexible density model should be seriously considered in architectures that attempt to capture segmentation ambiguity through density modeling. The benefit of this improved modeling will increase human confidence in annotation and segmentation, and enable eager adoption of the technology in practice.

Keywords: Segmentation · Uncertainty · Computer vision · Imaging

1 Introduction

As a result of the considerable advances in machine learning research over the past decade, computer-aided diagnostics (CAD) using deep learning has rapidly been gaining attention. The outcome from these deep learning-based CAD systems has to be highly accurate, since it is often connected to critical designs resulting in a potentially large impact on patient care. As such, conclusions

C. H. Sudre et al. (Eds.): UNSURE 2021/PIPPI 2021, LNCS 12959, pp. 75–88, 2021.
https://doi.org/10.1007/978-3-030-87735-4_8

drawn from these CAD systems should be interpreted with care and by experts. A convolutional neural network (CNN)-based approach has been adopted in a large number of CAD applications and especially in semantic segmentation. This approach segments the objects of interest by assigning class probabilities to all pixels of the image. In the medical domain and especially in the context of lesion segmentation, the exact edges or borders of these lesions are not always easily defined and delineated by radiologists and endoscopists. In the case of multiple annotators, clinicians can also disagree on the boundaries of the localized lesions based on their understanding of the surrounding anatomy. However, the exact edges or borders of these areas of interests often play a critical role in the diagnostic process. For example, when determining whether to perform surgery on a patient or the surgical planning thereof, the invasion of a tumour into local anatomical structures derived from a CT scan, is crucial. Thus, multiple forms of uncertainties come into play with semantic segmentation-based approach for CAD. As such, accurately quantifying these uncertainties have become an essential addition in CAD. Specialized doctors provide ground-truth segmentation for models to be trained, based on their knowledge and experience. When this is done by multiple individuals per image, often discrepancies in the annotations arise, resulting in ambiguities in the ground-truth labels.

Recent work [5] suggests two types of uncertainties exist in deep neural networks. First, epistemic uncertainty, which refers to the lack of knowledge and can be minimized with information gain. In the case of multi/single-annotated data, these are the preferences, experiences, knowledge (or lack thereof) and other biases of the multiple/single annotators. This epistemic uncertainty from the annotator(s) manifests into aleatoric uncertainty when providing annotations. *Aleatoric uncertainty* is the variability in the outcome of an experiment, due to the inherent ambiguity that exists in the data, captured through the multiple ground truths. By using a probabilistic segmentation model, we attempt to learn this as a distribution of possible annotations. It is important to enable expressiveness of the probability distributions to sufficiently capture the variability. In multi-annotator settings, the adoption of rich and multi-modal distributions may be more appropriate. We aim to show that by using invertible bijections, also known as Normalizing Flows (NFs), we can obtain more expressive distributions to adequately deal with the disagreement in the ground-truth information. Ultimately, this improves the ability to quantify the aleatoric uncertainty for segmentation problems.

In this work, we use the Probabilistic U-Net (PU-Net) [10] as the base model and subsequently improve on it by adding a planar and radial flow to render a more expressive learned posterior distribution. For quantitative evaluation, we use the Generalized Energy Distance (GED), as is done in previous work (see related work). We hypothesize that this commonly used metric is prone to some biases, such that it rewards sample diversity rather than predictive accuracy. Therefore, we also evaluate on the average and Hungarian-matched IoU for the single- and multi-annotated data, respectively, as is also done by Kohl *et al.* [9].

To qualitatively evaluate the ability to model the inter-variability of the annotations, we present the mean and standard deviation of the segmentation samples reconstructed from the model. In this paper, we make use of the multi- and single-annotated LIDC-IDRI (LIDC) and Kvasir-SEG datasets, thereby handling limited dataset size and giving insights on the effects of the complex posterior on hard-to-fit datasets.

2 Related Work

Kohl *et al.* [10] introduced the PU-Net for image segmentation, a model that combines the cVAE [15] and a U-Net [13]. Here, the uncertainty is captured by a down-sampled axis-aligned Gaussian prior that is updated through the KL divergence of the posterior. These distributions contain several low-dimensional representations of the segmentation, which can be reconstructed by sampling. We use this model as our baseline model.

The concept of using NFs has been presented in earlier literature. For an extensive introduction to NFs we suggest the paper from Kobyzev *et al.* [8]. The planar and radial NFs have been used for approximating flexible and complex posterior distributions [12].

Selvan *et al.* [14] used an NF on the posterior of a cVAE-like segmentation model and showed that this increases sample diversity. The increased sample diversity resulted in a better score on the GED metric and a slight decrease in DICE score. The authors reported significant gains in performance. However, we argue that this claim requires more evidence to confirm this positive effect, such as training with K-fold cross-validation and evaluating using other metrics. Also, insight into the reasons for their improvements are not provided and critical details of the experiments are missing, such as the number of samples used for the GED evaluation. We aim to provide a more complete argumentation and show clear steps towards improving the quantification of aleatoric uncertainty.

3 Methods

3.1 Model Architecture

We use a PU-Net extended with an NF, as is shown in Fig. 1. A key element of the architecture is the posterior network Q, which attempts to encapsulate the distribution of possible segmentations, conditioned on the input image $\mathbb{X}$ and ground truth $\mathbb{S}$ in the base distribution. The flexibility of the posterior is enhanced through the use of an NF, which warps it into a more complex distribution. During training, the decoder is sampled by the posterior and is constructing, based on the encoded input image, a segmentation via the proposed reconstruction network. The prior P is updated with the evidence lower bound (ELBO [7]), which is based on two components: first, the KL divergence between the distributions Q and P and second, the reconstruction loss between the predicted and ground-truth segmentation. The use of NFs is motivated by the fact that a Gaussian distribution is too limited to fully model the input-conditional latent distribution of annotations. An NF can introduce complexity to Q, e.g. multi-modality, in order to more accurately describe the characteristics of this relationship.

We proceed by extending the PU-Net objective (see Appendix A) and explain the associated parameters in detail. We make use of the NF-likelihood objective (see Appendix B) with transformation $f : \mathbb{R} \mapsto \mathbb{R}$ to define our posterior as

$$\log q(\mathbf{z}|\mathbf{s}, \mathbf{x}) = \log q_0(\mathbf{z}_0|\mathbf{s}, \mathbf{x}) - \sum_{i=1}^{K} \log \left(\left| \det \frac{df_i}{d\mathbf{z}_{i-1}} \right| \right). \tag{1}$$

to obtain the objective

$$\mathcal{L} = -\mathbb{E}_{q_\phi(\mathbf{z}_0|\mathbf{s},\mathbf{x})} \left[\log p(\mathbf{s}|\mathbf{z}, \mathbf{x}) \right]$$

$$+ \mathrm{KL}\left(q_\phi(\mathbf{z}_0|\mathbf{s},\mathbf{x}) \| p_\psi(\mathbf{z}|\mathbf{x}) \right) - \mathbb{E}_{q_\phi(\mathbf{z}_0|\mathbf{s},\mathbf{x})} \left[\sum_{i=1}^{K} \log \left(\left| \det \frac{df_i}{d\mathbf{z}_{i-1}} \right| \right) \right]. \tag{2}$$

The input-dependent context vector $\mathbf{c}$, is used to obtain the posterior flow parameters. During training, the posterior flow is used to capture the data distribution with the posterior network $Q(\boldsymbol{\mu}, \boldsymbol{\sigma}, \mathbf{c}|\mathbb{X}, \mathbb{S})$, followed by sampling thereof to reconstruct the segmentation predictions $\mathbb{Y}$. At the same time, a prior network $P(\boldsymbol{\mu}, \boldsymbol{\sigma}|\mathbb{X})$ only conditioned on the input image is also trained through constraining its KL divergence with the posterior distribution. The first term in Eq. (2) entails the reconstruction loss, in our case the cross-entropy function as mentioned earlier. At test time, the prior network produces latent samples to construct the segmentation predictions.

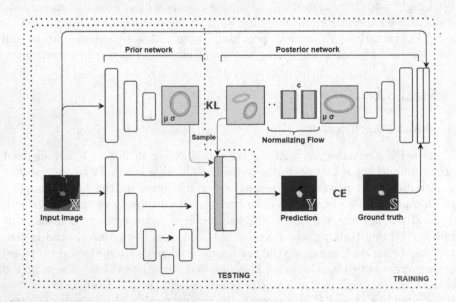

Fig. 1. Diagram of the PU-Net with a flow posterior.

3.2 Data and Baseline Experiments

We perform extensive experimental validation using the vanilla Probabilistic U-Net with 2-/4-step planar and radial flow variants on processed versions of the LIDC-IDRI (LIDC) [1] and the Kvasir-SEG [4] datasets. The preprocessed LIDC dataset [14] transforms the 1,018 thoracic CT scans with four annotators into 15,096 128 × 128-pixel patches according to the method of [2,10]. Each image has 4 annotations. The Kvasir-SEG dataset contains 1,000 polyp images of the gastrointestinal tract from the original Kvasir dataset [11]. We resize the images to be 128 × 128 pixels as well and convert the images to gray-scale. Example images of the datasets can be found in Appendix C. As for the NFs, we used the planar flow conform related work [3,14] and also experiment with the radial flow. These flows are usually chosen because they are computationally the cheapest transformations that possess the ability to expand and contract the distributions along a direction (planar) or around a specific point (radial).

3.3 Performance Evaluation

For evaluation, we deploy the *Generalized Energy Distance* (GED) (also known as the *Maximum Mean Discrepancy*), which is defined as

$$D^2_{GED}(P_{pr}, P_{out}) = 2\mathbb{E}\left[d(\mathbb{S}, \mathbb{Y})\right] - \mathbb{E}\left[d(\mathbb{S}, \mathbb{S}')\right] - \mathbb{E}\left[d(\mathbb{Y}, \mathbb{Y}')\right], \qquad (3)$$

where $\mathbb{Y}$, $\mathbb{Y}'$ and $\mathbb{S}$, $\mathbb{S}'$ are independent samples from the predicted distribution and ground truth distributions P_{pr} and P_{gt}, respectively. Here, d is a distance metric, in our case, one minus the Intersection over Union (1-IoU). When the predictions poorly match the ground truth, the GED is prone to simply reward diversity in samples instead of accurate predictions because the influence of the $\mathbb{E}\left[d(\mathbb{Y}, \mathbb{Y}')\right]$ term becomes dominant. Therefore, we also evaluate the Hungarian-matched IoU, using the average IoU of all matched pairs for the LIDC dataset. We duplicate the ground-truth set, hence matching it with the sample size. Since the Kvasir-SEG dataset only has a single annotation per sample, we simply take the average IoU from all samples. Furthermore, when the model correctly predicts the absence of a lesion (i.e. no segmentation), the denominator of the metric is zero and thus the IoU becomes undefined. In previous work, the mean excluding undefined elements was taken over all the samples. However, since this is a correct prediction, we award this with full score (IoU= 1) and compare this approach with the method of excluding undefined elements for the GED.

To qualitatively depict the model performance, we calculate the mean and standard deviation with Monte-Carlo simulations (i.e. sampling reconstructions from the prior). All evaluations in this paper are based on 16 samples to strike a right balance between sufficient samples and a justifiable approximation, while maintaining minimal computational time.

3.4 Training Details

The training procedure entails tenfold cross-validation using a learning rate of 10^{-4} with early stopping on the validation loss based on a patience of 20 epochs.

The batch size is chosen to be 96 and 32 for the LIDC and the Kvasir-SEG dataset, respectively. The dimensionality of the latent space is set to $L = 6$. We split the dataset as 90-10 (train/validation and test) and evaluate the test set on the proposed metrics. All experiments are done on an 11-GB RTX 2080TI GPU.

4 Results and Discussion

4.1 Quantitative Evaluation

We refer to the models by their posterior, either unaugmented (vanilla) or with their n-step Normalizing Flow (NF). The results of our experiments are presented in Table 1. In line with literature, it shown that the GED improves with the addition of an NF. This hypothesis is tested using both a planar and radial NF and observe that both have a similar effect. Furthermore, both the average and Hungarian-matched IoU improve with the NF. We find that the 2-step radial (2-radial) NF is slightly better than other models for the LIDC dataset, while for the Kvasir-SEG dataset the planar models tend to perform better. The original PU-Net introduced the variability capturing of annotations into a Gaussian model. However, this distribution is not expressive enough to efficiently capture this variability. The increase in GED and average IoU performance from our experiments confirm our hypothesis that applying NF to the posterior distribution of the PU-Net improves the accuracy of the probabilistic segmentation. This improvement occurs because the posterior becomes more complex and can thus provide more meaningful updates to our prior distribution.

Including/excluding correct empty predictions did not result in a significant difference in the metric value when comparing the vanilla models with the posterior NF models. Our results show that the choice in NF has minimal impact on the performance and suggest practitioners to experiment with both NFs. Another publication in literature [14] has experimented with more complex posteriors such as GLOW [6], where no increase in performance was obtained. In our research, we have found that even a 4-step planar or radial NF (which are much simpler in nature) can already be too complex for our datasets. A possible explanation is that the variance in annotations captured in the posterior distribution only requires a complexity that manifests from two NF steps. This degree of complexity is then most efficient for the updates of the prior distribution. More NF steps would then possibly introduce unnecessary model complexity as well parameters for training, thereby reducing the efficiency of the updates. Another explanation could be that an increase in complexity of the posterior distribution does in fact model the annotation variability in a better way. Nevertheless, not all information can be captured by the prior, as it is still a Gaussian. In this case, a 2-step posterior is close enough to a Gaussian for meaningful updates yet complex enough to be preferred over a Gaussian distribution. We consider that for similar problems, it is better to adopt simple NFs with a few steps only. However, we can imagine the need for a more complex NF for other scenarios,

Table 1. Test set evaluations on the GED and IoU based on 16 samples. Further distinction in the GED is made on whether the correct empty predictions are included. The IoU is evaluated with the Hungarian-matching algorithm and averaged with the LIDC and Kvasir-SEG dataset, respectively.

Dataset	Posterior	GED		IoU	
		Excl.	*Incl.*	*Avg.*	*Hungarian*
LIDC	Vanilla	0.33 ± 0.02	0.39 ± 0.02	—	0.57 ± 0.02
	2-planar	0.29 ± 0.02	0.35 ± 0.03	—	0.57 ± 0.01
	2-radial	$\mathbf{0.29 \pm 0.01}$	$\mathbf{0.34 \pm 0.01}$	—	$\mathbf{0.58 \pm 0.01}$
	4-planar	0.30 ± 0.02	0.35 ± 0.04	—	0.57 ± 0.02
	4-radial	0.29 ± 0.02	0.34 ± 0.03	—	0.57 ± 0.01
Kvasir-SEG	Vanilla	0.68 ± 0.18	0.69 ± 0.17	0.62 ± 0.07	—
	2-planar	0.62 ± 0.05	0.63 ± 0.05	$\mathbf{0.71 \pm 0.01}$	—
	2-radial	$\mathbf{0.63 \pm 0.03}$	$\mathbf{0.64 \pm 0.03}$	0.66 ± 0.06	—
	4-planar	0.63 ± 0.06	0.67 ± 0.05	0.71 ± 0.04	—
	4-radial	0.65 ± 0.04	0.67 ± 0.05	0.65 ± 0.07	—

where the varying nature in the ground truth follows different characteristics, e.g. encompassing non-linearities.

We have also compared the vanilla, 2-planar and 2-radial models by depicting their GEDs based on sample size (Appendix D). As expected, the GEDs decrease as the number of samples increase. It is also evident that the variability in metric evaluations is less for models with an NF posterior. The NF posteriors consistently outperform the vanilla PU-Net for the LIDC and Kvasir-SEG datasets.

4.2 Qualitative Evaluation

The mean and standard deviation based on 16 segmentation reconstructions from the validation set is shown in Figs. 2 and 3. Ideally, it is expected to obtain minimal uncertainty at the center of the segmentation, because annotations agree on the center area most of the time for our datasets. This also implies that the mean of the center should be high because of the agreement of the annotators. For both datasets, the means of our sampled segmentations match well with the ground truths and have high values in the center areas corresponding to good predictions. Furthermore, the PU-Net without an NF shows uncertainty at both edges and segmentation centers. In contrast, for all NF posterior PU-Net models, the uncertainty is mostly on the edges alone. A high uncertainty around the edges is also expected, since at those areas the annotators almost always disagree. From this, we can conclude that NF posterior models are better at quantifying the aleatoric uncertainty of the data. Even though there is no significant quantitative performance difference between the NF models, there is a well distinguishable difference in the visual analysis. In almost all cases, it can

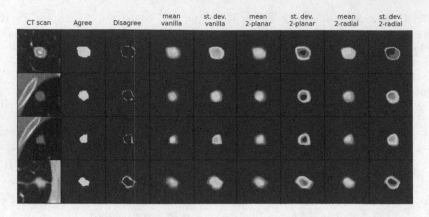

Fig. 2. Reconstructions of the LIDC test set.

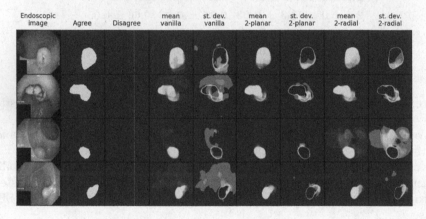

Fig. 3. Reconstructions of the Kvasir-SEG test set.

be observed that the planar is better than the radial NF posterior in learning the agreement between the segmentation centers. We also investigated the prior distribution to determine if it captures the ambiguity that exist in the input image. In Appendix E, we show the prior distribution variance for different test set input images. We qualitatively observe that with increasing variance, the subjective assessment of the annotation difficulty increases. This suggests the possibility of obtaining an indication of the uncertainty in a test input image without sampling and evaluating the segmentation reconstructions.

The prior distribution is an area that needs to be further explored, since this is still assumed Gaussian. We hypothesize that augmenting the prior with an NF could result in further improvements. Future work will also include an investigation into the correlation between the prior and segmentation variance. A limiting factor of the proposed model is the use of only a single distribution.

We consider that when using flexible distributions at multiple scales, the overall model will further improve.

5 Conclusion

Quantifying uncertainty in image segmentation is very important for decision-making in the medical domain. In this paper, we propose to use the broader concept of Normalizing Flows (NFs) for modeling both single- and multi-annotation data. This concept allows more complex modeling of aleatoric uncertainty. We consider modeling of the posterior distribution by Gaussians too restrictive to model variability. By augmenting the model posterior with a planar or radial NF, we attain up to 14% improvement in GED and 13% in IoU, resulting in a better quantification of the aleatoric uncertainty. We propose that density modeling with NFs is something that should be experimented with throughout other ambiguous settings in the medical domain, since we are confident this will result in valuable information for further research. A significant improvement has been found through only augmenting the posterior distribution with NFs, whereas little-to-none investigations have been made into the effect of additionally augmenting the prior distribution and is suggested for future work. Moreover, we suggest augmenting other architectures that aim to capture uncertainty and variability through a learnt probability distribution with Normalizing Flows.

Appendices

A Probabilistic U-Net Objective

The loss function of the PU-Net is based on the standard ELBO and is defined as

$$\mathcal{L} = -\mathbb{E}_{q_\phi(\mathbf{z}|\mathbf{s},\mathbf{x})}[\log p(\mathbf{s}|\mathbf{z},\mathbf{x})] + \mathrm{KL}\left(q_\phi(\mathbf{z}|\mathbf{s},\mathbf{x})||p_\psi(\mathbf{z}|\mathbf{x})\right), \tag{4}$$

where the latent sample $\mathbf{z}$ from the posterior distribution is conditioned on the input image $\mathbf{x}$, and ground-truth segmentation $\mathbf{s}$.

B Planar and Radial Flows

Normalizing Flows are trained by maximizing the likelihood objective

$$\log p(\mathbf{x}) = \log p_0(\mathbf{z}_0) - \sum_{i=1}^{K} \log\left(\left|\det \frac{df_i}{d\mathbf{z}_{i-1}}\right|\right). \tag{5}$$

In the PU-Net, the objective becomes

$$\log q(\mathbf{z}|\mathbf{s}, \mathbf{x}) = \log q_0(\mathbf{z}_0|\mathbf{s}, \mathbf{x}) - \sum_{i=1}^{K} \log \left(\left| \det \frac{df_i}{d\mathbf{z}_{i-1}} \right| \right), \tag{6}$$

where the i-th latent sample $\mathbf{z}_i$ from the Normalizing Flow is conditioned on the input image $\mathbf{x}$, and ground-truth segmentation $\mathbf{s}$.

The planar flow expands and contracts distributions along a specific directions by applying the transformation

$$f(\mathbf{x}) = \mathbf{x} + \mathbf{u}h(\mathbf{w}^T\mathbf{x} + \mathbf{b}), \tag{7}$$

while the radial flow warps distributions around a specific point with the transformation

$$f(\mathbf{x}) = \mathbf{x} + \frac{\beta}{\alpha\,|\mathbf{x} - \mathbf{x}_0|}(\mathbf{x} - \mathbf{x}_0). \tag{8}$$

C   Dataset Images

Here example images from the datasets used in this work can be seen. Figure 4 depicts four examples from the LIDC dataset. On the left in the figure the 2D CT image containing the lesion, followed by the four labels made by four independent annotators is shown. In Fig. 5, eight examples from the Kvasir-SEG dataset is depicted. An endoscopic image with its ground truth label can be seen.

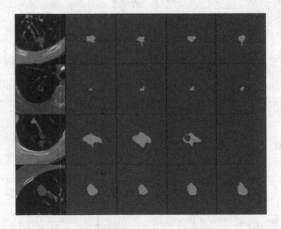

Fig. 4. Example images from the LIDC dataset.

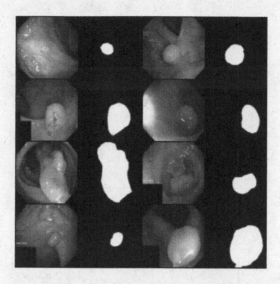

Fig. 5. Example images from the Kvasir-SEG dataset.

D   Sample Size Dependent GED

The GED evaluation is dependent on the number of reconstructions sampled from the prior distribution. Figure 6 depicts this relationship for the vanilla, 2-planar and 2-radial posterior models. The uncertainty in the values originate from the changing results when training with ten-fold cross validation. One can observe that with increasing sample size, the GED as well as the associated uncertainty decrease. This is also the case when the posterior is augmented with a 2-planar or 2-radial flow. Particularly, the uncertainty in the GED evaluation significantly decreases.

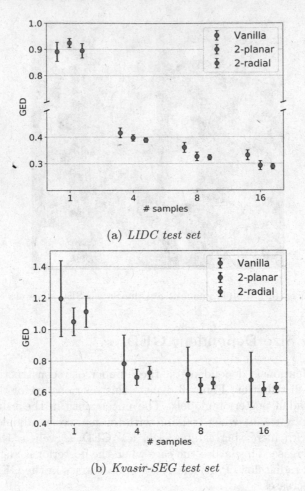

(a) *LIDC test set*

(b) *Kvasir-SEG test set*

Fig. 6. The GED based on sample size evaluated on the vanilla, 2-planar and 2-radial models.

E   Prior Distribution Variance

We investigated whether the prior distribution captures the degree of ambiguity in the input images. For every input image $\mathbb{X}$, we obtain a latent L-dimensional mean and standard deviation vector of the prior distribution $P(\boldsymbol{\mu}, \boldsymbol{\sigma}|\mathbb{X})$. The mean of the latent prior variance vector μ_{LV}, is obtained from the input images in an attempt to quantify this uncertainty. Figure 7 shows this for several different input images of the test set. As can be seen, the mean variance over the latent prior increases along with a subjective assessment of the annotation difficulty.

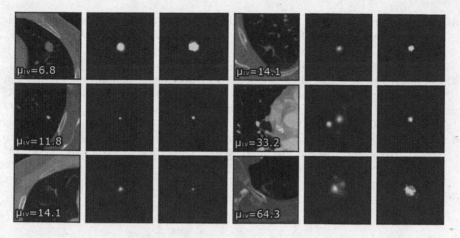

Fig. 7. Depicted in the CT image is the mean of the prior distribution variance of the 2-planar model. We show the input CT image, its average segmentation prediction (16 samples) and ground truth from four annotators.

References

1. Armato, S., III., et al.: The lung image database consortium (LIDC) and image database resource initiative (IDRI): a completed reference database of lung nodules on CT scans. Med. Phys. **38**, 915–931 (2011). https://doi.org/10.1118/1.3528204
2. Baumgartner, C.F., et al.: PHiSeg: capturing uncertainty in medical image segmentation (2019)
3. van den Berg, R., Hasenclever, L., Tomczak, J.M., Welling, M.: Sylvester normalizing flows for variational inference (2019)
4. Jha, D., et al.: Kvasir-SEG: a segmented polyp dataset. In: Ro, Y.M., et al. (eds.) MMM 2020. LNCS, vol. 11962, pp. 451–462. Springer, Cham (2020). https://doi.org/10.1007/978-3-030-37734-2_37
5. Kendall, A., Gal, Y.: What uncertainties do we need in Bayesian deep learning for computer vision? CoRR (2017). http://arxiv.org/abs/1703.04977
6. Kingma, D.P., Dhariwal, P.: Glow: generative flow with invertible 1x1 convolutions (2018)
7. Kingma, D.P., Welling, M.: Auto-encoding variational bayes (2014)
8. Kobyzev, I., Prince, S., Brubaker, M.: Normalizing flows: an introduction and review of current methods. IEEE Trans. Pattern Anal. Mach. Intell. p. 1 (2020). https://doi.org/10.1109/tpami.2020.2992934
9. Kohl, S.A.A., et al.: A hierarchical probabilistic U-Net for modeling multi-scale ambiguities (2019)
10. Kohl, S.A., et al.: A probabilistic u-net for segmentation of ambiguous images. arXiv preprint arXiv:1806.05034 (2018)
11. Pogorelov, K., et al.: KVASIR: a multi-class image dataset for computer aided gastrointestinal disease detection (2017). https://doi.org/10.1145/3083187.3083212
12. Rezende, D.J., Mohamed, S.: Variational inference with normalizing flows (2016)

13. Ronneberger, O., Fischer, P., Brox, T.: U-Net: convolutional networks for biomedical image segmentation. In: Navab, N., Hornegger, J., Wells, W.M., Frangi, A.F. (eds.) MICCAI 2015. LNCS, vol. 9351, pp. 234–241. Springer, Cham (2015). https://doi.org/10.1007/978-3-319-24574-4_28

14. Selvan, R., Faye, F., Middleton, J., Pai, A.: Uncertainty quantification in medical image segmentation with normalizing flows. In: Liu, M., Yan, P., Lian, C., Cao, X. (eds.) MLMI 2020. LNCS, vol. 12436, pp. 80–90. Springer, Cham (2020). https://doi.org/10.1007/978-3-030-59861-7_9

15. Sohn, K., Lee, H., Yan, X.: Learning structured output representation using deep conditional generative models. Adv. Neural. Inf. Process. Syst. **28**, 3483–3491 (2015)

UNSURE 2021: Domain Shift Robustness and Risk Management in Clinical Pipelines

Task-Agnostic Out-of-Distribution Detection Using Kernel Density Estimation

Ertunc Erdil[⊠], Krishna Chaitanya, Neerav Karani, and Ender Konukoglu

Computer Vision Lab., ETH Zurich, Sternwartstrasse 7, 8092 Zurich, Switzerland
ertunc.erdil@vision.ee.ethz.ch

Abstract. In the recent years, researchers proposed a number of successful methods to perform out-of-distribution (OOD) detection in deep neural networks (DNNs). So far the scope of the highly accurate methods has been limited to image level classification tasks. However, attempts for generally applicable methods beyond classification did not attain similar performance. In this paper, we address this limitation by proposing a simple yet effective task-agnostic OOD detection method. We estimate the probability density functions (pdfs) of intermediate features of a pre-trained DNN by performing kernel density estimation (KDE) on the training dataset. As direct application of KDE to feature maps is hindered by their high dimensionality, we use a set of lower-dimensional marginalized KDE models instead of a single high-dimensional one. At test time, we evaluate the pdfs on a test sample and produce a confidence score that indicates the sample is OOD. The use of KDE eliminates the need for making simplifying assumptions about the underlying feature pdfs and makes the proposed method task-agnostic. We perform experiments on classification task using computer vision benchmark datasets. Additionally, we perform experiments on medical image segmentation task using brain MRI datasets. The results demonstrate that the proposed method consistently achieves high OOD detection performance in both classification and segmentation tasks and improves state-of-the-art in almost all cases. Our code is available at https://github.com/eerdil/task_agnostic_ood. Longer version of the paper and supplementary materials can be found as preprint in [8].

Keywords: Out-of-distribution detection · Kernel density estimation

1 Introduction

Deep neural networks (DNNs) can perform predictions on test images with very high accuracy when the training and testing data come from the same distribution. However, the prediction accuracy decreases rapidly when the test image is sampled from a different distribution than the training one [16,33]. Furthermore, in such cases, DNNs can make erroneous predictions with very high confidence [11]. This creates a major obstacle when deploying DNNs for real applications, especially for the ones with a low tolerance for error, such as autonomous driving and medical diagnosis. Therefore,

© Springer Nature Switzerland AG 2021
C. H. Sudre et al. (Eds.): UNSURE 2021/PIPPI 2021, LNCS 12959, pp. 91–101, 2021.
https://doi.org/10.1007/978-3-030-87735-4_9

Input   Prediction   Ground Truth

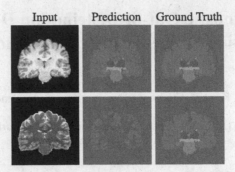

Fig. 1. A visual example demonstrating the importance of OOD detection on a segmentation task. A network trained on HCP_T1w images works well on a HCP_T1w test image (first row) while it produces a poor segmentation on the HCP_T2w image of the same patient (second row).

it is crucial to improve the robustness of DNN-based methods and prevent them from making big mistakes [1].

Recently, to improve the robustness of DNNs, substantial advances have been made for OOD detection in DNNs trained for image level classification tasks [12,15,22,24]. Although, OOD detection is equally crucial for non-classification tasks (e.g. segmentation), so far, attempts for developing more generic OOD detection methods did not attain similar performance [25]. For instance, a UNet [27] architecture trained on HCP_T1w images achieves average Dice score of 0.853 on the test set of the same dataset while it achieves 0.588 on ABIDE_Caltech_T1w and 0.107 on HCP_T2w datasets [16] where the upper bound (training and testing on the same dataset) is 0.896 for ABIDE_Caltech_T1w and 0.867 for HCP_T2w. In Fig. 1, we show visual results of test images from HCP_T1w and HCP_T2w datasets on a UNet model trained on HCP_T1w dataset. The example demonstrates that a DNN trained on T1w brain images (in-distribution (InD)) produces poor segmentation results for T2w images (OOD) of the same patient. Such results can be catastrophic in an automated image analysis pipeline and detecting OOD samples is a way to prevent such mistakes.

1.1   Related Work

One of the earlier methods proposed by Hendrycks et al. [12] uses maximum predicted class probability as a confidence score that the sample is OOD. ODIN [23] extended the baseline by applying an adversarial perturbation to the input image (referred to as input pre-processing) and temperature scaling before softmax to increase the difference between the prediction probabilities of InD and OOD samples. Hsu et al. [15] further extended ODIN, referred to as Generalized ODIN (G-ODIN), by introducing an additional output that indicates whether the input sample belongs to InD or OOD. The penultimate layer of a DNN is decomposed into two branches to model the conditional distribution of this indicator variable and its joint distribution with the class label. The conditional probability of the indicator variable is used as the confidence score while the ratio of the joint probability and the conditional probability serve as the final class prediction for an image after applying input processing. Yu et al. [36] propose a DNN with two classification heads, where one aims to minimize classification loss and the other aims to maximize the discrepancy between the two classifiers. The method, named as MCD, uses a subset of OOD samples along with the InD samples in the discrepancy loss. At test time, the samples with higher discrepancy are labeled as OOD. Liu et al. [24] proposed an energy-based method (EBM) which interprets softmax probabilities as energy scores and use for OOD detection. Lee et al. [22] proposed a method named as Mahalanobis, which models the class conditional pdfs of the features at intermediate layers of a DNN with Gaussian densities for InD samples. The parameters of each class conditional Gaussian are estimated by computing the empirical mean and co-variance using InD training samples belonging to that class. At test time, ODIN-style input preprocessing is applied before evaluating the estimated densities to obtain a confidence score, which is expected to be higher for InD samples and lower for OOD samples. More related work can be found in [6, 13, 21, 32].

Despite their successful performance, most of the aforementioned methods are designed for OOD detection for classification tasks and their extension to non-classification tasks usually is not trivial. Task-agnostic networks that do not share the same drawback have also been proposed. Hendryks et al. [14] proposed a self-supervised learning (SSL) based OOD detection method. The method trains an auxiliary rotation network, which predicts angle of rotation in discrete categories, on the InD dataset and computes a confidence score for a test image as the maximum of the softmax activation, expecting higher activations for InD samples compared to OOD samples. Kim et al. [18] proposed a method, referred to as RaPP, that is based on the observation that in an autoencoder the internal feature representations of an input image and its reconstructed version are very similar for InD samples and the similarity decreases for OOD samples that are not used for the training of the autoencoder. RaPP defines a confidence score based on this observation for OOD detection. Venkatakrishnan et al. [31] combine the ideas in SSL and RaPP, and propose a method (Multitask_SSL) by jointly training a network for both rotation prediction and reconstruction tasks for OOD detecion in brain images. As both SSL [14], RaPP [18], and Multitask_SSL [31] operate on auxiliary networks that are detached from the main network, they are task-agnostic and therefore can be applied to both classification and non-classification tasks.

1.2   Contribution

In this paper, we propose a simple yet effective task-agnostic OOD detection method. In the proposed method, we estimate feature pdfs of each channel in a DNN using KDE and InD training images. We evaluate the pdfs using a new test sample and obtain a confidence score for each channel. We combine all the scores into a final confidence score using a logistic regression model, which we train using channel-wise confidence scores of training images as InD samples and their adversarially perturbed versions as OOD samples.

We take our motivation from Mahalanobis [22] for developing the proposed method but extend it in multiple ways crucial for building a task-agnostic method that achieves improved detection accuracy. (1) Mahalanobis estimates class conditional densities while the distribution approximation in the proposed method is not conditioned on the class, making it task-agnostic. (2) Direct use of Mahalanobis in the task-agnostic setting is feasible using unconditioned Gaussian distributions to approximate layer-wise feature distributions of InD samples. However, the Gaussian assumption may be too restrictive to model unconditioned feature densities and lead to lower accuracy when the assumption does not hold. Using a nonparametric density estimation method KDEs, we extend the flexibility of the density approximation in the proposed method. (3) Layer-wise approximation is prone to the curse-of-dimensionality. Even though Mahalanobis takes channel-wise mean to reduce the dimension of a channel from $C \times H \times W$ to $C \times 1$ before density estimation, the resulting vector can still be high dimensional in modern architectures. We approximate 1D channel-wise distributions in the proposed method, which are simpler to estimate. This approach ignores dependencies between channels of a layer in the density estimation part but takes them into account in the logistic regression model that combines channel-wise scores.

The use of KDEs for OOD detection is not new. The first use dates back to 1994 when Bishop [2] applied KDE in the input space. In that work, the input was only 12-dimensional and application of KDE was feasible. In modern architectures, the input dimensions are often much larger thereby, making the direct application of Bishop's method infeasible. The application of a modified version of the method in high dimensional spaces is still possible by applying KDE over the distances between the test image and the training images [4,17]. Bishop's method as well as its modified version differs from the proposed method. In our method, we use multiple channel-wise KDEs and aggregate results.

Using KDE and estimating channel-wise pdfs are conceptually simple extensions that are very effective and yield substantial gains. We performed extensive comparisons on DNNs trained for classification and segmentation tasks. In the classification experiments, we use the common benchmark that contains 2 different classification networks trained on CIFAR-10 and CIFAR-100 datasets [20] and 6 other OOD datasets. We compare the proposed method with 6 methods in the literature most of which are either very recent or common baselines used in the literature. In the segmentation experiments, we use datasets for brain MRI segmentation and compare with 5 methods. In total, we compare with 10 different OOD detection methods to correctly position the proposed method within the current literature.

2 Method

Let us denote a set of training images with $X_{tr} = \{x_1, x_2, \ldots x_M\} \sim P_{in}$ and corresponding labels with $y_{tr} = \{y_1, y_2, \ldots, y_M\}$, where P_{in} denotes the InD. Let us also denote a DNN with f, trained using (X_{tr}, y_{tr}). f is more likely to perform good predictions on a test image x_{test} if $x_{test} \sim P_{in}$ and incorrect predictions if $x_{test} \sim P_{out}$, where $P_{out} \neq P_{in}$. In this section, we present the proposed task-agnostic KDE-based approach that identifies test images sampled from P_{out}.

The main output of the proposed method is a confidence score that indicates how likely a given sample belongs to OOD for the given DNN f. In this section, we describe how we compute this score. Let us assume that f consists of L layers and the feature map in a layer l for a given image x is denoted as $f_l(x)$ and it has dimensions $C_l \times H_l \times W_l$, where C_l, H_l, and W_l are the number of channels, height, and width of the feature map, respectively. We take the channel-wise mean of the feature map and reduce the dimensionality to $C_l \times 1$, as also done in [22]. We denote the resulting C_l-dimensional feature vector by $f'_l(x)$. We then estimate the marginal feature pdfs *for each channel* c using KDE:

$$p_{lc}(x) \approx \hat{p}_{lc}(x) = \frac{1}{M} \sum_{i=1}^{M} \mathcal{K}(f'_{lc}(x) - f'_{lc}(x_i); \sigma_{lc}) \tag{1}$$

where p_{lc} is the true marginal pdf of the features f'_{lc} in channel c of layer l, $\hat{p}_{lc}$ is the estimate of the pdf, and $\mathcal{K}(u, v; \sigma_{lc}) = e^{-(u-v)^2/\sigma_{lc}^2}$ is a 1D squared exponential kernel with σ_{lc} being the kernel size. We estimate σ_{lc} using Silverman's rule of thumb [29]. When using KDE, we use samples $x_i \in X_{tr}$ and thus model InD channel-wise pdfs with $\hat{p}_{lc}$. For a given sample x, $\hat{p}_{lc}(x)$ is the confidence score of channel c in layer l.

The advantage of estimating channel-wise pdfs over estimating layer-wise pdfs, as was done in [22], is performing density estimation in 1D space instead of C_l−D space. Typically, C_l can be very large in modern networks, and density estimation becomes less accurate in high-dimensions [28], channel-wise estimation avoids this.

In order to evaluate p_{lc} in Eq. (1) for a new sample, ideally we need to store all the InD training images in X_{tr}. In real-world applications where M is very large, storing the entire X_{tr} may not be feasible and the summation over M images in Eq. (1) can take very long. Improving the computational and memory efficiency of KDE-based methods are possible by defining an unbiased estimator [9], which simply uses a random subset of X_{tr} such that $\hat{X}_{tr} = \{x_{u_1}, x_{u_2}, \ldots, x_{u_N}\} \subset X_{tr}$ where $\{u_1, u_2, \ldots, u_N\} \subset \{1, 2, \ldots, M\}$ is a random subset of indices generated by sampling from a Uniform density, $\mathcal{U}(1, M)$, without replacement and $N << M$. Using the random subset, we replace Eq. (1) with the computationally more efficient unbiased estimator

$$p_{lc}(x) \approx \hat{p}_{lc}(x) = \frac{1}{N} \sum_{i=1}^{N} \mathcal{K}(f'_{lc}(x) - f'_{lc}(x_{u_i}); \sigma_{lc}). \tag{2}$$

In our experiments, we set $N = 5000$. In [8], we demonstrate results with different choices of N.

Estimating marginal pdfs using Eq. (2) does not model dependencies between channels. In the proposed method, we take into account such dependencies and compute the final confidence score using a logistic regression classifier $\mathcal{M}_x = \sum_{l=1}^{L} \sum_{c=1}^{C_l} \alpha_{lc} \hat{p}_{lc}(x)$ where α_{lc} are the weights that are learned as described next.

The role of the logistic regression model is to distinguish between InD and OOD samples given the channel-wise confidence scores. Training for the weights α_{lc} requires having access to both InD and OOD images. Although the InD images, X_{tr}, are already available, it is difficult to capture all possible images in P_{out}. Lee et al. [22] propose using adversarial examples obtained by FGSM [10] as samples from P_{out} for hyperparameter tuning. We use adversarial examples as OOD samples to train logistic regression in the proposed method. After obtaining OOD samples X_{tr}^{adv} by applying adversarial perturbation to the images in X_{tr} using FGSM, the logistic regression classifier is trained by using the confidence scores $\hat{p}_{lc}$ of X_{tr} and X_{tr}^{adv} as inputs, and the output labels are provided as positive for InD images (X_{tr}) and negative for the OOD (X_{tr}^{adv}) ones. Note that FGSM can work with any type of label that is for the task, e.g., image-level label for classification, ground truth mask for segmentation and so on. Therefore, using FGSM does not affect the task-agnostic nature of the proposed method. We present further details on the FGSM method in [8].

3 Experiments and Results

3.1 Experimental Details

In the classification experiments, we use two InD datasets: CIFAR-10 and CIFAR-100 [20][1]. We tested OOD detection performance of these models on four computer vision datasets (SVHN [26], TinyImageNet (TIN) [5], LSUN [35], iSUN [34][2]) as well as two datasets obtained by random Gaussian and Uniform noise, respectively.

In the segmentation experiments, we use images from 2 publicly available datasets for brain segmentation: Human Connectome Project (HCP) [30] and Autism Brain Imaging Data Exchange (ABIDE) [7]. HCP dataset contains both T1w and T2w images for each subject, while ABIDE dataset consists of T1w image from different imaging sites. HCP_T1w and HCP_T2w datasets contain images from 47 patients and we split 21 for training, 5 for validation, and 21 for testing. There are T1w images from 37 patients in both ABIDE_Caltech_T1w and ABIDE_Stanford_T1w datasets and we split 11, 5, 21 images for train, validation and test.

Using HCP and ABIDE datasets, we design 2 different experiments to evaluate OOD detection performance on segmentation task. In the first experiment, we train a UNet [27] architecture on ABIDE_Caltech_T1w images and use ABIDE_Stanford_T1w, HCP_T1w, and HCP_T2w images as OOD. In the second experiment, we train the UNet on HCP_T1w images and use ABIDE_Caltech_T1w, ABIDE_Stanford_T1w, and HCP_T2w as OOD. We choose UNet as the network architecture since it is the most common choice for medical image segmentation [3,19,27].

We use two quantitative measurements to evaluate the performance of OOD detecion methods: FPR at 95% TPR and the Area Under the Receiver Operating Characteristic curve (AUROC). In all evaluations, we take the InD as the positive class and OOD as the negative class. The proposed method is implemented in PyTorch and we run all experiments on a Nvidia GeForce Titan X GPU with 12 GB memory.

3.2 Results and Analysis

Results in Classification Tasks. We compare the proposed method with the Baseline method proposed by Hendryks et al. [12], ODIN [23], Mahalanobis [22], MCD [36], G-ODIN [15], and EBM [24] which are primarily designed for OOD detection in classification tasks. We took the results of G-ODIN from the corresponding experimental setting in the original paper (see Table 6 in [15]) since the code is not available.

---

[1] Pretrained models: https://github.com/pokaxpoka/deep_Mahalanobis_detector.

[2] TIN, LSUN, and iSUN are available at https://github.com/facebookresearch/odin.

Table 1. Quantitative results on CIFAR-10 InD dataset.

OOD	FPR at 95% TPR	AUROC
	Baseline/ODIN/Mahalanobis/MCD/G-ODIN/EBM/Proposed	
SVHN	25.77/16.65/8.37/60.61/10.50/6.86/**6.49**	89.88/95.42/98.12/72.86/97.80/98.19/**98.48**
TIN	28.37/11.24/18.89/40.44/18.60/35.88/**8.41**	90.53/96.78/96.73/89.75/96.10/86.21/**98.31**
LSUN	28.31/10.30/19.61/34.46/9.10/21.62/**3.80**	91.09/97.06/96.77/91.15/98.00/92.50/**99.01**
iSUN	28.02/12.37/22.46/37.72/11.20/22.52/**7.31**	91.01/96.03/96.34/89.89/97.60/92.03/**98.55**
Gaussian	6.44/2.69/**0.0**/4.21/**0.0**/0.13/**0.0**	97.11/98.45/**100.0**/97.14/**100.0**/99.96/**100.0**
Uniform	9.24/4.16/**0.0**/13.17/**0.0**/ 0.0/**0.0**	96.04/97.78/**100.0**/92.69/**100.0**/**100.0**/**100.0**

Table 2. Quantitative results on CIFAR-100 InD dataset.

OOD	FPR at 95% TPR	AUROC
	Baseline/ODIN/Mahalanobis/MCD/G-ODIN/EBM/Proposed	
SVHN	55.73/24.76/**15.53**/73.33/44.90/45.49/17.46	79.34/92.13/**97.01**/64.92/93.20/88.93/95.44
TIN	58.97/33.74/24.33/56.95/23.50/70.04/**7.64**	77.01/88.32/95.04/85.53/95.90/75.11/**98.38**
LSUN	64.71/37.09/28.68/58.40/23.20/67.99/**3.73**	75.58/87.70/94.66/84.97/96.10/76.45/**99.13**
iSUN	63.26/38.21/29.46/64.32/24.70/70.11/**6.07**	75.68/86.73/94.02/83.46/95.70/76.57/**98.75**
Gaussian	58.43/39.41/**0.0**/10.78/**0.0**/**0.0**/**0.0**	55.85/72.04/**100.0**/94.02/**100.0**/**100.0**/**100.0**
Uniform	32.04/18.49/**0.0**/15.99/**0.0**/**0.0**/**0.0**	85.13/89.81/**100.0**/92.34/**100.0**/**100.0**/**100.0**

ODIN, Mahalanobis, MCD, and EBM use a validation set from target OOD samples to tune hyperparameters and/or to build OOD detector. In our experiments, we use X^{adv} instead of target OOD when running these methods for a fair comparison since target OOD is usually not available in a real application. More details on how we run these methods can be found in Sect. 3.2 in [8].

We present the OOD detection results when CIFAR-10 and CIFAR-100 datasets are InD in Tables 1 and 2, respectively. The results in CIFAR-10 experiments demonstrate that the proposed methods achieves better OOD detection performance than the existing methods on all OOD datasets. In the experiments on CIFAR-100 dataset, our method produces the best OOD detection results on all datasets except SVHN where it achieves the second best results. The results on the classification tasks suggest that the the proposed method improves the state-of-the-art OOD detection methods in almost all cases.

Results in Segmentation Tasks. We present the OOD detection results when InD datasets are ABIDE_Caltech_T1w and HCP_T1w dataset in Tables 3 and 4, respectively. The results demonstrate that the proposed method improves the existing methods in all cases. Since Bishop [2] works on high-dimensional input space, it cannot achieve accurate density estimation and produces poor OOD detection results as expected. Here, the results of the self-supervised methods: SSL, RaPP, and Multitask_SSL, were lower than we expect, and we investigated further to interpret the results better. These methods exhibit diminished performance because the self-supervised networks generalize surprisingly well to OOD images. For example, the network trained on HCP_T1w images

Table 3. Quantitative results on ABIDE_Caltech_T1w InD dataset.

OOD	FPR at 95% TPR	AUROC
	Baseline/Bishop/SSL/ RaPP/Multitask_SSL/Proposed	
ABIDE_Stanford_T1w	78.90/76.21/49.60/69.57/66.48/**44.25**	48.26/81.71/63.45/52.44/54.30/**89.27**
HCP_T1w	88.06/79.72/63.51/87.20/76.58/**42.93**	39.30/75.83/55.13/40.11/45.35/**93.96**
HCP_T2w	80.37/41.77/81.28/57.85/70.39/**40.27**	42.19/92.93/43.82/52.78/47.88/**94.62**

Table 4. Quantitative results on HCP_T1w InD dataset.

OOD	FPR at 95% TPR	AUROC
	Baseline/Bishop/SSL/ RaPP/Multitask_SSL /Proposed	
ABIDE_Stanford_T1w	59.25/100.0/62.96/67.01/45.06/**44.78**	71.34/39.02/84.43/67.07/83.78/**90.42**
ABIDE_Caltech_T1w	83.26/100.0/58.94/99.68/63.88/**11.71**	68.41/17.21/87.22/59.38/79.56/**96.77**
HCP_T2w	47.55/94.98/76.89/47.94/61.62/**18.77**	72.88/62.12/57.79/70.56/73.39/**95.60**

for the SSL rotation task predicts the rotation angles with $\approx 75\%$ accuracy for both InD and OOD datasets. This holds for the case when we use ABIDE_Caltech_T1w as InD. Analogously, the autoencoder trained for RaPP successfully reconstructs OOD images and results diminished OOD detection accuracy. Note that the OOD detection accuracy are lower in segmentation experiments compared to classification ones for the proposed methods. We argue that this difference is due to having very large number of common pixels in InD and OOD images from background for the dataset in the segmentation experiments.

Table 5. Comparison between different combinations of density estimation methods (Gaussian and KDE) with feature spaces (layer-wise and channel-wise) in terms of FPR at 95% TPR in CIFAR-100 dataset.

	Layer-wise		Channel-wise	
	Gaussian (Mahalanobis)	KDE	Gaussian	KDE (Proposed)
SVHN	15.53	24.09	12.00	17.46
TIN	24.33	34.08	13.90	7.64
LSUN	28.68	28.47	5.10	3.73
iSUN	29.46	33.49	7.50	6.07

3.3   Channel-Wise vs Layer-Wise and KDE vs Parametric Estimation

In the proposed method, we perform channel-wise KDE. Compared to the closest work Mahalanobis [22], this introduces two changes in the density estimation, one in feature selection (layer-wise vs channel features) and the other in estimation methodology (KDE vs Gaussian). In this section, we quantify the contribution of each change. To this end, we perform OOD detection with all possible combinations. The results in Table 5 demonstrate that performing channel-wise density estimation leads to a large improvement on OOD detection accuracy compared to layer-wise density estimation. We argue that this improvement is due to achieving more accurate density estimation in 1D space with the channel-wise features. Dependencies between channels are taken into account in the logistic regression model. We also observe that performing KDE on the channel-wise features yields further improvements over using Gaussian in most cases. This is expected since KDE is more flexible and can lead to more accurate density estimations. We present further experiments to compare channel-wise vs layer-wise KDE in [8].

4 Conclusion

In this paper, we presented a task-agnostic OOD detection method that estimates feature densities for each channel of a DNN using KDE. Features of a test image are evaluated at the corresponding KDEs to obtain a confidence score per channel, which is expected to be higher for InD images than OOD ones. These scores are combined into a final score using logistic regression classifier, that is pre-trained using InD training images and their adversarially perturbed versions. Being task-agnostic, the proposed method can be applied to both classification and non-classification DNNs. We performed extensive experiments on both classification and segmentation networks and compare them with the state-of-the-art methods. The results demonstrate that the proposed method that uses channel-wise KDE improves state-of-the-art in majority of the cases. Possible research direction include performing experiments on 3D models and extending this work for pixel-wise OOD detection.

Acknowledgement. The presented work was partly funding by: 1. Personalized Health and Related Technologies (PHRT), project number 222, ETH domain, 2. Clinical Research Priority Program Grant on Artificial Intelligence in Oncological Imaging Network, University of Zurich, 3. Swiss Data Science Center (DeepMicroIA), 4. Swiss Platform for Advanced Scientific Computing (PASC), coordinated by Swiss National Super-computing Centre (CSCS). We also thank Nvidia for their GPU donation.

References

1. Amodei, D., Olah, C., Steinhardt, J., Christiano, P., Schulman, J., Mané, D.: Concrete problems in AI safety. arXiv preprint arXiv:1606.06565 (2016)
2. Bishop, C.M.: Novelty detection and neural network validation. IEE Proc.-Vis. Image Sig. Process. **141**(4), 217–222 (1994)
3. Chaitanya, K., Erdil, E., Karani, N., Konukoglu, E.: Contrastive learning of global and local features for medical image segmentation with limited annotations. In: Advances in Neural Information Processing Systems, vol. 33 (2020)
4. Cremers, D., Osher, S.J., Soatto, S.: Kernel density estimation and intrinsic alignment for shape priors in level set segmentation. Int. J. Comput. Vision **69**(3), 335–351 (2006). https://doi.org/10.1007/s11263-006-7533-5
5. Deng, J., Dong, W., Socher, R., Li, L.J., Li, K., Fei-Fei, L.: ImageNet: a large-scale hierarchical image database. In: 2009 IEEE Conference on Computer Vision and Pattern Recognition, pp. 248–255. IEEE (2009)

6. DeVries, T., Taylor, G.W.: Learning confidence for out-of-distribution detection in neural networks. arXiv preprint arXiv:1802.04865 (2018)
7. Di Martino, A., et al.: The autism brain imaging data exchange: towards a large-scale evaluation of the intrinsic brain architecture in autism. Mol. Psychiatry **19**(6), 659–667 (2014)
8. Erdil, E., Chaitanya, K., Karani, N., Konukoglu, E.: Task-agnostic out-of-distribution detection using kernel density estimation. arXiv preprint arXiv:2006.10712 (2020). https://arxiv.org/pdf/2006.10712.pdf
9. Erdil, E., Yildirim, S., Tasdizen, T., Cetin, M.: Pseudo-marginal MCMC sampling for image segmentation using nonparametric shape priors. IEEE Trans. Image Process. **28**(11), 5702–5715 (2019)
10. Goodfellow, I.J., Shlens, J., Szegedy, C.: Explaining and harnessing adversarial examples. arXiv preprint arXiv:1412.6572 (2014)
11. Guo, C., Pleiss, G., Sun, Y., Weinberger, K.Q.: On calibration of modern neural networks. In: Proceedings of the 34th International Conference on Machine Learning, vol. 70, pp. 1321–1330. JMLR. org (2017)
12. Hendrycks, D., Gimpel, K.: A baseline for detecting misclassified and out-of-distribution examples in neural networks. In: Proceedings of International Conference on Learning Representations (2017)
13. Hendrycks, D., Mazeika, M., Dietterich, T.: Deep anomaly detection with outlier exposure. In: International Conference on Learning Representations (2018)
14. Hendrycks, D., Mazeika, M., Kadavath, S., Song, D.: Using self-supervised learning can improve model robustness and uncertainty. In: Advances in Neural Information Processing Systems, pp. 15637–15648 (2019)
15. Hsu, Y.C., Shen, Y., Jin, H., Kira, Z.: Generalized ODIN: detecting out-of-distribution image without learning from out-of-distribution data. In: Proceedings of the IEEE/CVF Conference on Computer Vision and Pattern Recognition, pp. 10951–10960 (2020)
16. Karani, N., Erdil, E., Chaitanya, K., Konukoglu, E.: Test-time adaptable neural networks for robust medical image segmentation. Med. Image Anal. **68**, 101907 (2021)
17. Kim, J., Çetin, M., Willsky, A.S.: Nonparametric shape priors for active contour-based image segmentation. Signal Process. **87**(12), 3021–3044 (2007)
18. Kim, K.H., Shim, S., Lim, Y., Jeon, J., Choi, J., Kim, B., Yoon, A.S.: Rapp: novelty detection with reconstruction along projection pathway. In: International Conference on Learning Representations (2020)
19. Kohl, S.A., et al.: A probabilistic U-Net for segmentation of ambiguous images. arXiv preprint arXiv:1806.05034 (2018)
20. Krizhevsky, A., Hinton, G., et al.: Learning multiple layers of features from tiny images. Technical report, Citeseer (2009)
21. Lee, K., Lee, H., Lee, K., Shin, J.: Training confidence-calibrated classifiers for detecting out-of-distribution samples. In: ICLR 2018 (2018)
22. Lee, K., Lee, K., Lee, H., Shin, J.: A simple unified framework for detecting out-of-distribution samples and adversarial attacks. In: Advances in Neural Information Processing Systems, pp. 7167–7177 (2018)
23. Liang, S., Li, Y., Srikant, R.: Enhancing the reliability of out-of-distribution image detection in neural networks. arXiv preprint arXiv:1706.02690 (2017)
24. Liu, W., Wang, X., Owens, J., Li, Y.: Energy-based out-of-distribution detection. In: Advances in Neural Information Processing Systems, vol. 33 (2020)
25. Nalisnick, E., Matsukawa, A., Teh, Y.W., Gorur, D., Lakshminarayanan, B.: Do deep generative models know what they don't know? In: International Conference on Learning Representations (2018)
26. Netzer, Y., Wang, T., Coates, A., Bissacco, A., Wu, B., Ng, A.Y.: Reading digits in natural images with unsupervised feature learning (2011)

27. Ronneberger, O., Fischer, P., Brox, T.: U-Net: convolutional networks for biomedical image segmentation. In: Navab, N., Hornegger, J., Wells, W.M., Frangi, A.F. (eds.) MICCAI 2015. LNCS, vol. 9351, pp. 234–241. Springer, Cham (2015). https://doi.org/10.1007/978-3-319-24574-4_28

28. Scott, D.W.: Multivariate Density Estimation: Theory, Practice, and Visualization. Wiley, Hoboken (2015)

29. Silverman, B.W.: Density Estimation for Statistics and Data Analysis, vol. 26. CRC Press, Boco Raton (1986)

30. Van Essen, D.C., et al.: The WU-MINN human connectome project: an overview. Neuroimage **80**, 62–79 (2013)

31. Venkatakrishnan, A.R., Kim, S.T., Eisawy, R., Pfister, F., Navab, N.: Self-supervised out-of-distribution detection in brain CT scans. arXiv preprint arXiv:2011.05428 (2020)

32. Vyas, A., Jammalamadaka, N., Zhu, X., Das, D., Kaul, B., Willke, T.L.: Out-of-distribution detection using an ensemble of self supervised leave-out classifiers. In: Ferrari, V., Hebert, M., Sminchisescu, C., Weiss, Y. (eds.) ECCV 2018. LNCS, vol. 11212, pp. 560–574. Springer, Cham (2018). https://doi.org/10.1007/978-3-030-01237-3_34

33. Wang, D., Shelhamer, E., Liu, S., Olshausen, B., Darrell, T.: Fully test-time adaptation by entropy minimization. arXiv preprint arXiv:2006.10726 (2020)

34. Xu, P., Ehinger, K.A., Zhang, Y., Finkelstein, A., Kulkarni, S.R., Xiao, J.: TurkerGaze: crowdsourcing saliency with webcam based eye tracking. arXiv preprint arXiv:1504.06755 (2015)

35. Yu, F., Seff, A., Zhang, Y., Song, S., Funkhouser, T., Xiao, J.: LSUN: construction of a large-scale image dataset using deep learning with humans in the loop. arXiv preprint arXiv:1506.03365 (2015)

36. Yu, Q., Aizawa, K.: Unsupervised out-of-distribution detection by maximum classifier discrepancy. In: Proceedings of the IEEE International Conference on Computer Vision, pp. 9518–9526 (2019)

Out of Distribution Detection for Medical Images

Oliver Zhang[✉], Jean-Benoit Delbrouck, and Daniel L. Rubin

Stanford University, Stanford, CA, USA
{ozhang,jeanbenoit.delbrouck,rubin}@stanford.edu

Abstract. Neural network architectures behave in unpredictable ways when testing on inputs which do not resemble their training data. It is valuable to detect any out-of-distribution (OOD) inputs to make any overseers aware of the limitations of the model's output. To address this need, a large number of methods for detecting OOD inputs have been proposed and tested on small datasets such as CIFAR10, SVHN, or LSUN. The purpose of this study is to determine the effectiveness of different methods for OOD detection on the domain of medical images. We investigate three common OOD detection methods (Maximum Softmax Probability, Confidence Branch, and Outlier Exposure) and report their effectiveness on widely used medical image datasets. We find that OOD detection metrics are volatile and can have large changes in performance in a short amount of training steps. Moreover, we also observe that OOD detection is sensitive to the choice of hyperparameters. Our code is reproducible at this link (https://github.com/oliverzhang42/ood_medical_images).

Keywords: Out of distribution detection · Medical image processing · Deep learning

1 Introduction

Deep learning models [16] can provide high performance in a variety of applications, so long as the data seen at test time is similar to the training data. However, when there is a distribution mismatch, deep neural network classifiers tend to give high confidence predictions on anomalous test examples [23]. In the field of medical imaging, identifying such distributional shifts is an essential building block for safely deploying machine learning models for the medical community.

Several previous works seek to address these problems by giving deep neural network classifiers the ability to assign anomaly scores to inputs [1,4,9,10,19, 21]. So far, most investigations have been carried out on toy datasets such as

Electronic supplementary material The online version of this chapter (https://doi.org/10.1007/978-3-030-87735-4_10) contains supplementary material, which is available to authorized users.

C. H. Sudre et al. (Eds.): UNSURE 2021/PIPPI 2021, LNCS 12959, pp. 102–111, 2021.
https://doi.org/10.1007/978-3-030-87735-4_10

CIFAR-10 and CIFAR-100 [15], TinyImageNet [27], LSUN [28], MNIST [18], Gaussian Noise, and Uniform Noise as out-of-distribution data. Yet we argue that OOD detection performance on these smaller, simpler datasets does not necessarily reflect OOD detection performance on medical images. Importantly, medical imaging data differs from the aforementioned datasets in four ways. In contrast to natural image analysis, medical analysis must often deal with orientation invariance, high variance in feature scale (in x-ray images), and local specific features. Furthermore, medical images usually require larger image sizes for useful predictions (e.g., 224 pixels by 224 pixels rather than 32 pixels by 32 pixels).

In this work, we analyze the performance of several common OOD detection methods on medical images. First, we analyze the performance of three popular methods on large-scale (i.e., 224 by 224) medical datasets. We show that both Confidence Branch and Outlier Exposure outperforms the baseline Maximum Softmax Prediction. Outlier Exposure and Confidence Branch each have datasets where they outperform the other. Second, we show that each method's OOD detection performance is sensitive to hyperparameter tuning. Third, we demonstrate that the OOD detection metrics are moderately volatile and that volatility depends on both the dataset and the OOD detection method. Finally, we point out the need for a larger investigation of OOD volatility which covers more algorithms than our own.

2   Related Works

2.1   OOD Detection on Medical Images

In recent years, there has been limited work on applying existing OOD methods to medical images. Most papers [1,4,9,10,19,21] on OOD detection propose their own methods and test them on toy datasets such as CIFAR-10, CIFAR-100, TinyImageNet, LSUN, iSUN, MNIST, Gaussian Noise, and Uniform Noise. Of the other papers which consider medical images, most also seek to propose their own methods, some on specific domains (skin lesions [20], brain tumors [26], or breast cancer lymphoma [22]) and others more generally [6]. As a result, their papers only analyze their own methods rather than evaluating common methods' performance or training dynamics on common datasets.

An exception to this rule is Cao et al. [3], which reports the results of Classifier-only methods (such as Gaussian mixture, KNN classifier, and ODIN) and Auxiliary Models (which use autoencoders) on some medical datasets. This paper can be viewed as a broad comparison of many different approaches to OOD detection. Our paper, in contrast, analyzes a few approaches in greater depth and provides insight into the training dynamics (i.e., the volatility) of OOD detection methods. Moreover, this paper covers two methods not present in their comparison: Confidence Branch and Outlier Exposure.

2.2   OOD Methods Analysis

Additionally, our work relates to some analysis done elsewhere. Roady *et al.* [25] analyze the effectiveness of open-set methods on large-scale datasets, specifically ImageNet. We, like them, increase the difficulty and size of the task. However, the way in which our tasks become more difficult are different. ImageNet is difficult because it requires understanding many different classes and modalities. Medical images, on the other hand, requires making precise and accurate predictions when the signal is difficult to reconstruct.

The second work by Henriksson *et al.* [11] analyzes the training dynamics of OOD metrics on CIFAR10, TinyImageNet, FakeData, and SVHN. They identify the same volatility that we do, and we expand on their analysis in two ways. First, we provide metrics to quantify the volatility of OOD detection, and second, we show that this volatility exists on additional datasets and methods.

3   Materials and Methods

3.1   OOD Detection Methods

We chose to investigate three OOD detection methods on medical images– Maximum Softmax Probability, Confidence Branch, and Outlier Exposure. Maximum Softmax Probability, is a baseline method which involves taking the maximum from the final softmax classification layer as the model's confidence [9]. Confidence Branch, leverages an auxiliary neural network branch to output its confidence level [4]. In particular, the model is penalized but is given a hint if it outputs low confidence. Outlier Exposure, involves exposing a classifier to an auxiliary OOD dataset and training the classifier to be uncertain on this dataset [10]. During testing time, a separate OOD dataset is used to test whether our classifier can generalize to unknown anomalies as well.

We chose these methods because they are widely used, and in future work we hope to expand our analysis to other methods, temperature scaling methods [21], deep ensembles [17], Monte-Carlo dropout [5], and many of the domain-specific algorithms [20, 22, 26].

3.2   Medical Image Datasets

We use four datasets in our experiments. First, Diabetic Retinopathy (DR) provides a set of around 89,000 high-resolution retina images [2]. A clinician has rated the presence of diabetic retinopathy in each image on a scale of 0 to 4. Second, Musculoskeletal Radiographs (MURA) is a dataset of around 40,000 bone X-rays [24]. Algorithms are tasked with detecting medical abnormalities in musculoskeletal X-rays, including seven body parts: elbow, finger, forearm, hand, humerus, shoulder and wrist. The task is binary only asking whether the provided X-ray is normal or abnormal. Third, MIMIC Chest X-Ray (MIMIC-CXR) is a large publicly available dataset of chest radiographs in DICOM format [13]. The dataset contains 377,110 images corresponding to 227,835 radiographic studies performed. We make this task binary: findings or no findings. Fourth,

RNSA BoneAge provides hands x-rays of children, both male and female [7]. The target is to correctly identify the age of a child from the x-ray in months. Around 12,000 samples are available for 228 possible answers. Usually seen as a regression problem, we only use it as out-of-domain. To visualize the data, we include a figure in the appendix which displays example images from each data.

3.3 Evaluation Metrics

To be consistent with the previous works [4,9,10,21], we use the following out-of-distribution metrics to evaluate our trained detectors. We follow previous work [9] and use in-domain images as positives.

FPR95: the false positive rate of OOD examples when true positive rate of in-distribution examples is at 95%.

Detection Error: the misclassification probability when TPR is 95%, given by $0.5 \times (1 - \text{TPR}) + 0.5 \times \text{FPR}$, where positive and negative examples have equal probability of appearing in the test set.

AUROC: the Area Under the Receiver Operating Characteristic curve, which is a threshold-independent metric. The ROC curve depicts the relationship between TPR and FPR. The AUROC can be interpreted as the probability that a positive example is assigned a higher detection score than a negative example.

4 Experiments

To understand the performance of the OOD detection methods on real-world datasets, we trained each method on each of the Diabetic Retinopathy, MIMIC-CXR, and MURA datasets [2,13,24]. The BoneAge dataset was used exclusively for OOD detection [7]. For each training dataset, we balanced the labels such that each class contained the same number of training examples. In total, we keep 5,000, 28,000 and 50,000 training/validation samples for the Diabetic Retinopathy, MURA and MIMIC-CXR dataset. When acting as out-of-distribution, each dataset has 5000 samples.

4.1 Common Details

In each run, our network consists of a default ResNet50 [8] as a convolutional base, one dropout layer after the global average pooling with $p = 0.5$, and one final feedforward layer with softmax. We train the network from scratch using an Adam optimizer [14] on a cross-entropy loss with learning rate plateau scheduler and data augmentation. For further details, see the appendix.

4.2 Confidence Branch

For the Confidence Branch models, we used a hint rate of 0.5. Moreover, we did not use the budget hyperparameter but rather had a fixed lambda throughout training. To determine what would be a fitting lambda, we did hyperparameter tuning on lambda, with options 0.05, 0.1, and 0.5. In total, we ran 45 models with five models for each dataset hyperparameter pair. We found that $\lambda = 0.1$ worked best for the Diabetic Retinopathy dataset, while $\lambda = 0.5$ worked best for MIMIC-CXR and MURA. Check Sect. 5.2 for more details.

4.3 Outlier Exposure

For the Outlier Exposure models, we also performed some hyperparameter tuning. We tried out lambda equals 0.05, 0.1, and 0.5 and found that lambda equals 0.5 worked the best. Further details can be found in Sect. 5.2. In each of our Outlier Exposure experiments, the model was given images from the BoneAge dataset as examples of out-samples. For that reason, OOD detection performance on DR, MIMIC-CXR, and MURA are more representative of its real-world performance. Finally, for Outlier Exposure, we found that the method worked only when batch normalization [12] was not actively tracking running statistics.

5   Results

Table 1 displays the results averaged across five runs. We compare between the baseline model, confidence branch, and outlier exposure.

Table 1. OOD detection averaged over five runs

In-distribution dataset	Out-distribution dataset	In-domain accuracy	FPR at 95% TPR ↓	Detection error ↓	AUROC ↑
Baseline/Confidence Branch/Outlier Exposure					
DR (5 classes)	MIMIC-CXR	**0.45**/0.38/0.41	0.95/**0.08**/0.34	0.12/**0.03**/0.07	0.61/**0.99**/0.93
	MURA		0.98/**0.12**/0.21	0.12/**0.03**/0.05	0.44/**0.96**/0.96
	BoneAge		0.97/**0.08**/0.10	0.16/**0.02**/0.04	0.35/0.97/**0.98**
MIMIC-CXR (2 classes)	DR	**0.77**/0.77/0.73	0.83/0.82/**0.22**	0.73/0.73/**0.20**	0.83/0.73/**0.96**
	MURA		0.91/0.58/**0.08**	0.48/0.31/**0.06**	0.72/0.77/**0.98**
	BoneAge		0.91/0.68/**0.04**	0.54/0.41/**0.04**	0.70/0.75/**0.99**
MURA (2 classes)	MIMIC-CXR	**0.68**/0.67/0.61	0.98/**0.93**/0.95	0.50/**0.48**/0.50	0.33/**0.65**/0.55
	DR		1.00/**0.27**/0.96	0.88/**0.25**/0.85	0.08/**0.85**/0.54
	BoneAge		0.94/1.00/**0.29**	0.56/0.57/**0.19**	0.57/0.35/**0.95**

5.1 Performance

Accuracy. In every case, the baseline model was more accurate compared to Confidence Branch or Outlier Exposure. Outlier exposure was consistently four to seven percentage points behind the baseline. On the other hand, Confidence Branch's performance varied from dataset to dataset. On MIMIC-CXR and MURA, it only lagged behind the baseline model at most one percentage point, whereas on Diabetic Retinopathy, it performed worse by seven percentage points.

OOD Detection. Conversely, both the Confidence Branch and Outlier Exposure models consistently outperformed the baseline model in OOD detection, often by very large margins. Confidence Branch and Outlier Exposure therefore make a sacrifice in accuracy to gain much better OOD detection. When compared to each other, Confidence Branch and Outlier Exposure each do better on different in-distribution datasets. In these experiments, Confidence Branch outperforms Outlier Exposure on Diabetic Retinopathy and MURA, while Outlier Exposure outperforms Confidence Branch on MIMIC-CXR. Neither can outperform the other; instead, both are competitive options worth testing.

5.2 Effects of Hyper Parameters

In our analysis, we analyze the performance of both Confidence Branch and Outlier Exposure when their newly introduced hyperparameter, λ was equal to $0.05, 0.1$, and 0.5. All other hyperparameters remain constant.

Accuracy. We found that tuning each algorithm's newly introduced hyperparameter made a small difference on the average accuracy across the five runs. Given an algorithm and a dataset, tuning the hyperparameter only provided gains of 1–2 percentage points in accuracy.

OOD Detection. Instead, we found that tuning the new hyperparameters made a much bigger impact on the OOD detection. For the Confidence Branch models, $\lambda = 0.5$ outperformed the second-best option by 0.15 on the MIMIC-CXR dataset and 0.32 on the MURA dataset. On the retina dataset, $\lambda = 0.1$ was much more stable, outperforming $\lambda = 0.05$ by 0.2 and $\lambda = 0.5$ by 0.76. For Outlier Exposure, $\lambda = 0.5$ was optimal for MIMIC-CXR and DR, outperforming the second-best option by 0.34 and 0.46. Outlier Exposure did not solve MURA, with all runs having FPR95 above 0.9. Figure 1 helps visualize these results.

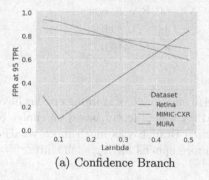
(a) Confidence Branch

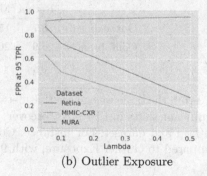
(b) Outlier Exposure

Fig. 1. Out of distribution detection changes significantly as lambda varies.

5.3   Volatility

We observe that OOD detection metrics can be volatile, especially with Confidence Branch. For instance, Fig. 2 shows a run using Confidence Branch and $\lambda = 0.1$ trained on the retina dataset. We can see that FPR95 ranges from 1.0 to 0.0 to 1.0 within the span of five epochs (epochs 12–16). Likewise, these patterns are inverted with AuROC: when the FPR at 95 TPR rises, the AuROC falls and vice versa. These graphs demonstrate that the OOD detection metrics are volatile. One epoch change an exemplary OOD detector to a poor one.

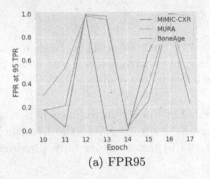
(a) FPR95

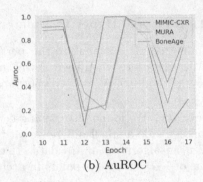
(b) AuROC

Fig. 2. OOD detection metrics can be volatile.

Volatility Across Methods and Datasets. We try to quantify this volatility. In particular, for each method and in-distribution dataset pair, we calculate the proportion of training epochs which change FPR95 by 25% (Table 2).

Table 2. Proportion of epochs which change FPR95 by 25%

	Retina	Mimic-CXR	MURA	Average
Out. exposure	8.7	15.8	4.0	9.5
Conf. branch	26.9	20.7	9.3	19.0

From the table we can see that between 4% and 26.9% of training epochs have significant shifts in FPR95. Moreover, volatility was different based on choice of method and dataset. For instance, Confidence Branch was much more volatile compared to Outlier Exposure, with 9.5% more epochs having such shifts.

Analysis of Volatility. Intuitively, we expect OOD metrics to be moderately volatile, as the model's OOD performance does not factor directly into its training. We hypothesize that a lack of consistent optimization pressure on OOD performance leads to volatility.

Our hypothesis may explain why Outlier Exposure has less volatile OOD detection than Confidence Branch. Outlier Exposure attempts to incorporate OOD performance directly into its loss function by providing an example class of OOD images. The consistent optimization pressure may help stabilize OOD performance. In constrast, Confidence Branch only trains on in-domain examples, with the hope that the model will recognize its uncertainty on OOD images. With Confidence Branch, uncertainty on OOD images is therefore a byproduct of the training process and is not directly accounted for in the loss function. This may account for the volatility in Confidence Branch's OOD performance.

Further Study. The volatility of the OOD detection methods warrants further study. We only found one other paper discussing this volatility [11]. We hypothesize that the volatility of OOD detection depends both on the complexity of the dataset as well as the OOD detection technique. In future work, we hope to expand this analysis to cover more datasets and more OOD detection methods.

6   Conclusion

In this paper, we have applied three common out-of-distribution detection techniques to medical images. We show that both Confidence Branch and Outlier Exposure outperform the Maximum Softmax Probability baseline model in OOD detection, but suffer small losses in accuracy. We also show that hyperparameters have a large effect on the OOD detection performance of both Confidence Branch and Outlier Exposure. Finally, we demonstrate that OOD detection performance can be volatile. In future work, we hope to continue to study the application of OOD detection to medical images, especially the volatility which arises from training on medical images.

References

1. Bendale, A., Boult, T.: Towards open set deep networks. arXiv:1511.06233 [cs], November 2015. http://arxiv.org/abs/1511.06233
2. California Healthcare Foundation, EyePACS: Diabetic Retinopathy Detection (2015). https://www.kaggle.com/c/diabetic-retinopathy-detection/overview
3. Cao, T., Huang, C.W., Hui, D.Y.T., Cohen, J.P.: A benchmark of medical out of distribution detection (2020)
4. DeVries, T., Taylor, G.W.: Learning confidence for out-of-distribution detection in neural networks. arXiv preprint arXiv:1802.04865 (2018)
5. Gal, Y., Ghahramani, Z.: Dropout as a bayesian approximation: representing model uncertainty in deep learning, October 2016. https://arxiv.org/abs/1506.02142
6. Gao, L., Wu, S.: Response score of deep learning for out-of-distribution sample detection of medical images. J. Biomed. Inform. **107**, 103442 (2020). https://doi.org/10.1016/j.jbi.2020.103442
7. Halabi, S.S., et al.: The RSNA pediatric bone age machine learning challenge. Radiology **290**(2), 498–503 (2019)

8. He, K., Zhang, X., Ren, S., Sun, J.: Deep residual learning for image recognition. arXiv preprint arXiv:1512.03385 (2015)
9. Hendrycks, D., Gimpel, K.: A baseline for detecting misclassified and out-of-distribution examples in neural networks. In: International Conference on Learning Representations (2016)
10. Hendrycks, D., Mazeika, M., Dietterich, T.: Deep anomaly detection with outlier exposure. In: International Conference on Learning Representations (2018)
11. Henriksson, J., Berger, C., Borg, M., Tornberg, L., Raman Sathyamoorthy, S., Englund, C.: Performance analysis of out-of-distribution detection on trained neural networks. Inform. Softw. Technol. **130**, 106409 (2021). https://doi.org/10.1016/j.infsof.2020.106409
12. Ioffe, S., Szegedy, C.: Batch normalization: accelerating deep network training by reducing internal covariate shift. arXiv:1502.03167 [cs], March 2015. http://arxiv.org/abs/1502.03167
13. Johnson, A.E., et al.: MIMIC-CXR-JPG, a large publicly available database of labeled chest radiographs. arXiv preprint arXiv:1901.07042 (2019)
14. Kingma, D.P., Ba, J.: Adam: a method for stochastic optimization, January 2017. https://arxiv.org/abs/1412.6980
15. Krizhevsky, A.: Learning multiple layers of features from tiny images. University of Toronto (2009)
16. Krizhevsky, A., Sutskever, I., Hinton, G.E.: ImageNet classification with deep convolutional neural networks. In: NIPS (2012)
17. Lakshminarayanan, B., Pritzel, A., Blundell, C.: Simple and scalable predictive uncertainty estimation using deep ensembles, November 2017. https://arxiv.org/abs/1612.01474
18. Lecun, Y., Bottou, L., Bengio, Y., Haffner, P.: Gradient-based learning applied to document recognition. Proc. IEEE **86**(11), 2278–2324 (1998). https://doi.org/10.1109/5.726791
19. Lee, K., Lee, H., Lee, K., Shin, J.: Training confidence-calibrated classifiers for detecting out-of-distribution samples. In: International Conference on Learning Representations (2018)
20. Li, X., Lu, Y., Desrosiers, C., Liu, X.: Out-of-distribution detection for skin lesion images with deep isolation forest. In: Liu, M., Yan, P., Lian, C., Cao, X. (eds.) MLMI 2020. LNCS, vol. 12436, pp. 91–100. Springer, Cham (2020). https://doi.org/10.1007/978-3-030-59861-7_10
21. Liang, S., Li, Y., Srikant, R.: Enhancing the reliability of out-of-distribution image detection in neural networks. In: International Conference on Learning Representations (2018)
22. Linmans, J., van der Laak, J., Litjens, G.: Efficient out-of-distribution detection in digital pathology using multi-head convolutional neural networks. In: Arbel, T., Ben Ayed, I., de Bruijne, M., Descoteaux, M., Lombaert, H., Pal, C. (eds.) Proceedings of the Third Conference on Medical Imaging with Deep Learning. Proceedings of Machine Learning Research, vol. 121, pp. 465–478. PMLR, 06–08 July 2020. http://proceedings.mlr.press/v121/linmans20a.html
23. Nguyen, A., Yosinski, J., Clune, J.: Deep neural networks are easily fooled: high confidence predictions for unrecognizable images. In: Proceedings of the IEEE Conference on Computer Vision and Pattern Recognition, pp. 427–436 (2015)
24. Rajpurkar, P., et al.: MURA: large dataset for abnormality detection in musculoskeletal radiographs. arXiv preprint arXiv:1712.06957 (2017)

25. Roady, R., Hayes, T.L., Kemker, R., Gonzales, A., Kanan, C.: Are open set classification methods effective on large-scale datasets? Plos One **15**(9) (2020). https://doi.org/10.1371/journal.pone.0238302

26. Wang, N., Chen, C., Xie, Y., Ma, L.: Brain tumor anomaly detection via latent regularized adversarial network. CoRR abs/2007.04734 (2020). https://arxiv.org/abs/2007.04734

27. Wu, J., Zhang, Q., Xu, G.: Tiny ImageNet challenge. Technical report, Stanford University (2017)

28. Yu, F., Seff, A., Zhang, Y., Song, S., Funkhouser, T., Xiao, J.: LSUN: construction of a large-scale image dataset using deep learning with humans in the loop. arXiv preprint arXiv:1506.03365 (2015)

Robust Selective Classification of Skin Lesions with Asymmetric Costs

Jacob Carse[1], Tamás Süveges[1], Stephen Hogg[1], Emanuele Trucco[1],
Charlotte Proby[2,3], Colin Fleming[3], and Stephen McKenna[1(✉)]

[1] CVIP, School of Science and Engineering, University of Dundee, Scotland, UK
{j.carse,t.suveges,s.c.z.hogg,e.trucco,s.j.z.mckenna}@dundee.ac.uk
[2] School of Medicine, Ninewells Hospital and Medical School, Dundee, UK
c.proby@dundee.ac.uk
[3] Department of Dermatology, Ninewells Hospital and Medical School, Dundee, UK
colin.fleming@nhs.scot

Abstract. Automated image analysis of skin lesions has potential to improve diagnostic decision making. A clinically useful system should be selective, rejecting images it is ill-equipped to classify, for example because they are of lesion types not represented well in training data. Furthermore, lesion classifiers should support cost-sensitive decision making. We investigate methods for selective, cost-sensitive classification of lesions as benign or malignant using test images of lesion types represented and not represented in training data. We propose EC-SelectiveNet, a modification to SelectiveNet that discards the selection head at test time, making decisions based on expected costs instead. Experiments show that training for full coverage is beneficial even when operating at lower coverage, and that EC-SelectiveNet outperforms standard cross-entropy training, whether or not temperature scaling or Monte Carlo dropout averaging are used, in both symmetric and asymmetric cost settings.

1 Introduction

Automated image analysis of skin lesions has great potential to improve diagnostic decision making and efficiency of clinical workflows in dermatology and primary care. Lesion classifiers that produce class probability distributions could be used to estimate the expected costs of clinical decisions such as whether or not to refer a patient, and thus inform effective decision making. Costs associated with mis-classification are usually asymmetric: deciding that a skin lesion is benign when it is really malignant is more costly than deciding it is malignant when it is benign. Optimal decision making requires predicted class probabilities to be well-calibrated. In addition, a clinically useful system should ascertain

Electronic supplementary material The online version of this chapter (https://doi.org/10.1007/978-3-030-87735-4_11) contains supplementary material, which is available to authorized users.

whether it has been sufficiently well trained to deal with the image under inspection. This is important for robustness and clinical usability. Classifiers should be selective, rejecting images they are ill-equipped to deal with; in particular, not all lesion types will be represented well in training data. Here we investigate methods for selective, cost-sensitive skin lesion classification. We focus on binary classification of malignant versus benign lesions using an experimental setup with test data from disease types represented in the training data as well as types not represented in the training data. Images were sourced from the ISIC 2019 data set [3,4,23].

We use empirical coverage and selective cost to evaluate performance and stress that selection and classification decisions must take into account asymmetry of mis-classification costs in the diagnostic setting (Sect. 3). We propose a modification to SelectiveNet [9] which we call EC-SelectiveNet (Sect. 4). SelectiveNet learns representations targeting expected image rejection rates by using two additional heads, a selection head and an auxiliary head, in addition to the usual predictive head; EC-SelectiveNet discards these additional heads at test time and makes selection decisions based on expected costs instead.

We provide empirical evidence that training selective networks for full coverage works well on skin lesion images, even when the desired coverage is lowered, somewhat counter to expectation (Sect. 5). We show that EC-SelectiveNet outperforms corresponding cross-entropy trained networks in both asymmetric and symmetric cost settings, whether or not temperature scaling [10] or Monte Carlo dropout averaging [8] are used (Sect. 5).

2   Background

AI systems for telediagnosis with performance comparable to human dermatologists have been demonstrated in some settings [1,7,11–13,16], providing evidence that deep learning can, if appropriately designed and integrated, assist diagnostic decision making effectively. However, deep learning models often overfit, resulting in over-confident predictions, and can struggle to decide which lesion images they are equipped to classify reliably [17]. Nevertheless, it has been noted that simply thresholding the maximum softmax response can be effective for rejecting images and reducing mis-classifications [14].

MC-Dropout [8] can be used to quantify uncertainty. It has been used in medical image analysis [18] including estimation of lesion segmentation quality [6] and provision of selection scores for active learning [2]. It uses dropout [15] at inference time, performing M forward passes of the model f on an image, x. Each pass is treated as a sample in a Bayesian approximation of a Gaussian process. Predictions are averaged to give an expected prediction $\hat{y} = \frac{1}{M} \sum_{m=1}^{M} f^m(x)$. Measures of uncertainty such as sample variance can also be calculated.

Temperature scaling can be used to improve calibration of class probabilities predicted by a network [10]. This can be important when making cost-sensitive decisions. Mozafari et al. [19] used temperature scaling with skin lesions and indicated potential hazards when working with noisy validation data. Nixon et al. [20] investigated calibration metrics and used temperature scaling. Temperature scaling [10] applies a scaling factor T to the output logits: $\hat{y}_i = \frac{\exp(\frac{z_i}{T})}{\sum_j \exp(\frac{z_j}{T})}$. The value of T is calculated by minimizing calibration error on a validation set.

SelectiveNet [9] jointly learns a classifier and selection function so that the deep representation can be learned with the expectation that some proportion of images should be rejected. We describe this in Sect. 4.

3   Robust Selective Classification

Selective classification is performed using a *selection function* and a *prediction function*. The selection function, $\sigma(x)$, indicates whether or not an image x should be rejected, in which case $\sigma(x) = 0$, or selected, in which case $\sigma(x) = 1$. Given a data set, S, of N images, the empirical *coverage*, $\phi(\sigma|S)$, is the proportion of images selected for classification, i.e., $\phi(\sigma|S) = \frac{1}{N}\sum_{i=1}^{N}\sigma(x_i)$. A classification decision is made for each selected image based on the classifier's prediction function; each such decision incurs a cost. The empirical *selective cost*, is this cost averaged over the selected images.

In general, mis-classification costs can be specified as a matrix C, where C_{jk} is the cost of assigning class k when the true class is j. These costs depend on the deployment setting and specifically on factors such as health economics, quality of life considerations, and available treatments. Many reported experiments on classification of dermatology images implicitly use $C = 1 - I$ where 1 is a matrix of ones and I is the identity matrix. This is unrealistic, as costs are in fact far from symmetric. Indeed, many medical classification tasks have highly asymmetric costs.

In this paper, we consider binary classification with class labels *malignant* (class 1) and *benign* (class 0). In a setting where mis-classifying a malignant lesion as benign is an order of magnitude more costly than mis-classifying a benign lesion as malignant, we have $C_{1,0} = 10.0, C_{0,1} = 1.0, C_{1,1} = 0.0, C_{0,0} = 0.0$. These values for the asymmetric costs were deemed reasonable through discussion with dermatolgists. The values used should however vary depending on the clinical setting, and further work should be done to investigate this in consultation with general practitioners, patient representative groups and health economists. Here we run experiments under different settings with the cost of mis-classifying a benign lesion as malignant, $C_{1,0}$, set to 1 (symmetric costs), 10, and 50 (highly asymmetric costs).

We use cost-coverage curves, showing the trade-off between cost incurred and coverage achieved, to characterize the performance of selective classifiers. A strongly performing selective classifier will have low cost and high coverage. As well as benign and malignant lesions from disease types present during training, we also test using images of benign and malignant lesions of disease types not represented in the training data. A robust system should either reject such data or maintain low selective cost on it.

4 SelectiveNet and EC-SelectiveNet

Deep representations can be learned specifically for a situation in which some proportion of data are expected to be rejected. SelectiveNet [9] trains a network end-to-end for a specific target coverage. This is enabled by adding two extra heads to the encoder, in addition to the standard predictive head p: a selective head g that outputs a selection score, and an auxiliary head a that outputs predictions used within the loss function. At test time, select/reject decisions are based on the output of the selective head. Here we propose the Expected-Costs SelectiveNet (EC-SelectiveNet) which modifies the network at test time. Specifically, the additional heads are discarded after training and select/reject decisions are based instead on expected costs computed using predicted class probabilities.

4.1 SelectiveNet

The SelectiveNet loss function (Eq. 1), is a combination of two functions ($L_{p,g}$ and L_a) weighted with a hyper-parameter α to control the relative importance of coverage optimization [9]:

$$L = \alpha L_{p,g} + (1 - \alpha)L_a \tag{1}$$

The first term uses predictive and selective heads (Eq. 2) and combines cross-entropy loss, l, with coverage. It uses hyper-parameter t as the target coverage for the model and λ to control the importance of this target coverage. The auxiliary head uses a standard cross-entropy loss for L_a, and is used to encourage the model to learn robust features from the training data.

$$L_{p,g} = \frac{1}{N\phi(g)} \sum_{i=1}^{N} l(p(x_i), y_i)g(x_i) + \lambda \max(t - \phi(g|S), 0)^2 \tag{2}$$

4.2 Selective Classification Based on Expected Costs

Given any trained classifier that outputs a (calibrated) posterior distribution $P(c|x)$ over classes given an image x, the expected costs of classification and rejection decisions can be used to decide *whether to select* and *how to classify* the image. Specifically, in the case of two classes, $c = 0$ (benign) and $c = 1$ (malignant), the expected cost of deciding benign is $R_0 = C_{10}P(c = 1|x)$ and the expected cost of deciding malignant is $R_1 = C_{01}P(c = 0|x)$. We should decide that x is in class 1 if $R_1 < R_0$, otherwise x is in class 0. Suppose that by rejecting an image we incur a cost θ. An optimal decision rule is then to reject x if $\min(R_0, R_1) > \theta$ and otherwise to decide the class with the lower expected cost. Note that a cost coverage plot can be generated by varying θ.

4.3   EC-SelectiveNet

Although SelectiveNet directly outputs a selection score, we propose to base selection instead on expected costs computed from the predictive head. We refer to this method as EC-SelectiveNet. The selective head is used during training to guide representation learning but, unlike [9], we discard the selective head along with the auxiliary head at test time.

Optionally, we apply temperature scaling to improve calibration to assist reliable estimation of expected costs. Temperature scaling was applied to the logit outputs of the predictive head p.

5   Experiments

Dataset and Implementation Setup. We used data from the ISIC Challenge 2019 [3,4,23] which consists in total of 25,331 images covering 8 classes: melanoma, melanocytic nevus, basal cell carcinoma, actinic keratosis, benign keratosis, dermatofibroma, vascular lesion, and squamous cell carcinoma. We compiled two datasets which we refer to as S_{in} and $S_{unknown}$.

S_{in}: These data were the melanoma, melanocytic nevus and basal cell carcinoma (BCC) images from the ISIC 2019 data. They were assigned to two classes for the purposes of our experiments: *malignant* (melanoma, BCC) and *benign* (melanocytic nevus). S_{in} was split into training, validation, and test sets consisting of 12432, 3316, and 4972 images respectively.

$S_{unknown}$: These data consisted of 4,360 ISIC 2019 images from classes that were not present in S_{in}, namely benign keratosis, dermatofibroma, actinic keratosis, and squamous cell carcinoma. They were assigned to *malignant* or *benign*. $S_{unknown}$ was not used for training but for testing selective classification performance on images from disease types not represented in the training data.

We refer to the union of the S_{in} and $S_{unknown}$ test sets as $S_{combined}$. Figure 1 shows example images.

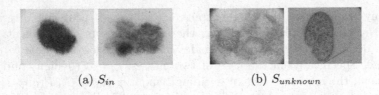

(a) S_{in} (b) $S_{unknown}$

Fig. 1. Example images from the test data sets S_{in} and $S_{unknown}$.

All code used for experiments can be downloaded from the project Github repository[1] along with reproduction instructions, trained models and expanded

---

[1] GitHub Repository: https://github.com/UoD-CVIP/Selective_Dermatology.

testing metrics. For all experiments we use an EfficientNet [22] encoder with compound coefficient 7, pre-trained on ImageNet [5]. Models were trained using stochastic gradient descent. Cross-entropy loss was used with a two-output softmax. SelectiveNet hyperparameters were $\alpha = 0.5$ and $\lambda = 32$ as recommended in [9]. MC-Dropout used a dropout rate of 50% and $M = 100$ samples. Learning rates were adjusted using a cyclical scheduler [21] that cycled between 10^{-4} and 0.1. Batch size was 8 to enable each batch to fit on our Nvidia RTX 2080TI GPU. Each model was trained for a total of 25 epochs with the weights from the model with the lowest validation loss used for testing.

SelectiveNet: Effect of Target Coverage. We examined the effect of the SelectiveNet target-coverage parameter, t, when SelectiveNet's selection head is used to make selection decisions. Figure 2 shows cost-coverage curves for values of t ranging from 0.7 to 1.0. These are plotted for S_{in}, $S_{unknown}$, and $S_{combined}$.

We expected to find, in accordance with the intended purpose of this parameter, that lower values of t would be effective at lower coverage. On the contrary, training with $t = 1.0$ incurred the lowest test cost on S_{in} for coverage values as low as 0.2. Costs incurred on $S_{unknown}$ are higher as expected, and curves show no clear ordering; the $t = 1.0$ curve, however, does show a clear reduction in cost as coverage is reduced.

Does SelectiveNet Training Help? The extent to which the target coverage t is enforced is controlled by the weighting parameter λ. Even when set to target full coverage ($t = 1.0$), the model can trade off coverage for cost in extreme cases during training. For this reason, results obtained by SelectiveNet with $t = 1.0$ will differ from those obtained by training a network without selective and auxiliary heads. We trained such a network using cross-entropy loss, retaining only the softmax predictive head. It made selection decisions at test time based on the maximum softmax output. The resulting cost-coverage curve is plotted in Fig. 2 (labelled 'softmax'). SelectiveNet trained with a target coverage of 1.0 performed better than a standard CNN with softmax for any coverage above 0.4.

MC-Dropout, Temperature Scaling, and EC-SelectiveNet. We investigated the effect of MC-Dropout on selective classification, using the mean and variance of the Monte Carlo iterations as selection scores, respectively. Figure 3 compares the resulting cost-coverage curves with those obtained using a network with no dropout at test time ('softmax response'). On S_{in}, using the MC-Dropout average had negligible effect whereas MC variance performed a little worse than simply using the maximum softmax response. In contrast, gains in cost were obtained by MC variance on $S_{unknown}$ for which model uncertainty should be high.

Figure 4 plots curves for a softmax network using temperature scaling (trained with cross-entropy loss). Although temperature scaling improved calibration it had negligible effect on cost-coverage curves. Figure 4 also shows curves

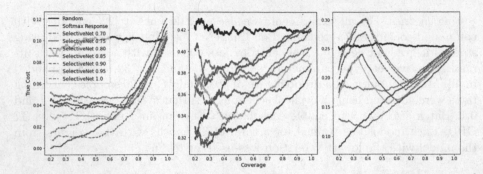

Fig. 2. Cost-coverage curves for SelectiveNets trained with different target coverages. From left to right: S_{in}, $S_{unknown}$ and $S_{combined}$.

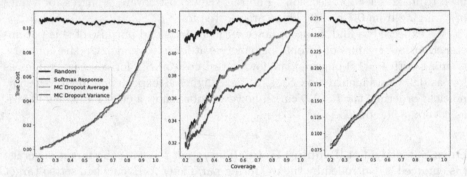

Fig. 3. Cost-coverage curves using MC-Dropout on S_{in}, $S_{unknown}$, and $S_{combined}$

obtained using EC-SelectiveNet in which the selection head is dropped at test time. EC-SelectiveNet showed a clear benefit on both S_{in} and $S_{unknown}$ compared to training a softmax network without the additional heads.

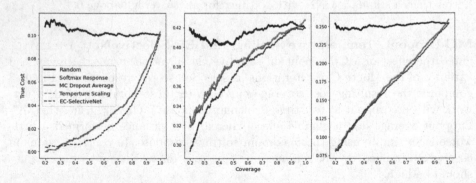

Fig. 4. Cost-coverage curves. From left to right: S_{in}, $S_{unknown}$ and $S_{combined}$.

Asymmetric Costs. We investigated the effect of asymmetric mis-classification costs. Figure 5 compares SelectiveNet with EC-SelectiveNet ($t = 1.0$). They performed similarly when costs were symmetric with SelectiveNet achieving a small cost reduction (approximately 0.015) at middling coverage. However, in the more realistic asymmetric settings, EC-SelectiveNet achieved cost reductions of approximately 0.1 at all coverages below about 0.8.

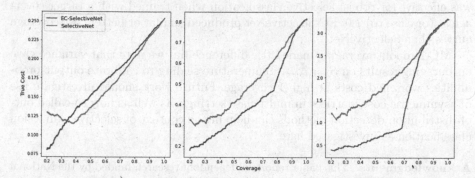

Fig. 5. Cost-coverage curves for SelectiveNet and EC-SelectiveNet. From left to right: $C_{1,0} = 1$ (symmetric costs), 10, and 50 (highly asymmetric costs)

Figure 6 plots the effect of temperature scaling. Both the softmax response and temperature scaling selection methods are based on the expected costs. The effect of temperature scaling was negligible with symmetric costs. In the asymmetric settings it had a small effect on selective classification. This effect was similar whether using EC-SelectiveNet ($t = 1.0$) or standard network training with cross-entropy loss. In both cases, temperature scaling increased costs at high coverage and reduced costs at low coverage. Figure 6 also makes clear the relative advantage of EC-SelectiveNet.

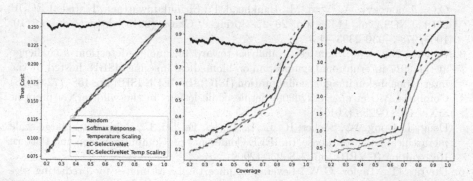

Fig. 6. Cost-coverage curves for cross-entropy training and EC-SelectiveNet combined with temperature scaling. From left to right: $C_{1,0} = 1$ (symmetric costs), 10, and 50 (highly asymmetric costs)

6 Conclusion

This study set out to better understand selective classification of skin lesions using asymmetric costs. In a primary care setting, for example, the cost of mis-classifying a life-threatening melanoma is clearly greater than that of misclassifying a benign lesion. We also investigated selective classification with lesion types not adequately represented during training. Generally, EC-SelectiveNet was effective for robust selective classification when trained with a target coverage at (or close to) 1.0. EC-SelectiveNet produced similar or better cost-coverage curves than SelectiveNet.

MC-Dropout averaging made little difference but we note that variance gave encouraging results on $S_{unknown}$. Temperature scaling to calibrate output probabilities worsened costs at higher coverage. Future work should investigate use of asymmetric cost matrices in multi-class settings, as well as how so-called out-of-distribution detection methods can help in the context of selective skin lesion classification as investigated here.

Acknowledgments. This paper reports independent research funded by the National Institute for Health Research (Artificial Intelligence, Deep learning for effective triaging of skin disease in the NHS, AI_AWARD01901) and NHSX. The views expressed in this publication are those of the authors and not necessarily those of the National Institute for Health Research, NHSX or the Department of Health and Social Care. This research was also funded by the Detect Cancer Early programme, and the Discovery Institute of Dermatology.

References

1. Brinker, T.J., et al.: Deep learning outperformed 136 of 157 dermatologists in a head-to-head dermoscopic melanoma image classification task. Eur. J. Cancer **113**, 47–54 (2019)
2. Carse, J., McKenna, S.: Active learning for patch-based digital pathology using convolutional neural networks to reduce annotation costs. In: Reyes-Aldasoro, C.C., Janowczyk, A., Veta, M., Bankhead, P., Sirinukunwattana, K. (eds.) ECDP 2019. LNCS, vol. 11435, pp. 20–27. Springer, Cham (2019). https://doi.org/10. 1007/978-3-030-23937-4_3
3. Codella, N., et al.: Skin lesion analysis toward melanoma detection: a challenge at the 2017 international symposium on biomedical imaging (ISBI), hosted by the international skin imaging collaboration (ISIC). In: IEEE ISBI, pp. 168–172 (2018)
4. Combalia, M., et al: BCN20000: dermoscopic lesions in the wild. arXiv preprint arXiv:1908.02288 (2019)
5. Deng, J., Dong, W., Socher, R., Li, L., Li, K., Fei-Fei, L.: ImageNet: a large-scale hierarchical image database. In: IEEE Conference on Computer Vision and Pattern Recognition (CVPR), pp. 248–255 (2009)
6. DeVries, T., Taylor, G.W.: Leveraging uncertainty estimates for predicting segmentation quality. In: Conference on Medical Imaging with Deep Learning (MIDL) (2018)
7. Esteva, A., et al.: Dermatologist-level classification of skin cancer with deep neural networks. Nature **542**, 115–8 (2017)

8. Gal, Y., Ghahramani, Z.: Dropout as a Bayesian approximation: representing model uncertainty in deep learning. In: Proceedings of the 33rd International Conference on Machine Learning (ICML), vol. PMLR 48, pp. 1050–1059 (2016)

9. Geifman, Y., El-Yaniv, R.: SelectiveNet: a deep neural network with an integrated reject option. In: Proceedings of the 36th International Conference on Machine Learning (ICML), vol. PMLR 97, pp. 2151–2159 (2019)

10. Guo, C., Pleiss, G., Sun, Y., Weinberger, K.Q.: On calibration of modern neural networks. In: Proceedings of the 34th International Conference on Machine Learning (ICML), vol. PMLR 70, pp. 1321–1330 (2017)

11. Haenssle, H.A., Fink, C., Schneiderbauer, R., et al.: Man against machine: diagnostic performance of a deep learning convolutional neural network for dermoscopic melanoma recognition in comparison to 58 dermatologists. Ann. Oncol. 29(8), 1836–1842 (2018)

12. Han, S.S., Kim, M.S., Lim, W., Park, G.H., Park, I., Chang, S.E.: Classification of the clinical images for benign and malignant cutaneous tumors using a deep learning algorithm. J. Inv. Dermatol. 138(7), 1529–1538 (2018)

13. Han, S.S., et al.: Augmented intelligence dermatology: deep neural networks empower medical professionals in diagnosing skin cancer and predicting treatment options for 134 skin disorders. J. Inv. Dermatol. 140(9), 1753–1761 (2020)

14. Hendrycks, D., Gimpel, K.: A baseline for detecting misclassified and out-of-distribution examples in neural networks. In: ICLR (2017)

15. Hinton, G., Srivastava, N., Krizhevsky, A., Sutskever, I., Salakhutdinov, R.: Improving neural networks by preventing co-adaptation of feature detectors. arXiv:1207.0580 (2012)

16. Kawahara, J., Hamarneh, G.: Visual diagnosis of dermatological disorders: human and machine performance. arxiv:1906.01256, 6 (2019)

17. Mårtensson, G., et al.: The reliability of a deep learning model in clinical out-of-distribution MRI data: a multicohort study. Med. Image Anal. 66, 101714 (2020)

18. Mobiny, A., Singh, A., Van Nguyen, H.: Risk-aware machine learning classifier for skin lesion diagnosis. J. Clin. Med. 8(8), 1241 (2019)

19. Mozafari, A.S., Gomes, H.S., Leão, W., Janny, S., Gagné, C.: Attended temperature scaling: a practical approach for calibrating deep neural networks. arXiv preprint arXiv:1810.11586 (2018)

20. Nixon, J., Dusenberry, M.W., Zhang, L., Jerfel, G., Tran, D.: Measuring calibration in deep learning. In: CVPR Workshops, vol. 2 (2019)

21. Smith, L.: Cyclical learning rates for training neural networks. In: IEEE Winter Conference on Applications of Computer Vision (WACV), pp. 464–472. IEEE (2017)

22. Tan, M., Le, Q.V.: EfficientNet: rethinking model scaling for convolutional neural networks. In: Proceedings of the 36th International Conference on Machine Learning (ICML), vol. PMLR 97, pp. 6105–6114 (2019)

23. Tschandl, P., Rosendahl, C., Kittler, H.: The HAM10000 dataset, a large collection of multi-source dermatoscopic images of common pigmented skin lesions. Sci. Data 5, 180161 (2018)

Confidence-Based Out-of-Distribution Detection: A Comparative Study and Analysis

Christoph Berger[1]([✉]), Magdalini Paschali[1], Ben Glocker[2], and Konstantinos Kamnitsas[2]

[1] Computer Aided Medical Procedures, Technical University Munich, Munich, Germany
c.berger@tum.de
[2] Department of Computing, Imperial College London, London, UK

Abstract. Image classification models deployed in the real world may receive inputs outside the intended data distribution. For critical applications such as clinical decision making, it is important that a model can detect such out-of-distribution (OOD) inputs and express its uncertainty. In this work, we assess the capability of various state-of-the-art approaches for confidence-based OOD detection through a comparative study and in-depth analysis. First, we leverage a computer vision benchmark to reproduce and compare multiple OOD detection methods. We then evaluate their capabilities on the challenging task of disease classification using chest X-rays. Our study shows that high performance in a computer vision task does not directly translate to accuracy in a medical imaging task. We analyse factors that affect performance of the methods between the two tasks. Our results provide useful insights for developing the next generation of OOD detection methods.

1 Introduction

Supervised image classification has produced highly accurate models, which can be utilized for challenging fields such as medical imaging. For the deployment of such models in critical applications, their raw classification accuracy does not suffice for their thorough evaluation. Specifically, a major flaw of modern classification models is their overconfidence, even for inputs beyond their capacity. For instance, a model trained to diagnose pneumonia in chest X-rays may have only been trained and tested on X-rays of healthy controls and patients with pneumonia. However, in practice the model may be presented with virtually infinite variations of patient pathologies. In such cases, overly confident models may give a false sense of their competence. Ideally, a classifier should know its capabilities and signal to the user if an input lies out of distribution.

In this work, we first explore confidence- and distance-based approaches for out-of-distribution (OOD) detection on a standard computer vision (CV) task

© Springer Nature Switzerland AG 2021
C. H. Sudre et al. (Eds.): UNSURE 2021/PIPPI 2021, LNCS 12959, pp. 122–132, 2021.
https://doi.org/10.1007/978-3-030-87735-4_12

and afterwards evaluate the best OOD detection methods on a medical bench-mark dataset. Moreover, we provide a set of useful insights for leveraging OOD approaches from computer vision to challenging medical datasets.

Related Work: OOD detection methods can be divided in two categories. The first consists of methods that build a **dedicated model for OOD detection** [25]. Some works accomplish this via estimating density $p(x)$ of 'normal' in-distribution (ID) data and then classify samples with low $p(x)$ as OOD [10]. However, learning $p(x)$ accurately can be challenging. An alternative is to learn a decision boundary between ID and OOD samples. Methods [27] attempt this in an unsupervised fashion using only 'normal' data. Nonetheless, supervised alternatives have also been introduced for CV and medical imaging [6,24,29], exposing the OOD classifier to OOD data during training. Such OOD data can originate from another database or be synthesized. However, collecting or synthesising samples that capture the heterogeneity of OOD data is challenging. Another approach for creating OOD detection neural networks (NNs) is *reconstruction-based* models [8,21]. A model, such as an auto-encoder, is trained with a reconstruction loss using ID data. Then, it is assumed that the reconstruction of unseen OOD samples will fail, thus enabling their detection. This approach is especially popular in medical imaging research [1,22,23,26,32], likely because it produces a per-pixel OOD score, allowing its use for unsupervised segmentation. It has shown promise for localisation of salient abnormalities but does not reach the performance of supervised models in more challenging tasks.

The second category of OOD detection methods, which this study focuses on, enhances a task-specific model to detect when an input is OOD. These approaches are commonly based on **confidence of model predictions**. They are compact, integrated straight into an existing model, and operate in the task-specific feature or output space. Their biggest theoretical advantage in comparison to training a dedicated OOD detector is that if the main model is unaffected by a change in the data, the OOD detector also remains unaffected. A subset of confidence-based methods has a probabilistic motivation, exploring the use of the predictive uncertainty of a model, such as Maximum Class Probability (MCP) [5], MCDropout [2] or ensembling [12]. Others derive confidence-scores based on distance in feature space [31], or learn spaces that better separate samples via confidence maximization [14] or contrastive losses [28,31]. In medical imaging, related work is mostly focused on improving uncertainty estimates by DNNs [17,30], or analysing quality of uncertainty estimates in *ID* settings [9,19]. In contrast, investigation of OOD detection based on model confidence is limited. A recent study compared MCDropout and ensembling [16] for medical imaging, finding the latter more beneficial. The potential of other OOD detection methods for medical imaging is yet to be assessed adequately, despite their importance for the field.

Contributions: This study assesses confidence-based methods for OOD detection. To this end, we re-implement and compare approaches, shown in Fig. 1, in a common test-bed to accomplish a fair and cohesive comparison.[1] We first evaluate those approaches on a CV benchmark to gain insights for their performance. Then, we benchmark all approaches on real-world chest X-rays [7]. We find that the performance of certain methods varies drastically between OOD detection tasks, which raises concerns about their reliability for real-world use, and we identify ODIN as consistently high performing in our tasks. Finally, we conduct an empirical analysis to identify the factors that influence the OOD performance of these methods, providing useful insights towards building the next generation of OOD detection methods.

2   Out-of-Distribution Detection Methods

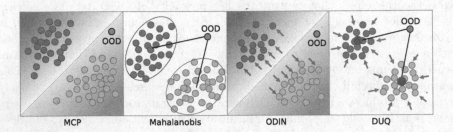

Fig. 1. Overview of the OOD detection methods studied: maximum class probability (baseline), Mahalanobis distance, ODIN and DUQ.

We study the following methods for OOD detection in image classification:

Maximum Class Probability (MCP) [5]: Any softmax-based model produces an estimate of confidence in its predictions via its class posteriors. Specifically, the probability $\max_y p(y|x)$ of the most likely class is interpreted as an ID score and, conversely, low probability indicates possible OOD input. Even though modern NNs have been shown to often produce over-confident softmax outputs [3], this method is a useful baseline for OOD detection.

Mahalanobis Distance [13]: Lee et al. propose the Mahalanobis distance as OOD metric in combination with NNs. The method can be integrated to any pre-trained classifier. It assumes that the class-conditional distributions of activations $z(x; \theta) \in \mathbb{R}^Z$ in the last hidden layer of the pre-trained model follow multivariate Gaussian distributions. After training model parameters θ, the model is applied to all training data to compute for each class c, the mean $\hat{\mu}_c \in \mathbb{R}^Z$ of activations z over all training samples x of class c, and the covariance matrix $\hat{\Sigma}$ of the class-conditional distributions of z. To perform OOD detection, the

---

[1] The framework is available online: https://github.com/christophbrgr/ood_detection_framework.

method computes the Mahalanobis distance between a test sample x and the closest class-conditional distribution as follows:

$$M(x) = max_c - (z(x; \theta) - \hat{\mu}_c)^T \hat{\Sigma}^{-1} (z(x; \theta) - \hat{\mu}_c) \tag{1}$$

The threshold to decide whether an input is OOD or ID is then set as a certain distance from the closest distribution.

Out-of-Distribution Detector for Neural Networks (ODIN) [14]: This method is also applicable to pre-trained classifiers which output class-posteriors using a softmax. Assume $f(x; \theta) \in \mathbb{R}^C$ are the logits for C classes. We write $S(x, \tau) = \text{softmax}(f(x; \theta), \tau) \in \mathbb{R}^C$ for the softmax output calculated for temperature τ ($\tau_{tr} = 1$ for training), and $S(x, \tau)_c$ is the value for class c. The method is based on the assumption that we can find perturbations of the input that increase the model's confidence more drastically for ID samples than for OOD samples. The perturbed version of input x is given by:

$$\tilde{x} = x - \varepsilon sign(-\nabla x \log max_c S(x; \theta, \tau_{tr})_c) \tag{2}$$

Here, a gradient is computed that maximizes the softmax probability of the most likely class. The model is then applied on the perturbed sample $\tilde{x}$ and outputs softmax probabilities $S(\tilde{x}; \tau') \in \mathbb{R}^C$. From this, the MCP ID score is derived as $max_c S(x; \tau')_c$. Since the perturbation forces over-confident predictions, it negatively affects calibration. To counteract this, ODIN proposes using a different softmax temperature τ' when predicting the perturbed samples, to re-calibrate its predictions. τ' is a hyperparameter that requires tuning. We assess the effect of the perturbation and τ in an ablation study.

Deep Ensembles [12]: This method trains multiple models from scratch, while initialisation and order of training data is varied. During inference, predicted posteriors of all models are averaged to compute the ensemble's posteriors. This in turn is used to compute MCP of the ensemble as an ID score. While deep ensembles have been shown to perform well for OOD detection, they come with high computational cost as training and inference times scale linearly with number of ensemble members. In our experiments, we also investigate an ensemble that uses a consensus Mahalanobis distance as OOD score instead of MCP.

Monte Carlo Dropout (MCDP) [2]: MCDP trains a model with dropout. At test time, multiple predictions are made per input with varying dropout masks. The predictions are averaged and MCP is used as ID score. The method interprets these predictions as samples from the model's posterior, where their average is a better predictive uncertainty estimate, improving OOD detection.

Deterministic Uncertainty Quantification (DUQ) [31]: This method trains a feature extractor without a softmax layer. Instead, it learns a centroid per class and attracts samples towards the centroids of their class, similar to contrastive losses [4]. It uses a Radial Basis Function (RBF) kernel to compute the distance between the input's embedding and the class centroids. The distance to the closest centroid defines classification, and is also used as the OOD score. Because RBF networks are prone to feature collapse, DUQ introduces a gradient penalty to regularize learnt embedding and alleviate the issue. Nonetheless, we still faced difficulties with DUQ convergence despite considerable attempts.

3 Benchmarking on CIFAR10 Vs SVHN

We first show results on a common computer-vision (CV) benchmark to gain insights about methods' performance, and validate our implementations by replicating results of original works before applying them to a biomedical benchmark.

Table 1. Out-of-distribution detection performance of WideResNet 28 × 10 trained on CIFAR10 with SVHN as OOD set. We report averages over 3 seeds.

Method	AUROC	AUCPR	ID acc.
MCP (baseline)	0.939	0.919	0.952
MCDP	0.945	0.919	**0.956**
Deep ensemble	0.960	0.951	0.954
Mahalanobis	0.984	0.960	0.952
Mahalanobis Ens.	**0.987**	**0.967**	0.954
ODIN	0.964	0.939	0.952
ODIN (pert. only)	0.968	0.948	0.952
ODIN (temp. only)	0.951	0.920	0.952
DUQ	0.833	–	0.890

3.1 Experimental Setup

Dataset: We use the training and test splits of CIFAR10 [11] as ID and SVHN [20] as OOD test set ($n_{\text{test ID}} = 10000$, $n_{\text{test OOD}} = 26032$). A random subset of 10% CIFAR training data is used as validation set, to tune method hyperparameters, such as temperature τ for ODIN.

Model: We use a WideResNet (WRN) [33] with depth 28 and widen factor 10 (WRN 28 × 10), trained with SGD using momentum 0.9, weight decay 0.0005, batch normalization and dropout of 0.3 for 200 epochs with early stopping. The Deep Ensemble uses 5 models and MCDP uses 10 samples.

Evaluation Metrics: We use the following metrics to assess the performance of a method in separating ID from OOD inputs: (1) area under the receiver operating characteristic (AUROC), (2) area under the precision-recall curve (AUCPR), (3) accuracy (Acc) on ID test set.

3.2 Results

In Table 1, we compare OOD detection performance for all studied methods. MCDP marginally improves over the baseline, with higher gains by Deep Ensembles. Interestingly, ODIN achieves comparable AUROC with Deep Ensembles and ODIN's input perturbation is the component responsible for the performance (see ODIN (pert. only)). The results of only applying temperature scaling and no input perturbation are listed under ODIN (temp. only). The highest AUROC over all methods is achieved by Mahalanobis distance both as a single model and an ensemble. Moreover, none of the OOD detection methods compromised the accuracy on the classification task. We reproduced the results of original implementation of DUQ with ResNet50. However, we faced unstable training of DUQ on our WRN and did not obtain satisfactory performance despite our efforts.

Table 2. Performance of different methods for separation of out-of-distribution (OOD) from in-distribution (ID) samples for CheXpert in two settings. **Setting 1:** Classifier trained to separate *Cardiomegaly* from *Pneumothorax* (ID) is given samples with *Fractures* (OOD). **Setting 2:** Classifier trained to separate *Lung Opacity* from *Pleural Effusion* (ID) is given samples with *Fracture* or *Pneumonia* (OOD). We report average over 3 seeds per experiment. Best in **bold**.

Method	Setting 1			Setting 2		
	OOD		ID	OOD		ID
	AUROC	AUCPR	Acc	AUROC	AUCPR	Acc
MCP (baseline)	0.678	0.695	0.888	0.458	0.586	0.758
MCDP	0.696	0.703	0.880	0.519	0.637	0.756
Deep ensemble	0.704	0.705	**0.895**	0.445	0.582	**0.769**
Mahalanobis	0.580	0.580	0.888	0.526	0.601	0.758
Mahalanobis Ens.	0.596	0.586	**0.895**	0.537	0.613	0.758
ODIN	**0.841**	**0.819**	0.888	0.862	0.856	0.758
ODIN (pert. only)	**0.841**	**0.819**	0.888	**0.865**	**0.856**	0.757
ODIN (temp. only)	0.678	0.695	0.888	0.444	0.575	0.757

4   Benchmarking on the X-Ray Lung Pathology Dataset

4.1   Experimental Setup

Dataset: To simulate a realistic OOD detection task in a clinical setting, we use subsets of the CheXpert X-ray lung pathology dataset [7] as ID and OOD data, in two different settings. Since CheXpert images are multi-labeled, we only used samples where ID and OOD classes were mutually exclusive. **Setting 1:** We train a classifier to distinguish *Cardiomegaly* from *Pneumothorax* (ID), and use images with *Fracture* as OOD ($n_{\text{test ID}} = 4300$, $n_{\text{test OOD}} = 7200$). **Setting 2:** We train a classifier to separate *Lung Opacity* and *Pleural Effusion* (ID), and use *Fracture* and *Pneumonia* as OOD classes ($n_{\text{test ID}} = 6000$, $n_{\text{test OOD}} = 8100$).

Model: We use WRN with depth 100 and a widen factor 2 (WRN 100x2). The Deep Ensemble uses 3 models and MCDP uses 10 samples. All other parameters remain the same as for the CIFAR10 vs SVHN benchmark.

Evaluation: We analyse performance based on the same metrics as in Sect. 3.

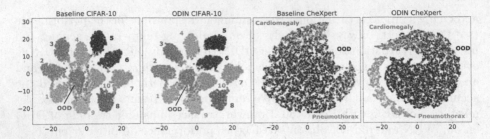

Fig. 2. T-SNE of embeddings for CIFAR10 vs SVHN and for CheXpert Setting 1. The OOD cluster is less separated for the latter, challenging benchmark. ODIN perturbations improve separation, which may explain its performance.

4.2 Results

Results for the two ID/OOD settings in CheXpert are shown in Table 2. The baseline performance indicates that the ID and OOD inputs are harder to separate for Setting 2, and much harder than the CIFAR vs SVHN task. MCDP improves OOD detection in both Settings. Interestingly, Deep Ensembles, often considered the most reliable method for OOD detection, do not improve Setting 2, although the Mahalanobis Ensemble does. Moreover, ODIN shows best performance in both settings with a considerable margin, even when only using the adversarial-inspired component of the method without softmax tempering (see ODIN (pert. only) in Fig. 2). Mahalanobis distance, which was the best method on the CIFAR10 vs SVHN task, performs worse than the Baseline on Setting 1 and only yields modest improvements in Setting 2. Reliability of OOD methods is crucial. Thus, the next section further analyses ODIN and Mahalanobis, to gain insights in the consistent performance of ODIN and the difference between the CV benchmark and CheXpert Setting 1 that may be causing the inconsistency of Mahalanobis distance.

4.3 Further Analysis

Mahalanobis: Our first hypothesis to explain the poor performance of Mahalanobis on the medical OOD detection task in comparison to the CV task was that the Mahalanobis distance may be ineffective in higher dimensional spaces. In the CIFAR10 vs SVHN task, the Mahalanobis distance is calculated in a hidden layer with [640, 8, 8] (40960 total) activations, whereas the WRN 100x2 for CheXpert has a corresponding layer with shape [128, 56, 56] (401408 total) activations. To test this hypothesis, we reduce the number of dimensions on which we compute the distributions by applying strided max pooling before computing the Mahalanobis distance and report the results in Fig. 3a. We find that this dimensionality reduction is not effective and conclude that this is not the major cause of Mahalanobis ineffectiveness in CheXpert.

To further investigate, we visualize with T-SNE [15] the last layer activations when trained models process perturbed samples for the CIFAR10 vs SVHN task

Pooling	Layer shape	AUROC
None	[128,56,56]	**0.6018**
4x4, stride=2	[128,56,56]	0.5634
2x2, stride=4	[128,14,14]	0.5608
8x8, stride=1	[128,14,14]	0.5508
1x1, stride=4	[128,14,14]	0.545

(a) Results with dimensionality reduction

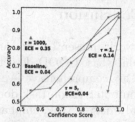

(b) Calibration curves

Fig. 3. (a) Results on CheXpert Setting 1 from experiments with dimensionality reduction in last hidden layer. Lower dimensionality did not improve OOD detection via Mahalanobis distance. (b) Calibration of baseline and ODIN for varying temperature τ and associated Expected Calibration Error (ECE, as a summary statistic for model calibration [18]), for CheXpert Setting 1. The baseline (green) is reasonably calibrated. Adding noise to the inputs with ODIN leads to highly overconfident model (purple, all samples very high confidence). For CheXpert, $\tau = 1000$ as used for CIFAR10 leads to under-confident model, whereas $\tau = 5$ restores good calibration. Interestingly, all ODIN settings achieve the same AUROC for OOD irrespective of τ value and calibration. (Color figure online)

and the CheXpert Setting 1. Figure 2 shows that activations for CIFAR10 classes are clearly separated and the OOD set is distinguishable from the ID clusters. For CheXpert, the baseline model achieves less clear separation of the two ID classes and the OOD class overlaps substantially with the ID classes. This suggests that fitting a Gaussian distribution to the ID embeddings is challenging, causing the Mahalanobis distance to not yield significant OOD detection benefits.

ODIN: We investigate how the perturbation that ODIN adds to inputs benefits OOD detection. For this, we also show T-SNE plots for both CIFAR10 and CheXpert Setting 1 in Fig. 2. The added perturbation results in a better separation of ID classes in both datasets, with the effect more pronounced for CheXpert. While there is still overlap between the *Fracture* OOD class and the *Pneumothorax* ID class, the clusters are more pronounced which ultimately leads to better OOD detection. Finally, we investigate the effect of temperature variation in ODIN. Following [14], temperature 1000 was used for CIFAR10 and CheXpert. By comparing baseline, ODIN (temp. only) and (pert. only) on Tables 1 and 2, we find that OOD detection is primarily improved by perturbation, not temperature scaling, especially on CheXpert. We note, however, that the perturbations lead to a completely over-confident model using training temperature 1, with all predictions having very high confidence (Fig. 3b). AUROC and AUCPR are calculated via ordering the OOD score (i.e. confidence) of predictions, so even slight differences between ID and OOD samples suffice to separate false and true detections. If only those metrics were taken into account, temperature scaling might have been considered redundant. However, to deploy an OOD system, a threshold on the confidence/OOD score needs to be chosen. Spreading the confidence estimates via temperature scaling ($\tau = 5$ in Fig. 3b) enables more reliable choice and deployment of a confidence threshold in practical settings.

5   Conclusion

This work presented an analysis of various state-of-the-art methods for confidence-based OOD detection on a computer vision and a medical imaging task. Our comprehensive evaluation showed that the performance of methods in a computer vision task does not directly translate to high performance on a medical imaging task, emphasized by the analysis of the Mahalanobis method. Therefore, care must be given when a method is chosen. We also identified ODIN as a consistently beneficial OOD detection method for both tasks. Our analysis showed that its effect can be attributed to its input perturbation, which enhances separation of ID and OOD samples. This insight could lead to further advances that exploit this property. Future work should further evaluate OOD detection methods across other datasets and tasks to better understand which factors affect their performance and reliability towards real-world deployment.

Acknowledgements. This work received funding from the European Research Council (ERC) under the European Union's Horizon 2020 research and innovation programme (grant agreement No. 757173, project MIRA, ERC-2017-STG), and the UKRI London Medical Imaging & Artificial Intelligence Centre for Value Based Healthcare.

References

1. Baur, C., Denner, S., Wiestler, B., Navab, N., Albarqouni, S.: Autoencoders for unsupervised anomaly segmentation in brain MR images: a comparative study. Med. Image Anal. 101952 (2021)
2. Gal, Y., Ghahramani, Z.: Dropout as a Bayesian approximation: representing model uncertainty in deep learning. In: Balcan, M.F., Weinberger, K.Q. (eds.) Proceedings of The 33rd International Conference on Machine Learning. Proceedings of Machine Learning Research, vol. 48, pp. 1050–1059. PMLR, New York, 20–22 June 2016. http://proceedings.mlr.press/v48/gal16.html
3. Guo, C., Pleiss, G., Sun, Y., Weinberger, K.Q.: On calibration of modern neural networks. In: Precup, D., Teh, Y.W. (eds.) Proceedings of the 34th International Conference on Machine Learning. Proceedings of Machine Learning Research, vol. 70, pp. 1321–1330. PMLR, International Convention Centre, Sydney, 06–11 August 2017. http://proceedings.mlr.press/v70/guo17a.html
4. Hadsell, R., Chopra, S., LeCun, Y.: Dimensionality reduction by learning an invariant mapping. In: 2006 IEEE Computer Society Conference on Computer Vision and Pattern Recognition (CVPR 2006), vol. 2, pp. 1735–1742. IEEE (2006)
5. Hendrycks, D., Gimpel, K.: A baseline for detecting misclassified and out-of-distribution examples in neural networks. In: Proceedings of International Conference on Learning Representations (2017)
6. Hendrycks, D., Mazeika, M., Dietterich, T.: Deep anomaly detection with outlier exposure. arXiv preprint arXiv:1812.04606 (2018)
7. Irvin, J., et al.: CheXpert: a large chest radiograph dataset with uncertainty labels and expert comparison. In: Proceedings of the AAAI Conference on Artificial Intelligence, vol. 33, pp. 590–597 (2019)
8. Japkowicz, N., Myers, C., Gluck, M., et al.: A novelty detection approach to classification. In: IJCAI, vol. 1, pp. 518–523. Citeseer (1995)

9. Jungo, A., Balsiger, F., Reyes, M.: Analyzing the quality and challenges of uncertainty estimations for brain tumor segmentation. Front. Neurosci. **14**, 282 (2020)
10. Kobyzev, I., Prince, S., Brubaker, M.: Normalizing flows: an introduction and review of current methods. IEEE Trans. Pattern Anal. Mach. Intell. (2020)
11. Krizhevsky, A., Nair, V., Hinton, G.: CIFAR-10 (Canadian institute for advanced research). http://www.cs.toronto.edu/~kriz/cifar.html
12. Lakshminarayanan, B., Pritzel, A., Blundell, C.: Simple and scalable predictive uncertainty estimation using deep ensembles. In: Advances in Neural Information Processing Systems, vol. 30 (2017)
13. Lee, K., Lee, K., Lee, H., Shin, J.: A simple unified framework for detecting out-of-distribution samples and adversarial attacks. In: Advances in Neural Information Processing Systems, vol. 31 (2018)
14. Liang, S., Li, Y., Srikant, R.: Enhancing the reliability of out-of-distribution image detection in neural networks. In: ICLR (2018)
15. van der Maaten, L., Hinton, G.: Visualizing data using t-SNE. J. Mach. Learn. Res. **9**(86), 2579–2605 (2008)
16. Mehrtash, A., Wells, W.M., Tempany, C.M., Abolmaesumi, P., Kapur, T.: Confidence calibration and predictive uncertainty estimation for deep medical image segmentation. IEEE Trans. Med. Imaging **39**(12), 3868–3878 (2020)
17. Monteiro, M., et al.: Stochastic segmentation networks: modelling spatially correlated aleatoric uncertainty. arXiv preprint arXiv:2006.06015 (2020)
18. Naeini, M.P., Cooper, G.F., Hauskrecht, M.: Obtaining well calibrated probabilities using Bayesian binning. In: Proceedings of the Twenty-Ninth AAAI Conference on Artificial Intelligence, AAAI 2015, pp. 2901–2907. AAAI Press (2015)
19. Nair, T., Precup, D., Arnold, D.L., Arbel, T.: Exploring uncertainty measures in deep networks for multiple sclerosis lesion detection and segmentation. Med. Image Anal. **59**, 101557 (2020)
20. Netzer, Y., Wang, T., Coates, A., Bissacco, A., Wu, B., Ng, A.Y.: Reading digits in natural images with unsupervised feature learning (2011)
21. Pang, G., Shen, C., Cao, L., Hengel, A.V.D.: Deep learning for anomaly detection: a review. arXiv preprint arXiv:2007.02500 (2020)
22. Pawlowski, N., et al.: Unsupervised lesion detection in brain CT using Bayesian convolutional autoencoders (2018)
23. Pinaya, W.H.L., et al.: Unsupervised brain anomaly detection and segmentation with transformers. arXiv preprint arXiv:2102.11650 (2021)
24. Roy, A.G., et al.: Does your dermatology classifier know what it doesn't know? Detecting the long-tail of unseen conditions. arXiv preprint arXiv:2104.03829 (2021)
25. Ruff, L., et al.: A unifying review of deep and shallow anomaly detection. arXiv preprint arXiv:2009.11732 (2020)
26. Schlegl, T., Seebӧck, P., Waldstein, S.M., Schmidt-Erfurth, U., Langs, G.: Unsupervised anomaly detection with generative adversarial networks to guide marker discovery. In: Niethammer, M., et al. (eds.) IPMI 2017. LNCS, vol. 10265, pp. 146–157. Springer, Cham (2017). https://doi.org/10.1007/978-3-319-59050-9_12
27. Schӧlkopf, B., Platt, J.C., Shawe-Taylor, J., Smola, A.J., Williamson, R.C.: Estimating the support of a high-dimensional distribution. Neural Comput. **13**(7), 1443–1471 (2001)
28. Tack, J., Mo, S., Jeong, J., Shin, J.: CSI: novelty detection via contrastive learning on distributionally shifted instances. In: NeurIPS (2020)
29. Tan, J., Hou, B., Batten, J., Qiu, H., Kainz, B.: Detecting outliers with foreign patch interpolation. arXiv preprint arXiv:2011.04197 (2020)

30. Tanno, R., et al.: Bayesian image quality transfer with CNNs: exploring uncertainty in dMRI super-resolution. In: Descoteaux, M., Maier-Hein, L., Franz, A., Jannin, P., Collins, D.L., Duchesne, S. (eds.) MICCAI 2017. LNCS, vol. 10433, pp. 611–619. Springer, Cham (2017). https://doi.org/10.1007/978-3-319-66182-7_70
31. Van Amersfoort, J., Smith, L., Teh, Y.W., Gal, Y.: Uncertainty estimation using a single deep deterministic neural network. In: III, H.D., Singh, A. (eds.) Proceedings of the 37th International Conference on Machine Learning. Proceedings of Machine Learning Research, vol. 119, pp. 9690–9700. PMLR, 13–18 July 2020
32. You, S., Tezcan, K.C., Chen, X., Konukoglu, E.: Unsupervised lesion detection via image restoration with a normative prior. In: International Conference on Medical Imaging with Deep Learning, pp. 540–556. PMLR (2019)
33. Zagoruyko, S., Komodakis, N.: Wide residual networks. In: BMVC (2016)

Novel Disease Detection Using Ensembles with Regularized Disagreement

Alexandru Țifrea[✉], Eric Stavarache, and Fanny Yang

ETH Zurich, Zurich, Switzerland
{tifreaa,ericst,fan.yang}@ethz.ch

Abstract. Automated medical diagnosis systems need to be able to recognize when new diseases emerge, that are not represented in the training data (ID). Even though current out-of-distribution (OOD) detection algorithms can successfully distinguish completely different data sets, they fail to reliably identify samples from novel classes that are similar to the training data. We develop a new ensemble-based procedure that promotes model diversity and exploits regularization to limit disagreement to only OOD samples, using a batch containing an unknown mixture of ID and OOD data. We show that our procedure significantly outperforms state-of-the-art methods, including those that have access, during training, to known OOD data. We run extensive comparisons of our approach on a variety of novel-class detection scenarios, on standard image data sets as well as on new disease detection on medical image data sets (Our code is publicly available at https://github.com/ericpts/reto).

Keywords: Novelty detection · Novel disease detection · Ensemble diversity · Regularization

1 Introduction

Modern machine learning (ML) systems are gaining popularity in many real-world applications, such as aiding medical diagnosis [2]. Despite achieving great test performance, many approaches have trouble dealing with out-of-distribution (OOD) data, i.e. test inputs that are unlike the data seen during training. For example, ML models often make incorrect predictions with high confidence when new unseen classes emerge over time (e.g. undiscovered bacteria [38], new diseases [18]), or when data suffers from distribution shift (e.g. corruptions [29], environmental changes [22]). If the OOD data consists of novel classes, then we must identify the OOD samples and bring them to the attention of human experts. This scenario is the focus of this paper and we use the terms OOD and novelty detection interchangeably.

Electronic supplementary material The online version of this chapter (https://doi.org/10.1007/978-3-030-87735-4_13) contains supplementary material, which is available to authorized users.

© Springer Nature Switzerland AG 2021
C. H. Sudre et al. (Eds.): UNSURE 2021/PIPPI 2021, LNCS 12959, pp. 133–144, 2021.
https://doi.org/10.1007/978-3-030-87735-4_13

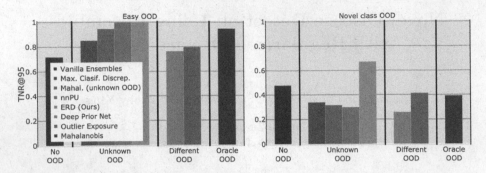

Fig. 1. Comparison of OOD detection methods ordered by the amount of information about the OOD distribution that they require. **Left:** On the easy settings usually reported in the literature, many methods achieve near-perfect detection. **Right:** On novel-class settings where ID and OOD data are difficult to distinguish, the baselines reach a much lower TNR@95 compared to our method.

Novelty detection aims to identify test samples that have a low probability under the marginal ID distribution P_X, i.e. x should be flagged as OOD if $P_X(x) \leq \alpha$ for some small constant α. If we could learn a model that estimates precisely the level sets of P_X, we would have perfect OOD detection. Unfortunately, when the input space is high-dimensional (e.g. high resolution medical images) and we only have access to limited data, this problem is intractable. In reality, however, we only need to detect outliers that actually appear in a test set, which makes the problem more amenable to statistical methods.

Apart from a labeled ID training set, state-of-the-art (SOTA) OOD detection methods often use some OOD data for training or calibration. We separate existing approaches into four different levels of access to OOD data: 1) *no OOD data* [23,39]; 2) an unlabeled set with an unknown mixture of ID and OOD data where OOD samples are not marked (*Unknown OOD*) [28,35,40,47]; 3) known OOD data, but from a different distribution than the test OOD (*Different OOD*) [17,30]; or 4) known OOD data from the same distribution as test OOD (*Oracle OOD*) [25,27].

Notably, prior work on OOD detection reports remarkably good detection performance: when 95% of the true OOD samples are correctly identified (i.e. the true positive rate is 95%), the ratio of ID samples correctly identified as ID (i.e. the true negative rate) is often larger than 80% (this metric is known as the TNR@95). However, these numbers are largely obtained when the in-distribution (ID) and the OOD data sets are vastly different (e.g. SVHN vs CIFAR10), while in real-world applications it is unlikely that the novel data is so easy to distinguish from ID samples (e.g. chest X-rays of a new disease may look quite similar to another pathology). When evaluating state-of-the-art (SOTA) methods on novel-class settings on standard image data sets (e.g. SVHN, CIFAR10), the TNR@95 for the best baseline drops below 40% (see Fig. 1 Right).

In this work, we adopt the *Unknown OOD* setting and introduce a principled method to obtain diverse ensembles by leveraging the unlabeled set and using early stopping regularization. We call our method **Ensembles with Regularized Disagreement** (ERD) and motivate it using a theoretical result on the dynamics of gradient descent training under label noise. Our method improves the state-of-the-art for novel-class detection, surpassing even approaches that assume oracle knowledge of OOD samples, as illustrated in Fig. 1. Moreover, we show that our algorithm consistently outperforms all baselines on a recently proposed medical image OOD detection benchmark.

2 Problem Setting

In this section we motivate the *Unknown OOD* setting that we adopt for our method and stress its practical relevance.

Problem statement. We consider a labeled data set $S = \{(x_i, y_i)\}_{i=1}^n \sim P$, where $x_i \in \mathcal{X}$ are the covariates and $y_i \in \mathcal{Y}$ are discrete labels. We assume that the labels are obtained as a deterministic function of the covariates, which we denote $y^* : \mathcal{X} \to \mathcal{Y}$. In this paper we focus on detecting samples from novel classes, unseen at training time. We define $\mathcal{X}_{ID} := \{x : P_X(x) > \alpha\}$ as ID points and $\mathcal{X}_{OOD} = \{x : P_X(x) \leq \alpha\}$ as the set of OOD points, where P_X is the marginal ID distribution.

The *Unknown OOD* settings has been proposed in prior work on OOD detection [28,40,47] and assumes that, apart from the ID training data with class labels, we also have access to a batch of unlabeled data U drawn from the same distribution P_{test} as the test data. This distribution consists of a mixture of ID and OOD data, with OOD proportion $\pi \in [0,1]$, that is $P_{\text{test}}[x \in \mathcal{X}_{OOD}] = \pi$. The goal is to use the set U to learn to distinguish between ID and OOD data drawn from P_{test}, without explicit knowledge of π nor which samples in U are OOD.

The *Unknown OOD* setting is relevant for many practical applications that would benefit from more effective novel-class detection. Consider, for instance, a medical center that uses an automated system for real-time diagnosis. In addition, the hospital may wish to run a novelty detection algorithm offline every week to check for possible new pathologies. A procedure based on the unknown OOD setting can use all the X-rays from the week as an unlabeled set U. If U contains X-rays exhibiting a new disease, the algorithm can be used to flag such novel classes both in the already collected unlabeled set and for future patients suffering from the same new disease. Furthermore, the flagged samples can be examined and labeled by experts.

3 Proposed Method

In this section we introduce our proposed algorithm, ERD, and provide a principled justification for the key ingredients that lead to the improved performance of our method.

Algorithm 1. Fine-tuning the ERD ensemble

Input: Train set S, Validation set V, Unlabeled set U, Weights W pretrained on S, Ensemble size K
Output: ERD ensemble $\{f_{y_i}\}_{i=1}^{K}$
Sample K different labels $\{y_1, ..., y_K\}$ from $\mathcal{Y}$
for all $c \in \{y_1, ..., y_K\}$ **do**
  $f_c \leftarrow Initialize(W)$
  $(U, c) \leftarrow \{(x, c) : x \in U\}$
  $f_c \leftarrow EarlyStoppedFinetuning(f_c, S \cup (U, c); V)$
end for
return $\{f_{y_i}\}_{i=1}^{K}$

Algorithm 2. OOD detection with ERD

Input: Ensemble $\{f_{y_i}\}_{i=1}^{K}$, Test set T, Threshold t_0, Disagreement metric ρ
Output: O, i.e. the OOD data in T
$O \leftarrow \emptyset$
for all $x \in U$ **do**
  if $(\text{Avg} \circ \rho)(f_{y_1}, ..., f_{y_K})(x) > t_0$ **then**
    $O \leftarrow O \cup \{x\}$
  end if
end for
return O

3.1 The Complete ERD Procedure

Recall that we have access to both a labeled training set S and an unlabeled set U that contains both ID and unknown OOD samples. Moreover, we initialize the models of the ensemble using weights pretrained on S.

In Algorithm 1 we show how to obtain an ERD ensemble. We begin by assigning an arbitrary label $c \in \mathcal{Y}$, to all the unlabeled samples in U, resulting in the c-labeled set $(U, c) := \{(x, c) : x \in U\}$. We then fine-tune a classifier f_c on the union $S \cup (U, c)$ of the correctly-labeled training set S and the unlabeled set (U, c). We perform early stopping by picking a model at an intermediate epoch, before the accuracy on a holdout ID validation set V starts to decrease. We repeat this procedure to create an ensemble of several classifiers f_c, for different choices of $c \in \mathcal{Y}$. Finally, during test time in Algorithm 2, we use this ensemble to flag as OOD all the points for which an aggregate disagreement measure surpasses a threshold value t_0, as we elaborate later in this section.

3.2 Role of Regularization

Recall that, in our approach, each member of the ensemble tries to fit a different label c to the entire unlabeled set U in addition to the correct labels of the ID training set S. We train the models to fit $S \cup (U, c)$, where we use the notation $(U, c) = (U_{\text{ID}}, c) \cup (U_{\text{OOD}}, c) = \{(x, c) : x \in U_{\text{ID}}\} \cup \{(x, c) : x \in U_{\text{OOD}}\}$. Moreover, we can partition the set (U_{ID}, c) into the subset of samples whose ground truth label differs from c and are thus incorrectly labeled with c, and the subset whose correct label is indeed c:

$$(U_{\text{ID}}^{\neg c}, c) := \{(x, c) : x \in U_{\text{ID}} \text{ with } y^*(x) \neq c\}$$
$$(U_{\text{ID}}^{c}, c) := \{(x, c) : x \in U_{\text{ID}} \text{ with } y^*(x) = c\}$$

We now explain why and how we can regularize the model complexity such that the classifier fits S and all of (U, c), except for $(U_{\text{ID}}^{\neg c}, c)$. The *key intuition* why regularization helps is that it is more difficult to fit the labels c on $(U_{\text{ID}}^{\neg c}, c)$ than on (U_{OOD}, c), since $(U_{\text{ID}}^{\neg c}, c)$ lies closer in covariate space to points in the correctly labeled training set S. Hence, we can exactly fit (U_{OOD}, c) but not the entire (U_{ID}, c) if we adequately limit the function complexity (e.g. by choosing a small model class, or through regularization), as illustrated in Fig. 2 Left. Moreover, since regularized predictors are smooth, it follows that the model generalizes well on ID data and also predicts the label c on holdout OOD samples similar to the ones in the U_{OOD}. On the other hand, if the models are too complex (e.g. deep neural networks [49]), then they can even fit the wrong labels on (U_{ID}, c) (see Fig. 2 Right), causing the models in the ensemble to disagree on the entire unlabeled set U.

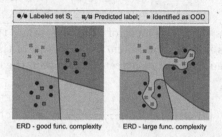

Fig. 2. Restricting model complexity is necessary to prevent from flagging the whole U as OOD. **Left:** Linear classifiers disagree on points in U_{OOD}, but agree to predict the correct label on samples from U_{ID}. **Right:** The models are too complex so they fit the arbitrary label on the entire U.

We use early stopping regularization, motivated by recent empirical and theoretical works that have found that early stopped neural networks are less vulnerable to label noise in the training data [26, 46]. In Proposition 1 from Appendix B we argue that there exists an optimal stopping time for gradient descent at which all points in (U, c) are fit, except for the wrongly labeled samples in (U_{ID}, c). To find the best stopping time in practice, we use a validation set of labeled ID points to select an intermediate checkpoint before convergence. As a model starts to fit $(U_{ID}^{\neg c}, c)$, i.e. the wrongly labeled ID samples in U_{ID}, it also predicts the label c on some validation ID points, leading to a decrease in validation accuracy, as shown in Fig. 3.

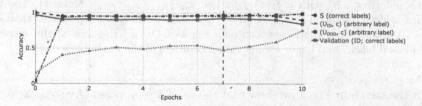

Fig. 3. Accuracy measured while fine-tuning a model pretrained on S (epoch 0 indicates values obtained with the initial weights). The samples in (U_{OOD}, c) are fit first, while the model reaches high accuracy on (U_{ID}, c) much later. We fine-tune for at least one epoch and then early stop when the validation accuracy starts decreasing after 7 epochs (vertical line). The model is trained on SVHN[0:4] as ID and SVHN[5:9] as OOD.

3.3   Ensemble Disagreement Score

We now motivate a novel ensemble aggregation technique tailored to exploit ensemble diversity that we use to detect OOD samples with ERD. Note that we can cast the OOD detection problem as a hypothesis test with null hypothesis $H_0 : x \in \mathcal{X}_{ID}$. Our procedure tests the null hypothesis by using an ensemble-based score: The null hypothesis is *rejected* and we report x as OOD (*positive*) if the score is larger than a threshold t_0 (see Algorithm 2).

Previous works [23,34] first average the softmax predictions of the models in the ensemble $\bar{f}(x) := \frac{1}{K} \sum_{i=1}^{K} f_i(x) \in [0,1]^{|\mathcal{Y}|}$ and then use the entropy of $\bar{f}(x)$ as a metric, i.e. $(\mathrm{H} \circ \mathrm{Avg})(f_1(x), ..., f_K(x)) := -\sum_{i=1}^{|\mathcal{Y}|} (\bar{f}(x))_i \log(\bar{f}(x))_i$ where $(\bar{f}(x))_i$ denotes the ith element of $\bar{f}(x)$. We argue that averaging model outputs first, discards information about the diversity of the ensemble. Instead, we propose the average pairwise *disagreement* between the outputs of K models in an ensemble:

$$(\mathrm{Avg} \circ \rho)(\{f_i(x)\}_{i=1}^{K}) := \frac{2}{K(K-1)} \sum_{i \neq j} \rho(f_i(x), f_j(x)),$$

where ρ is a measure of disagreement between the softmax outputs of two predictors, for example the total variation distance $\rho_{\mathrm{TV}}(f_i(x), f_j(x)) = \frac{1}{2}\|f_i(x) - f_j(x)\|_1$ used in our experiments.

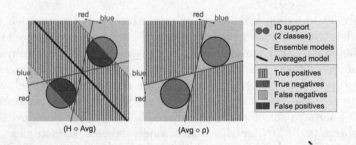

Fig. 4. Cartoon illustration showing a diverse ensemble of linear binary classifiers (solid purple). We compare OOD detection performance for two aggregation scores: $(\mathrm{H} \circ \mathrm{Avg})$ (**Left**) and $(\mathrm{Avg} \circ \rho)$ with $\rho(f_1(x), f_2(x)) = \mathbb{1}_{\mathrm{sgn}(f_1(x)) \neq \mathrm{sgn}(f_2(x))}$ (**Right**). The two metrics achieve similar TPRs, but $(\mathrm{H} \circ \mathrm{Avg})$ leads to more false positives than our score, $(\mathrm{Avg} \circ \rho)$, since the former can only flag as OOD a band around the averaged model (solid black) and hence cannot take advantage of the ensemble's diversity.

In the sketch in Fig. 4 we show that the score we propose, $(\mathrm{Avg} \circ \rho)$, achieves a higher TNR compared to $(\mathrm{H} \circ \mathrm{Avg})$, for a fixed TPR – a common way of evaluating statistical tests. Notice that the detection region for $(\mathrm{H} \circ \mathrm{Avg})$ is always limited to a band around the average model. In order for the $(\mathrm{H} \circ \mathrm{Avg})$ to have large TPR, this band needs to be wide, leading to many false positives. This example demonstrates how averaging softmax outputs relinquishes the benefits of a diverse ensemble that our disagreement score can exploit. In Appendix C we provide quantitative experimental evidence that reveals that, indeed, our disagreement score is crucial for good OOD detection performance with diverse ensembles, such as the one obtained with Algorithm 1.

4 Experimental Results

4.1 Standard Image Data Sets

ID vs OOD Settings. We report results on two broad types of OOD detection scenarios: *(1) Easy OOD data:* ID and OOD samples come from strikingly different data sets (e.g. CIFAR10 vs SVHN). These are the settings usually considered in the literature and on which most baselines perform well; and *(2) Hard OOD data:* The OOD data consists of "novel" classes that resemble the ID samples: e.g. the first 5 classes of CIFAR10 are ID, the last 5 classes are OOD. The similarities between the ID and the OOD classes make these settings significantly more challenging (see Appendix E for more details).

Table 1. AUROC and TNR@95 for different OOD detection scenarios (the numbers in squared brackets indicate the ID or OOD classes). We highlight the best ERD variant and the *best baseline* among prior work. The asterisk marks methods proposed in this paper. nnPU (†) assumes oracle knowledge of the OOD ratio in the unlabeled set.

ID data	OOD data	Other settings					Unknown OOD			
		Vanilla Ensembles	Gram	DPN	OE	Mahal.	nnPU†	MCD	Bin. Classif. *	ERD *
						AUROC ↑/TNR@95 ↑				
SVHN	CIFAR10	0.97/0.88	0.97/0.86	*1.00/1.00*	*1.00/1.00*	0.99/0.98	*1.00/1.00*	0.97/0.85	1.00/1.00	1.00/0.99
CIFAR10	SVHN	0.92/0.78	*1.00/0.98*	0.95/0.85	0.97/0.89	0.99/0.96	*1.00/1.00*	*1.00/0.98*	1.00/1.00	1.00/1.00
CIFAR100	SVHN	0.84/0.48	0.99/0.97	0.77/0.44	0.82/0.50	0.98/0.90	*1.00/1.00*	0.97/0.73	1.00/1.00	1.00/1.00
FMNIST [0,2,3,7,8]	FMNIST [1,4,5,6,9]	0.64/0.07	–/–	0.77/0.15	0.66/0.12	0.77/0.20	*0.95/0.71*	0.78/0.30	0.95/0.66	0.94/0.67
SVHN [0:4]	SVHN [5:9]	0.92/0.69	0.81/0.31	0.87/0.19	0.85/0.52	0.92/0.71	*0.96/0.73*	0.91/0.51	0.81/0.40	0.95/0.74
CIFAR10 [0:4]	CIFAR10 [5:9]	0.80/0.39	0.67/0.15	*0.82/0.32*	*0.82/0.41*	0.79/0.27	0.61/0.11	0.69/0.25	0.85/0.43	0.93/0.70
CIFAR100 [0:49]	CIFAR100 [50:99]	*0.78/0.35*	0.71/0.16	0.70/0.26	0.74/0.31	0.72/0.20	0.53/0.06	0.70/0.26	0.66/0.13	0.82/0.44
Average		0.84/0.52	0.86/0.57	0.84/0.46	0.84/0.54	*0.88/0.60*	0.86/*0.66*	0.86/0.55	0.89/0.66	0.95/0.79

Baselines. We compare ERD against previous methods that are applicable to the unknown OOD setting and also include well-known baselines that require different kinds of access to OOD data for training, as indicated in Table 4. In addition, we propose a novel simple approach that uses an unlabeled set: an early stopped binary classifier (*Bin. Classif.*) trained to distinguish between S and U. We include a detailed description of all the baselines together with precise hyperparameter choices in Appendix D.

Our method – ERD. For our method we train ensembles of 5 MLP models for FashionMNIST and ResNet20 [14] models for the other settings. The networks are initialized with weights pretrained on the ID training set. For each model in the ensemble we perform post-hoc early stopping: we train for 10 epochs and select the iteration with the lowest validation loss. In the appendix we also present results for a variant of ERD trained from random initializations.

Evaluation Metrics. We use two standard metrics common in the OOD detection literature: the area under the ROC curve (AUROC; larger values are better) and the TNR at a TPR of 95% (TNR@95; larger values are better).

Summary of Results. Table 1 summarizes the main empirical results. On the easy scenarios (top part of the table) most methods achieve near-perfect OOD detection with AUROC close to 1. However, on the novelty detection scenarios (bottom part), ERD has a clear edge over the other approaches and improves the average TNR@95 by 20% relative to the best baseline. Furthermore, on the novel-class setting on CIFAR10, the TPR@95 gains of our method compared to the best prior work go as high as 70%. A similar trend can be observed for AUROC as well. The substantial gap between ERD and other approaches, both in average AUROC and average TNR@95, indicates that our method lends itself well to practical situations when accurate OOD detection is critical.

Table 2. AUROC for some representative baselines from [5] on a medical OOD detection benchmark. We highlight the *best ERD* variant and the *best baseline* among prior work. See Appendix G for details about the baselines and the data sets.

Data set	Use case	Mahal.	kNN-8	VAE-BCE	AE-MSE	ERD	ERD++
PADChest	1, 2 (easy OOD)	0.94	0.97	0.95	*0.99*	<u>1.00</u>	<u>1.00</u>
NIHCC	1, 2 (easy OOD)	0.98	*0.99*	*0.99*	0.98	0.97	<u>0.99</u>
DRD	1, 2 (easy OOD)	0.82	0.96	0.97	*0.99*	0.98	<u>1.00</u>
PADChest	3 (hard OOD)	0.53	0.46	0.52	*0.55*	<u>0.77</u>	0.72
NIHCC	3 (hard OOD)	*0.52*	*0.52*	0.50	*0.52*	0.46	<u>0.50</u>
DRD	3 (hard OOD)	*0.70*	0.60	0.67	0.64	0.91	<u>1.00</u>
	Average AUROC	0.80	0.82	0.83	*0.85*	0.89	<u>0.91</u>

4.2 Medical Image OOD Detection Benchmark

Data Sets and Baselines. We use the benchmark proposed in [5] which comprises different kinds of medical image data from both healthy and unhealthy patients. For our comparison, we consider three modalities for the ID data: lateral (PADChest) and frontal (NIHCC) chest X-rays and retinal images (DRD). The authors annotate the training data with binary labels indicating whether the patient is healthy or unhealthy, thus discarding information about the condition. For each ID data set, the benchmark examines three categories of OOD detection problems:

- Use case 1: The OOD data set contains images from a completely different domain (e.g. X-rays as ID, and samples from CIFAR10 as OOD).

- Use case 2: The OOD data contains images captured incorrectly, e.g. lateral chest X-rays as ID and frontal X-rays as OOD.
- Use case 3: The OOD data set contains images that come from novel diseases, not present in the training set.

While the first two categories of outliers can be detected accurately by most methods, the third scenario turns out to be more challenging, due to how similar the ID and OOD samples are. We use the baselines considered in [5] and compare them to the performance of both ERD and ERD++. The latter is a variant of our method fine-tuned for more epochs and initialized from weights pretrained on ImageNet (see Appendix D for more details). Since the training labels are binary (healthy/unhealthy), we train ensembles of two models and note that the performance of our method could be improved if one provides finer-grained annotations (e.g. by assigning different labels to the various diseases in the training set, instead of collecting them all in the "unhealthy" class).

Summary of Results. Our method improves the average AUROC from 0.85 to 0.91, compared to the best performing baseline, with the gap being even more significant on novel disease scenarios, as indicated in Table 2. Appendix G contains more detailed results for then medical settings.

4.3   Limitations of Related OOD Detection Methods

We now discuss some shortcomings of existing OOD detection approaches closely related to ours and indicate how our method attempts to address them. Firstly, vanilla ensembles use only the stochasticity of the training process and the random initialization to obtain diverse models, but this often leads to similar classifiers, that predict the same incorrect label on OOD data [15]. Secondly, in the absence of proper regularization, optimizing the MCD objective leads to models that agree to a similar extent on both ID and OOD data so that one cannot distinguish them from one another (as indicated by low AUROC scores). Furthermore, nnPU does not exploit all the signal in the training set and discards the labels of the ID data, which leads to poor performance on hard OOD data.

ERD successfully diversifies an ensemble on OOD data by using the unlabeled set and without requiring additional information about the test distribution (e.g. unlike nnPU which requires the true OOD ratio). We identify the key reasons behind the good performance of our approach to be as follows: 1) utilizing the labels of the ID training data and the complexity of deep neural networks to diversify model outputs on OOD data; 2) choosing an appropriate disagreement score that draws on ensemble diversity; 3) employing early stopping regularization to prevent diversity on ID inputs.

5   Conclusions

Reliable OOD detection is essential in order to deploy classification systems in critical applications in the medical domain. We propose a procedure that results in an ensemble with selective disagreement only on OOD data, by successfully leveraging unlabeled data to fine-tune the models in the ensemble. It outperforms state-of-the-art methods that also have access to a mixture of ID and unknown OOD samples, and even surpasses approaches that use known OOD data for training.

References

1. Barbu, A., et al.: ObjectNet: a large-scale bias-controlled dataset for pushing the limits of object recognition models. In: Advances in Neural Information Processing, vol. 32, pp. 9453–9463 (2019)
2. Beede, E., et al.: A human-centered evaluation of a deep learning system deployed in clinics for the detection of diabetic retinopathy. In: Proceedings of the CHI Conference on Human Factors in Computing Systems, pp. 1–12 (2020)
3. Ben-David, S., Blitzer, J., Crammer, K., Pereira, F.: Analysis of representations for domain adaptation. In: Advances in Neural Information Processing Systems, vol. 19, pp. 137–144 (2007)
4. Blundell, C., Cornebise, J., Kavukcuoglu, K., Wierstra, D.: Weight uncertainty in neural networks. In: Proceedings of the 32th International Conference on Machine Learning (2015)
5. Cao, T., Huang, C.W., Hui, D.Y.T., Cohen, J.P.: A benchmark of medical out of distribution detection. arXiv preprint arXiv:2007.04250 (2020)
6. Chen, Y., Wei, C., Kumar, A., Ma, T.: Self-training avoids using spurious features under domain shift. arXiv preprint arXiv:2006.10032 (2020)
7. Choi, H., Jang, E., Alemi, A.A.: WAIC, but why? Generative ensembles for robust anomaly detection. arXiv preprint arXiv:1810.01392 (2018)
8. Deng, J., Dong, W., Socher, R., Li, L.J., Li, K., Fei-Fei, L.: ImageNet: a large-scale hierarchical image database. In: Proceedings of the IEEE Conference on Computer Vision and Pattern Recognition (CVPR) (2009)
9. Fu, Y., Hospedales, T.M., Xiang, T., Gong, S.: Transductive multi-view zero-shot learning. IEEE Trans. Pattern Anal. Mach. Intell. **37**, 2332–2345 (2015)
10. Gal, Y., Ghahramani, Z.: Dropout as a Bayesian approximation: representing model uncertainty in deep learning. In: Proceedings of Machine Learning Research, vol. 48, pp. 1050–1059 (2016)
11. Ganin, Y., et al.: Domain-adversarial training of neural networks. J. Mach. Learn. Res. **17**, 1–35 (2016)
12. Geifman, Y., El-Yaniv, R.: Selective classification for deep neural networks. In: Advances in Neural Information Processing Systems, vol. 30, pp. 4878–4887 (2017)
13. Graves, A.: Practical variational inference for neural networks. In: Advances in Neural Information Processing Systems, vol. 24, pp. 2348–2356 (2011)
14. He, K., Zhang, X., Ren, S., Sun, J.: Deep residual learning for image recognition. In: Proceedings of the IEEE Conference on Computer Vision and Pattern Recognition (CVPR) (2016)
15. Hein, M., Andriushchenko, M., Bitterwolf, J.: Why ReLU networks yield high-confidence predictions far away from the training data and how to mitigate the problem. In: Proceedings of the IEEE Conference on Computer Vision and Pattern Recognition (CVPR) (2019)
16. Hendrycks, D., Dietterich, T.: Benchmarking neural network robustness to common corruptions and perturbations. In: Proceedings of the International Conference on Learning Representations (2019)

17. Hendrycks, D., Mazeika, M., Dietterich, T.: Deep anomaly detection with outlier exposure. In: Proceedings of the International Conference on Learning Representations (2019)
18. Katsamenis, I., Protopapadakis, E., Voulodimos, A., Doulamis, A., Doulamis, N.: Transfer learning for COVID-19 pneumonia detection and classification in chest X-ray images. medRxiv (2020)
19. Kirichenko, P., Izmailov, P., Wilson, A.G.: Why normalizing flows fail to detect out-of-distribution data. In: Advances in Neural Information Processing Systems, vol. 33, pp. 20578–20589 (2020)
20. Kiryo, R., Niu, G., du Plessis, M.C., Sugiyama, M.: Positive-Unlabeled learning with non-negative risk estimator. In: Advances in Neural Information Processing Systems, vol. 30 (2017)
21. Krizhevsky, A.: Learning multiple layers of features from tiny images. Technical report (2009)
22. Kumar, A., Ma, T., Liang, P.: Understanding self-training for gradual domain adaptation. arXiv preprint arXiv:2002.11361 (2020)
23. Lakshminarayanan, B., Pritzel, A., Blundell, C.: Simple and scalable predictive uncertainty estimation using deep ensembles. In: Advances in Neural Information Processing Systems, vol. 30, pp. 6402–6413. Curran Associates, Inc. (2017)
24. Lecun, Y., Bottou, L., Bengio, Y., Haffner, P.: Gradient-based learning applied to document recognition. In: Proceedings of the IEEE, pp. 2278–2324 (1998)
25. Lee, K., Lee, K., Lee, H., Shin, J.: A simple unified framework for detecting out-of-distribution samples and adversarial attacks. In: Advances in Neural Information Processing Systems, vol. 31, pp. 7167–7177 (2018)
26. Li, M., Soltanolkotabi, M., Oymak, S.: Gradient descent with early stopping is provably robust to label noise for overparameterized neural networks, pp. 4313–4324. Proceedings of Machine Learning Research (2020)
27. Liang, S., Li, Y., Srikant, R.: Enhancing the reliability of out-of-distribution image detection in neural networks. In: Proceedings of the International Conference on Learning Representations (2018)
28. Liu, S., Garrepalli, R., Dietterich, T., Fern, A., Hendrycks, D.: Open category detection with PAC guarantees, pp. 3169–3178 (2018)
29. Lu, A.X., Lu, A.X., Schormann, W., Andrews, D.W., Moses, A.M.: The cells out of sample (COOS) dataset and benchmarks for measuring out-of-sample generalization of image classifiers. arXiv preprint arXiv:1906.07282 (2019)
30. Malinin, A., Gales, M.: Predictive uncertainty estimation via prior networks. In: Advances in Neural Information Processing Systems, vol. 32, pp. 7047–7058 (2018)
31. Nalisnick, E., Matsukawa, A., Teh, Y.W., Gorur, D., Lakshminarayanan, B.: Do deep generative models know what they don't know? In: Proceedings of the International Conference on Learning Representations (2019)
32. Neal, R.M.: Bayesian Learning for Neural Networks. Springer, Heidelberg (1996). https://doi.org/10.1007/978-1-4612-0745-0
33. Netzer, Y., Wang, T., Coates, A., Bissacco, A., Wu, B., Ng, A.Y.: Reading digits in natural images with unsupervised feature learning. In: NIPS Workshop on Deep Learning and Unsupervised Feature Learning (2011)
34. Ovadia, Y., et al.: Can you trust your model's uncertainty? Evaluating predictive uncertainty under dataset shift. In: Advances in Neural Information Processing Systems, vol. 32, pp. 13991–14002 (2019)
35. du Plessis, M.C., Niu, G., Sugiyama, M.: Analysis of learning from positive and unlabeled data. In: Advances in Neural Information Processing Systems, vol. 27 (2014)

36. Recht, B., Roelofs, R., Schmidt, L., Shankar, V.: Do CIFAR-10 classifiers generalize to CIFAR-10? arXiv preprint arXiv:1806.00451 (2018)
37. Recht, B., Roelofs, R., Schmidt, L., Shankar, V.: Do ImageNet classifiers generalize to ImageNet? arXiv preprint arXiv:1902.10811 (2019)
38. Ren, J., Liu, P.J., Fertig, E., Snoek, J., Poplin, R., Depristo, M., Dillon, J., Lakshminarayanan, B.: Likelihood ratios for out-of-distribution detection. In: Advances in Neural Information Processing Systems, vol. 32, pp. 14707–14718 (2019)
39. Sastry, C.S., Oore, S.: Detecting out-of-distribution examples with in-distribution examples and Gram matrices. arXiv preprint arXiv:1912.12510 (2019)
40. Scott, C., Blanchard, G.: Transductive anomaly detection. Technical report (2008)
41. Shimodaira, H.: Improving predictive inference under covariate shift by weighting the log-likelihood function. J. Stat. Plan. Infer. **90**, 227–244 (2000)
42. Torralba, A., Fergus, R., Freeman, W.T.: 80 million tiny images: a large data set for nonparametric object and scene recognition. IEEE Trans. Pattern Anal. Mach. Intell. **30**, 1958–1970 (2008)
43. Vapnik, V.N.: Statistical Learning Theory. Wiley, Hoboken (1998)
44. Wan, Z., Chen, D., Li, Y., Yan, X., Zhang, J., Yu, Y., Liao, J.: Transductive zero-shot learning with visual structure constraint. In: Advances in Neural Information Processing Systems, vol. 32, pp. 9972–9982 (2019)
45. Xiao, H., Rasul, K., Vollgraf, R.: Fashion-MNIST: A novel image dataset for benchmarking machine learning algorithms (2017)
46. Yilmaz, F.F., Heckel, R.: Image recognition from raw labels collected without annotators. arXiv preprint arXiv:1910.09055 (2019)
47. Yu, Q., Aizawa, K.: Unsupervised out-of-distribution detection by maximum classifier discrepancy. In: Proceedings of the IEEE/CVF International Conference on Computer Vision (ICCV) (2019)
48. Zagoruyko, S., Komodakis, N.: Wide residual networks. In: Proceedings of the British Machine Vision Conference (BMVC) (2016)
49. Zhang, C., Bengio, S., Hardt, M., Recht, B., Vinyals, O.: Understanding deep learning requires rethinking generalization. arXiv preprint arXiv:1611.03530 (2016)

PIPPI 2021

Automatic Placenta Abnormality Detection Using Convolutional Neural Networks on Ultrasound Texture

Zoe Hu[1,2(✉)], Ricky Hu[1,2], Ryan Yan[3], Chantal Mayer[3], Robert N. Rohling[2,4,5], and Rohit Singla[4]

[1] School of Medicine, Queen's University, Kingston, ON, Canada
zhu@qmed.ca
[2] Electrical and Computer Engineering, University of British Columbia, Vancouver, BC, Canada
[3] Obstetrics and Gynecology, University of British Columbia, Vancouver, BC, Canada
[4] School of Biomedical Engineering, University of British Columbia, Vancouver, BC, Canada
[5] Mechanical Engineering, University of British Columbia, Vancouver, BC, Canada

Abstract. Diseases related to the placenta, such as preeclampsia (PE) and fetal growth restriction (FGR), are major causes of mortality and morbidity. Diagnostic criteria of such diseases are defined by biomarkers, such as proteinuria, that appear in advanced gestational age. As placentally-mediated disease is often clinically unrecognized until later stages, accurate early diagnosis is required to allow earlier intervention, which is particularly challenging in low-resource areas without subspecialty clinicians. Proposed attempts at early diagnosis involve a combination of subjective and objective ultrasound placental assessments which have limited accuracy and high interobserver variability. Machine learning, particularly with convolutional neural networks, have shown potential in analyzing complex textural features in ultrasound imaging that may be predictive of disease. We propose a model utilizing a two-stage convolutional neural network pipeline to classify the presence of placental disease. The pipeline involves a segmentation stage to extract the placenta followed by a classification stage. We evaluated the pipeline on retrospectively collected placenta ultrasound scans and diagnostic outcomes of 321 patients taken by 18 sonographers and 3 ultrasound machines. Compared to existing clinical algorithms and neural networks, our classifier achieved significantly higher accuracy of 0.81 ± 0.02 ($p < 0.05$). Class activation maps were generated to identify potential abnormal regions of interest in placenta tissue. This study provides support that automated image analysis of ultrasound texture may assist physicians in early identification of placental disease, with potential benefits to low-resource environments.

Keywords: Placenta · Ultrasound · Convolutional neural networks · Classification · Preeclampsia · Fetal growth restriction

© Springer Nature Switzerland AG 2021
C. H. Sudre et al. (Eds.): UNSURE 2021/PIPPI 2021, LNCS 12959, pp. 147–156, 2021.
https://doi.org/10.1007/978-3-030-87735-4_14

1   Introduction

1.1   Background and Motivation

The placenta is essential for fetal development and maternal health, providing nutrients, waste exchange, and transfer of antibodies to the fetus. Preeclampsia (PE) and fetal growth restriction (FGR) are two clinical manifestations of placental insufficiency with much overlap in placental pathology. Both diseases hold a significant risk to the fetus and the mother. PE occurs in 3–10% of all pregnancies and is a leading cause of maternal mortality [1]. Placentally mediated FGR occurs up to 5% of all pregnancies and can result in fetal demise or several neonatal complications [2]. Hence, there is a motivation to identify the placental abnormalities early to study and prevent adverse outcomes.

A challenge is that clinical features of placental pathology, such as proteinuria, are not observable until an advanced gestational age [3]. While screening for risk factors such as diabetes or family history are strongly associated with PE, up to 70% of patients with PE do not have these risk factors [4]. Hence, the lack of early and accurate screening results in difficulty for effective interventions as placental insufficiency may have progressed before the disease is identified [5]. While there have been developments in multi-modality first trimester screening to improve risk assessment [6], overall patient access is limited. The specialized services for such screening is available only in tertiary care centers, which incur additional travel costs and barriers to care for patients in rural and low-resource settings.

The development of a placental assessment tool at the time of routine obstetrical ultrasound may address the need for early screening that can be automated for low-resource areas. Diseases such as PE and FGR are known to have a subclinical phase where the placenta tissue changes before signs and symptoms are observable [7]. Ultrasound assessment of the placenta is routinely performed and includes the analysis of the uterine Doppler waveform or subjectively assessing image texture to assess for PE [8]. Ultrasound contains complex features, such as speckle patterns due to scattering interaction of acoustic waves with tissue, that can be used to characterize placenta tissue microstructure [9]. O'Gorman et al. (2016) [10] assessed 10 different ultrasound assessment algorithms for PE, reporting a sensitivity range of 0.27–0.57 for all 10 algorithms and a sensitivity of 0.50–0.56 for the best performing algorithm. This variance may be due to how ultrasound is dependent on both the acquisition technique, resulting in interobserver error, as well as difficulty in subjective interpretation of ultrasound. Hence, there is a motivation to develop a higher accuracy, automatic, and quantifiable tool to assess placental disease. Automation may enable deployment in low-resource settings to be used by local healthcare providers and potentially triage accessing expensive specialized services.

1.2   Existing Work

There are several existing methods that attempt to automatically predict placental disease. Moreira et al. [11] utilized a Bayesian network with clinical variables (age and parity) and patient symptoms to identify the probability of preeclampsia and hypertension on 20 patients. Jhee et al. [12] compared predictions from popular classification algorithms of a decision tree, naive Bayes, support vector machine, random forest, and stochastic gradient boost. The input features used included maternal medical history, physical exam finding, and laboratory values. Sufriyana et al. [13] utilized maternal medical history and ultrasound uterine Doppler measurements as inputs to a logistic regression model to predict PE and FGR. The challenge with utilizing clinical variables such as hypertension is that PE is primarily a disease of poor placentation and blood flow, which progresses to clinically observable metrics of proteinuria and hypertension. When protein levels in the urine and blood pressure have increased near the threshold for clinical diagnosis, then the disease has likely progressed significantly [4]. There lies a gap in placental disease prediction without the reliance of late stage clinical markers.

Convolutional neural networks (CNNs) have been applied to placental imaging in the past mainly for feature detection, such as localizing lacunae [14] or segmentation of the placenta [15], indicating the potential for a CNN to identify complex textural features. To our knowledge, there has been no prior method utilizing a machine learning method to characterize ultrasound placenta features to predict placental disease, which we aim to achieve with this work.

1.3 Contributions

The main contribution of the work is the proposal and development of a pipeline that uses 2 separate CNNs to identify the placenta in an ultrasound image and then classify if the patient is predicted to have placental disease. We conducted a study acquiring 13,384 frames of 2D first trimester fetal ultrasound images from 321 patients across 5 years from different ultrasound machines, operators, settings, and view angles. The purpose of this study is specifically so that any model built will be tested on a diverse dataset to prevent overfitting and overestimation of accuracies from a single controlled use case as a proof of concept that a generalized signal can be detected by a CNN. To our knowledge, this is the largest study analyzing placental ultrasound for placental disease classification.

2 Methods

2.1 Dataset and Equipment

The dataset was acquired at the British Columbia Women's Hospital and Health Center (BCW). The dataset included ultrasound images and the patient's diagnostic outcome. From 321 patients, 321 2-dimensional (2D) second trimester placental ultrasound sweeps were obtained as per standard protocol at BCW (13,384 frames total). Age ranged from 17 to 50 years of age and gestational ages ranged from 59 days to 193 days (mean: 160 days). Of 321 patients, 93 (29%) were diagnosed with either PE or FGR at time of delivery and 9 (3%) were diagnosed with either PE or IUGR before time of scan.

Images were acquired by 18 different sonographers. The machines used during that time period include GE Voluson E8 (Boston, USA), GE Voluson E10 (Boston, USA), and Philips IU22 (Amsterdam, Netherlands). The Keras library was used to develop the CNN model and the model was trained with a NVIDIA Titan V graphics processing unit (Santa Clara, USA). The retrieval and analysis of the data was approved by the University of British Columbia Research Ethics Board (ID: H18-01199).

2.2   Classification Pipeline

Developing a placental screening tool from ultrasound poses as a binary classification problem. The 2 classes were either an abnormal placenta, defined as being diagnosed with placentally-mediated diseases of PE or FGR by a physician, or normal placenta in the absence of both conditions. We recognize there is likely not a perfect bijection between abnormal placental texture and the presence of PE of FGR, which is discussed in our limitations. The full system contains 3 consecutive stages: preprocessing, segmentation, then classification (Fig. 1).

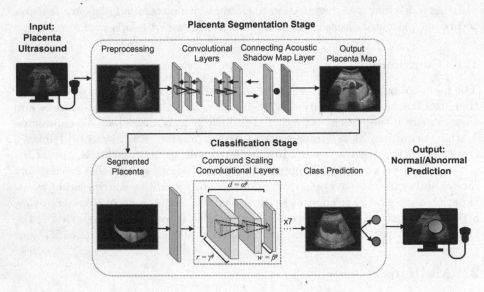

Fig. 1. The full classification pipeline. First, the image is preprocessed by resizing and normalizing in gray-level. The image is then segmented with an attention-based convolutional neural network, generating a segmented placenta. The segmentation is then classified to a placental disease label using a compound scaling convolutional neural network. In both networks, hyperparameters were optimized using gridsearch algorithms.

Preprocessing. The images were extracted frame-by-frame from an ultrasound sweep and each image first had identifying information and metadata removed. Frames without a placenta in view were manually removed. The images were resized to 600×600 and normalized by linearly scaling gray level intensities to $[0, 1]$. Images were labelled as abnormal if the patient was diagnosed with PE or FGR on retrospective chart data. The patients were randomly split with 80% of the data used for training of any subsequent CNNs and 20% used for testing.

Segmentation Neural Network. Segmentation utilized the attention-based CNN from on Hu et al. (2019) [15], which reported an accuracy range of $0.88 - 0.96$, which is within the 95% confidence interval of expert observers. The hyperparameters of the network include number of layers, learning rate, epochs, and batch size and were all optimized via gridsearch. The output of the segmentation is an image with only pixels of the placenta, which is the input to the classification stage.

Classification Neural Network. 4 different CNN architectures were tested (DenseNet [16], ResNet [17], ResNeXt [18], EfficientNet [19]) and the final architecture chose was EfficientNet developed by Tan et al. (2019) [19]. Placental ultrasound likely contains highly detailed texture features, which may be processed through a deep CNN. However, the increased complexity of the CNN may result in overfitting. EfficentNet was designed to have 7 intermediate sub-blocks which vary in width ($w = \alpha^\phi$), depth ($d = \alpha^\phi$), and resolution ($r = \gamma^\phi$) by a compound scaling parameter ϕ. The hyperparameters for this stage include dropouts, learning rate, and batch size, all of which were optimized via gridsearch. Our final model used $\phi = 7$, base constants $\alpha = \beta = \gamma = 2$ as proposed by Tan et al. (2019) [19], a dropout ratio of 0.20 at the top layer, a learning rate of 0.01, and a batch size of 8.

To visualize predictive regions of an image, class activation maps (CAM) were computed by computing the gradients of the final convolutional layer output, multiplied by the weights of the feature map, and summing all channels. This visualizes which kernels were activated for a particular class [20].

2.3 Validation

The classification pipeline was validated with 5 k-fold cross validation. The mean accuracy, sensitivity, and specificity were computed across the 5 k-folds for the test set. The gold standard used was a physician diagnosis of placental disease recorded in the patient's chart at the completion of the pregnancy.

To assess the performance of our system, we compared against the clinical assessment algorithm at BCW, testing on the same set of patients and evaluating with a Wilcoxon rank-sum test. The clinical assessment algorithm defines a placenta as abnormal if any of the following are seen on a second trimester ultrasound scan: a) bilateral uterine artery Doppler notch, b) unilateral or indeterminate uterinary artery notch and a pulsatility index >1.55, and c) any 2 of placental thickness greater than 5.0 cm, placental length less than 11.0 cm, abnormal echotexture, or greater than 2 infarcts.

An ablation analysis included repeated 5 k-fold cross validation with different variations of input data. The first variation ("No Segmentation") was removing the segmentation stage to assess accuracy if the region of interest was not isolated a priori. The second variation ("Balanced") forced class balance by randomly removing normal patients until normal samples equalled abnormal samples to assess if accuracy changes due to bias in the original unbalanced dataset. The third variation ("Sampled") was forcing equal sampling of the ultrasound sweep such that a patient contained no more than 20 frames of images (the minimum observed in the dataset) to assess if accuracies changed due to certain patients with more images created bias in training the classifier.

3   Results

The accuracies, sensitivities, and specificities of the different classifiers are displayed in Table 1. The results of the ablation analysis are displayed in Table 2. The accuracy of the final model (Segmentation + EfficientNetB7) performed significantly better than the existing clinical algorithm ($p < 0.05$). An example of a CAM for normal and abnormal placentas are visualized in Fig. 2.

Table 1. Classification accuracies of different architectures when compared to the gold standard of the diagnosis recorded in the patient chart.

Method	Accuracy	Sensitivity	Specificity
Segmentation + ResNet	0.65 ± 0.07	0.00 ± 0.00	0.93 ± 0.15
Segmentation + ResNeXt	0.66 ± 0.05	0.72 ± 0.04	0.50 ± 0.18
Segmentation + DenseNet	0.68 ± 0.05	0.00 ± 0.00	0.96 ± 0.04
Clinical Algorithm	0.63 ± 0.06	0.34 ± 0.07	0.88 ± 0.04
Segmentation + EfficientNetB7	**0.81 ± 0.02**	**0.88 ± 0.06**	**0.65 ± 0.06**

Table 2. The accuracies of the classification model when different components are manipulated. Variations included removing the segmentation stage, forcing class balance, and sampling to create an equal number of frames for each patient.

Method	Accuracy	Sensitivity	Specificity
No Segmentation + EfficientNetB7	0.66 ± 0.06	0.80 ± 0.03	0.29 ± 0.13
Segmentation + EfficientNetB7	0.81 ± 0.02	0.88 ± 0.06	0.65 ± 0.06
Segmentation + EfficientNetB7 + Balanced	0.81 ± 0.04	0.74 ± 0.11	0.89 ± 0.07
Segmentation + EfficientNetB7 + Sampled	0.73 ± 0.11	0.92 ± 0.02	0.43 ± 0.20

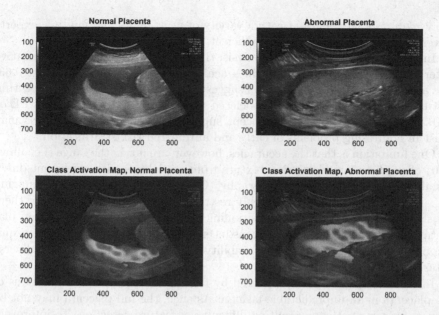

Fig. 2. The resulting class activation map of processing the segmented placenta through the convolutional kernels. The highly activated regions indicate that the classifier identifies certain regions to be predictive of a normal or abnormal class.

4 Discussion

Our dual-stage CNN pipeline performed with greater accuracy than the existing standard-of-care clinical algorithm at BCW. It additionally outperformed the reported accuracies of all manual ultrasound assessment algorithms analyzed by O'Gorman et al. (2016) [10]. Selecting the EfficientB7 architecture achieved significantly greater accuracy than other popular CNNs ($p < 0.05$). When segmentation was removed, the model performance decreased, indicating that a segmentation step may be necessary for a CNN to focus on a region of interest pertaining to placental texture. The accuracies remained similar when class balance was forced, providing support that the accuracies were not due to a biased dataset. The biases are seen with DenseNet and ResNet which overpredicts normal classes, likely due to a larger number of normal patients. This may also be the case for the existing clinical assessment algorithm where the high specificity and low sensitivity indicates overpredicting normal placenta, which would report a higher accuracy than in a balanced dataset due to 71% of patients being normal.

There are several implications and limitations of our study. We manually selected only ultrasound frames with the placenta in view before segmentation. We recognize that this may not be reproducible at other centres and may hinder the ability to generalize our method. A potential solution would be to utilize the segmentation network to extract placenta regions and filter images that include some minimum area of placenta. However, validation for this method likely requires a large dataset with a variety of view angles and multiple experts to verify that the images contain sufficient placenta for analysis.

In terms of placenta texture analysis, there are currently a few quantitative criteria, but no assessment guidelines achieving accuracies greater than 0.63. The accuracy achieved by our model suggests that there is a viable signal within texture analysis of the placenta associated with placental disease such as PE or FGR. This provides a proof-of-concept support that models using CNNs may assist in addressing problems of early and automated risk stratification.

One limitation is that the accuracies, however, are not in the range (i.e. above 0.90) of predictive models using texture from modalities with consistent directionality such as computed tomography (CT) or magnetic resonance imaging [21]. As fetal ultrasound is not highly restrained in its image acquisition, there are significant variations in the positioning of the fetus, amniotic fluid and placenta relative to the ultrasound transducers. These variations in image acquisition could impact the algorithm's ability to perform prediction on a diverse dataset.

In addition, ultrasound is reflected by bone, which may obscure portions of the placenta in an unpredictable fashion. As such, the full placenta may not be in view for some patients, resulting difficulty of feature detection. To improve a placental assessment model, future studies may assess *ex-vivo* placentas or full 3-dimensional volume reconstructions to reduce imaging variability and allow a CNN to focus on texture. A larger dataset would also provide more training samples for the classifier for future validation.

Another limitation is that abnormal microstructure and placental disease (by our classification definition) is not necessarily a bijection. That is, it is unknown to what extent does abnormal placentation and microstructure lead to PE or FGR. However, while FGR may be constitutionally solved and a result of maternal factors such as teratogens or genetic disease, this study focuses on patients with placentally-mediated FGR which overlaps significantly with PE in terms of pathology. Thus, the model provides insight to a correlative link, but future studies are required to be validated if the current diagnostic criteria for PE or FGR are the appropriate gold standard for placenta texture analysis. One such direction could be to analyze the class activation map produced by a CNN and investigate the identified regions of interest with a histological study to see what cellular changes are associated with PE or FGR. Combination with a larger dataset and a diverse patient population, such as patients undergoing regular screening, are required to satisfy criteria to conclude causality between CNN features and clinical outcomes [22].

5   Conclusion

We have developed a classifier model from a dual-stage CNN pipeline involving the segmentation of a placenta from ultrasound and predicting placenta-mediated disease. Our model achieved higher accuracies than existing manual assessment algorithms, though several limitations in the variability of the scans may reduce the ability to generalize the model. The model provides a proof-of-concept support that machine learning has the potential to address challenges in predicting placental disease, providing automated analysis and early identification.

Acknowledgments. Funding for this study was provided by The Natural Sciences and Engineering Research Council of Canada and the Canadian Institutes of Health Research.

References

1. Mackay, A., Berge, C., Atrash, H.: Pregnancy-related mortality from preeclampsia and eclampsia. Obstet. Gynecol. **97**(4), 533–538 (2001)
2. Garite, T.J., Clark, R., Thorpe, J.A.: Intrauterine growth restriction increases morbidity and mortality among premature neonates. Am. J. Obstet. Gynecol. **191**(2), 481–487 (2004)
3. Redman, C.W.: Latest advances in understanding preeclampsia. Science **308**(5728), 1592–1594 (2005)
4. Leslie, K., Thilaganathan, B., Papageorghiou, A.: Early prediction and prevention of pre-eclampsia. Best Pract. Res. Clin. Obstet. Gynaecol. **25**(3), 343–354 (2011)
5. Mol, B.W., Roberts, C.T., Thangarantinam, S., Magee, L.A., De Groot, C.J., Hofmeyr, G.J.: Pre-eclampsia. Lancet **387**(10022), 999–1011 (2016)
6. Rolnik, D.L., Wright, D., Poon, L.C., O'Gorman, N., Syngelaki, A., de Paco Matallana, C., Akolekar, R., Cicero, S., Janga, D., Singh, M., Molina, F.S.: Aspirin versus placebo in pregnancies at high risk for preterm preeclampsia. N. Engl. J. Med. **377**(7), 613–622 (2017)
7. Romero, R.: Prenatal medicine: the child is the father of the man. J. Matern. Neonatal Med. **22**(8), 636–639 (2009)
8. A. C. of Obstetricians and Gynecologists: CO638: first-trimester risk assessment for early-onset pre-eclampsia. Obstet. Gynecol. **126**(638), 25–274 (2015)
9. Deeba, F., et al.: Multiparametric QUS analysis for placental tissue characterization. In: 40th Annual International Conference of the IEEE Engineering in Medicine and Biology Society (EMBC), pp. 3477–3480 (2018)
10. O'Gorman, N., Nicolaides, K.H., Poon, L.C.Y.: The use of ultrasound and other markers for early detection of preeclampsia: Women's Health, pp. 197–207 (2012)
11. Moreira, M.W.L., Rorigues, J.J.P.C., Oliveira, A.M.B., Ramos, R.F., Saleem, K.: A preeclampsia diagnosis approach using Bayesian networks. In: 2016 IEEE International Conference on Communications (ICC), pp. 1–5 (2016)
12. Jhee, J., et al.: Prediction model development of late-onset preeclampsia using machine learning-based methods. PLoS ONE **14**(8), e0221202 (2019)
13. Sufriyana, H., Wu, Y., Su, E.: Prediction of preeclampsia and intrauterine growth restriction: development of machine learning models on a prospective cohort. JMIR Med. Inform. **8**(5), 215411 (2020)
14. Qi, H., Collins, S., Noble, J.A.: Automatic lacunae localization in placental ultrasound images via layer aggregation. In: Frangi, A.F., Schnabel, J.A., Davatzikos, C., Alberola-López, C., Fichtinger, G. (eds.) MICCAI 2018. LNCS, vol. 11071, pp. 921–929. Springer, Cham (2018). https://doi.org/10.1007/978-3-030-00934-2_102

15. Hu, R., Singla, R., Yan, R., Mayer, C., Rohling R.N.: Automated placenta seg-
    mentation with a convolutional neural network weighted by acoustic shadow detec-
    tion. In: Proceedings of the International Conference of the IEEE Engineering in
    Medicine and Biology Society (EMBC), pp. 6718–6723 (2019)
16. Huang, G., Liu, Z., Van Der Maaten, L., Weinberger, K.Q.: Densely connected
    convolutional networks. In: Proceedings of the IEEE Conference on Computer
    Vision and Pattern Recognition 2017, pp. 4700–4708 (2017)
17. He, K., Zhang, X., Ren, S., Sun, J.: Deep residual learning for image recognition. In:
    Proceedings of the IEEE Conference on Computer Vision and Pattern Recognition
    2016, pp. 770–778 (2016)
18. Xie, S., Girshick, R., Dollar, P., Tu, Z., He, K.: Aggregated residual transformations
    for deep neural networks. In: Proceedings of the IEEE Conference on Computer
    Vision and Pattern Recognition, pp. 1492–1500 (2017)
19. Tan, M., Le, Quoc.: Efficientnet: rethinking model scaling for convolutional neural
    networks. In: International Conference on Machine Learning, pp. 6105–6114 (2019)
20. Yang, W., Huang, H., Zhang, Z., Chen, X., Huang, K., Zhang, S.: Towards rich fea-
    ture discovery with class activation maps augmentation for person re-identification.
    In: Proceedings of the IEEE Conference on Computer Vision and Pattern Recog-
    nition 2019, pp. 1389–1398 (2019)
21. Anwar, S.M., Majid, M., Qayyum, A., Awais, M., Alnowami, M., Khan, M.K.:
    Medical image analysis using convolutional neural networks: a review. Image Signal
    Process. **42**(11), 1–13 (2018)
22. Castro, D.C., Walker, I., Glocker, B.: Causality matters in medical imaging. Nat.
    Commun. **11**(1), 1–10 (2020)

Simulated Half-Fourier Acquisitions Single-shot Turbo Spin Echo (HASTE) of the Fetal Brain: Application to Super-Resolution Reconstruction

Hélène Lajous[1,2(✉)] , Tom Hilbert[1,3,4] , Christopher W. Roy[1] ,
Sébastien Tourbier[1] , Priscille de Dumast[1,2] , Yasser Alemán-Gómez[1] ,
Thomas Yu[4], Hamza Kebiri[1,2] , Jean-Baptiste Ledoux[1,2] ,
Patric Hagmann[1] , Reto Meuli[1] , Vincent Dunet[1] , Mériam Koob[1] ,
Matthias Stuber[1,2] , Tobias Kober[1,3,4] , and Meritxell Bach Cuadra[1,2,4]

[1] Department of Radiology, Lausanne University Hospital (CHUV)
and University of Lausanne (UNIL), Lausanne, Switzerland
helene.lajous@unil.ch
[2] CIBM Center for Biomedical Imaging, Lausanne, Geneva, Switzerland
[3] Advanced Clinical Imaging Technology (ACIT), Siemens Healthcare,
Lausanne, Switzerland
[4] Signal Processing Laboratory 5 (LTS5), Ecole Polytechnique Fédérale de Lausanne
(EPFL), Lausanne, Switzerland

Abstract. Accurate characterization of *in utero* human brain maturation is critical as it involves complex interconnected structural and functional processes that may influence health later in life. Magnetic resonance imaging is a powerful tool complementary to the ultrasound gold standard to monitor the development of the fetus, especially in the case of equivocal neurological patterns. However, the number of acquisitions of satisfactory quality available in this cohort of sensitive subjects remains scarce, thus hindering the validation of advanced image processing techniques. Numerical simulations can mitigate these limitations by providing a controlled environment with a known ground truth. In this work, we present a flexible numerical framework for clinical T2-weighted Half-Fourier Acquisition Single-shot Turbo spin Echo of the fetal brain. The realistic setup, including stochastic motion of the fetus as well as intensity non-uniformities, provides images of the fetal brain throughout development that are comparable to real data acquired in clinical routine. A case study on super-resolution reconstruction of the fetal brain from synthetic motion-corrupted 2D low-resolution series further demonstrates the potential of such a simulator to optimize post-processing methods for fetal brain magnetic resonance imaging.

Keywords: Fetal brain Magnetic Resonance Imaging (MRI) ·
Numerical phantom · Half-Fourier Acquisition Single-shot Turbo spin
Echo (HASTE) sequence · Super-Resolution (SR) reconstruction

T. Hilbert, C. W. Roy—These authors contributed equally to this work.

© Springer Nature Switzerland AG 2021
C. H. Sudre et al. (Eds.): UNSURE 2021/PIPPI 2021, LNCS 12959, pp. 157–167, 2021.
https://doi.org/10.1007/978-3-030-87735-4_15

1   Introduction

Brain maturation involves complex intertwined structural and functional processes that can be altered by various genetic and environmental factors. As such, early brain development is critical and may impact health later in life [1–3].

Magnetic resonance imaging (MRI) may be required during pregnancy to investigate equivocal situations as a support for diagnosis and prognosis, but also for postnatal management planning [4]. In clinical routine, T2-weighted (T2w) fast spin echo sequences are used to scan multiple 2D thick slices that provide information on the whole brain volume with a good signal-to-noise ratio (SNR) while minimizing the effects of random fetal motion during the acquisition [5]. Contrary to periodic movements that can be directly related to physiological processes such as breathing or a heartbeat, and may therefore be compensated during post-processing, stochastic movements of the fetus in the womb cause various artifacts in the images and impede the repeatability of measurements [6], thus hindering retrospective motion correction. The difficulty of estimating such unpredictable movements results in the lack of any ground truth, yet necessary for the validation of new methods [7]. Post-processing approaches built on motion estimation and correction can compensate for motion artifacts. Especially, super-resolution (SR) reconstruction techniques take advantage of the redundancy between low-resolution (LR) series acquired in orthogonal orientations to reconstruct an isotropic high-resolution (HR) volume of the fetal brain with reduced intensity artifacts and motion sensitivity [8–11]. The development and validation of such advanced image processing strategies require access to large-scale data to account for the subject variability, but the number of good quality exploitable MR acquisitions available in this sensitive cohort remains relatively scarce. Therefore, numerical phantoms are an interesting alternative that offers a fully scalable and flexible environment to simulate a collection of data in various controlled conditions. As such, they make it possible to conduct accurate, robust and reproducible research studies [6,12], especially to evaluate post-processing techniques with respect to a synthetic ground truth.

In this work, we present a simulation framework for T2w Half-Fourier Acquisition Single-shot Turbo spin Echo (HASTE) of the fetal brain based on segmented HR anatomical images from a normative spatiotemporal MRI atlas of the fetal brain [4]. It relies on the extended phase graph (EPG) formalism [13,14] of the signal formation, a surrogate for Bloch equations to describe the magnetization response to various MR pulse sequences. EPG simulations are particularly relevant in the case of multiple radiofrequency pulses that are responsible for stimulated echoes [13], as in the HASTE acquisition scheme. The proposed pipeline is highly flexible and built on a realistic setup that accounts for intensity non-uniformities and stochastic fetal motion. A case study on SR fetal brain MRI further explores the value of such a numerical phantom to evaluate and optimize an SR reconstruction algorithm [11,15].

2   Methods

2.1   Numerical Implementation of HASTE Acquisitions

Fig. 1 provides an overview of the workflow implemented in MATLAB (Math-Works, R2019a) to simulate clinical HASTE acquisitions of the fetal brain. We have developed this numerical phantom with the idea of keeping the framework as general as possible to enable users a large flexibility in the type of simulated images. As such, multiple acquisition parameters can be set up with respect to the MR contrast (effective echo time, excitation/refocusing pulse flip angles, echo spacing, echo train length), the geometry (number of 2D slices, slice orientation, slice thickness, slice gap, slice position, phase oversampling), the resolution (field-of-view, matrix size), the resort to any acceleration technique (acceleration factor, number of reference lines), as well as other settings related to the gestational age (GA) of the fetus, the radiofrequency transmit field inhomogeneities, the amplitude of random fetal motion in the three main directions, and the SNR. The entire simulation pipeline is described in detail in the following.

Fetal Brain Model and MR Properties. Our numerical phantom is based on segmented 0.8-mm-isotropic anatomical images (Fig. 1-i) from the normative spatiotemporal MRI atlas of the developing brain built by Gholipour and colleagues from normal fetuses scanned between 19 and 39 weeks of gestation [4]. Due to the lack for ground truth relaxometry measurements in the fetal brain, all thirty-four segmented tissues are merged into three classes: gray matter, white matter and cerebrospinal fluid (Fig. 1-ii and Table 1). Corresponding T1 and T2 relaxation times at 1.5 T [16–20] are assigned to these anatomical structures to obtain reference T1 and T2 maps, respectively (Fig. 1-iii).

Table 1. Classification of segmented brain tissues [4] as gray matter, white matter and cerebrospinal fluid.

Gray matter	Amygdala, Caudate, Cortical plate, Hippocampus, Putamen, Subthalamic nuclei, Thalamus
White matter	Cerebellum, Corpus callosum, Fornix, Hippocampal commissure, Intermediate zone, Internal capsule, Midbrain, Miscellaneous, Subplate, Ventricular zone
Cerebrospinal fluid	Cerebrospinal fluid, Lateral ventricles

Intensity Non-Uniformities (INU). Non-linear slowly-varying INU fields are based on BrainWeb estimations from real scans to simulate T2w images [21]. The available 20% INU version is resized to fit the dimensions of the atlas images and normalized by 1.2 to provide multiplicative fields from 0.8 to 1.2 over the brain area.

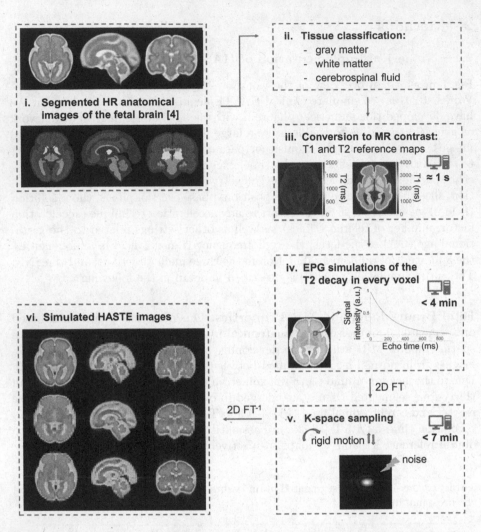

Fig. 1. Workflow for simulating HASTE images of the fetal brain (i) from segmented HR anatomical MR images [4], illustrated for a fetus of 30 weeks of GA. (ii) Brain tissues are organized into gray matter, white matter and cerebrospinal fluid. (iii) Anatomical structures are converted to the corresponding MR contrast to obtain reference T1 and T2 maps of the fetal brain at 1.5 T. (iv) The T2 decay over time is computed in every brain voxel by the EPG algorithm and subsequently used (v) to sample the Fourier domain of the simulated HASTE images of the moving fetus. After the addition of noise to match the SNR of real clinical acquisitions, (vi) HASTE images of the fetal brain are eventually recovered by 2D inverse Fourier transform.

EPG Formalism. From the HASTE sequence pulse design, the T1 and T2 maps of the fetal brain and the realistic INU, the EPG algorithm [14] computes the T2 decay in every voxel of the anatomical images over each echo train

(Fig. 1-iv). The resulting 4D matrix that combines information about both the anatomy and the magnetic relaxation properties of the fetal brain is hereafter referred to as the T2 decay matrix.

K-Space Sampling and Image Formation. The T2 decay matrix is Fourier-transformed in the spatial dimensions and subsequently used for k-space sampling of the simulated HASTE images. For a given echo time (TE), at most one line from the associated Fourier domain of the T2 decay matrix is used, with the central line corresponding to the effective TE. Forty-two reference lines are consecutively sampled around the center of k-space. Beyond, one line out of two is actually needed to simulate an acceleration factor of two resulting from the implementation of GRAPPA interpolation in the clinical HASTE acquisitions. As a first approximation, these sampled lines are copied to substitute the missing lines. As HASTE is a partial Fourier imaging technique, the properties of Hermitian symmetry in the frequency domain are used to fill the remaining unsampled part of k-space. While intra-slice motion can be neglected, inter-slice random 3D translation and rotation of the fetal brain are implemented during k-space sampling (Fig. 1-v). Complex Gaussian noise (mean, 0; standard deviation, 0.15) is also added to simulate thermal noise generated during the acquisition process and qualitatively match the SNR of clinical data. The simulated images are eventually recovered by 2D inverse Fourier transform (Fig. 1-vi).

With the aim of replicating the clinical protocol for fetal brain MRI, HASTE acquisitions are simulated in the three orthogonal orientations. Besides, the position of the field-of-view is shifted by ± 1.6 mm in the slice thickness orientation to produce additional partially-overlapping datasets in each orientation.

Fetal Motion. The amplitude of typical fetal movements is estimated from clinical data [22]. Three levels are defined accordingly for little, moderate and strong motion of the fetus. They are characterized by less than 5%, 10% and 25% of corrupted slices respectively, and simulated by a uniform distribution of $[-2, 2]$ mm, $[-3, 3]$ mm and $[-3, 3]$ mm for translation in every direction and $[-2, 2]°, [-4, 4]°$ and $[-4, 4]°$ for 3D rotation respectively (Fig. 1-v).

Computational Performance. Since the addition of 3D motion during k-space sampling is expensive in computing memory, the simulations are run on 16 CPU workers in parallel with 20 GB of RAM each. In this setup and for a fetus of 30 weeks of GA whose brain is covered by twenty-five slices, the computation time to convert segmented HR images of the fetal brain to MR contrast and to run EPG simulations in every voxel of the 3D HR anatomical images is in the order of one second, respectively less than four minutes. K-space sampling takes less than seven minutes for one axial series with the different levels of motion.

2.2   Clinical Protocol

Typical fetal brain acquisitions are performed on patients at 1.5 T (MAGNE-TOM Sola, Siemens Healthcare, Erlangen, Germany) with an 18-channel body coil and a 32-channel spine coil at our local hospital. At least three T2w series of 2D thick slices are acquired in three orthogonal orientations using an ultra-fast multi-slice HASTE sequence (TR/TE, 1200 ms/90 ms; excitation/refocusing pulse flip angles, 90°/180°; echo train length, 224; echo spacing, 4.08 ms; field-of-view, 360×360 mm^2; voxel size, $1.13 \times 1.13 \times 3.00$ mm^3; inter-slice gap, 10%). The position of the field-of-view is slightly shifted in the slice thickness orientation to acquire additional data with some redundancy. In clinical practice, six partially-overlapping LR HASTE series are commonly acquired for subsequent SR reconstruction of the fetal brain.

2.3   Datasets

Six subjects in the GA range of 21 to 33 weeks were scanned at our local hospital as part of a larger institutional research protocol with written consent approved by the local ethics committee.

  These clinical cases are used as representative examples of fetal brain HASTE acquisitions: the corresponding sequence parameters are replicated to simulate HASTE images of the fetal brain at various GA, and with realistic SNR. The amplitude of fetal movements in clinical acquisitions is assessed by an engineer expert in MR image analysis to ensure a similar level of motion in the simulated images. The original real cases are also used to visually compare the quality and realistic appearance of the synthetic images generated.

  A 3D HR 1.1-mm-isotropic HASTE image of the fetal brain is simulated without noise or motion to serve as a reference for the quantitative evaluation of SR reconstructions from simulated LR 1.1-mm-in-plane HASTE images.

2.4   Qualitative Assessment

Two medical doctors specialized in neuroradiology and pediatric (neuro) radiology respectively, provided qualitative assessment of the fetal brain HASTE images simulated in the GA range of 21 to 33 weeks, in the three orthogonal orientations with various levels of motion. Special attention was paid to the MR contrast between brain tissues, to the SNR, to the delineation and sharpness of the structures of diagnostic interest that are analyzed in clinical routine, as well as to characteristic motion artifacts.

2.5   Application Example: Parameter Fine-Tuning for Optimal SR Reconstruction

Implementation of SR Reconstruction. Orthogonal T2w LR HASTE series from clinical examinations, respectively simulated images, are combined into a motion-free 3D image $\hat{\mathbf{X}}$ using the Total Variation (TV) SR reconstruction algorithm [11,15] which solves:

$$\hat{\mathbf{X}} = \arg\min_{\mathbf{X}} \frac{\lambda}{2} \sum_{kl} \| \underbrace{\mathbf{D}_{kl}\mathbf{B}_{kl}\mathbf{M}_{kl}}_{\mathbf{H}_{kl}} \mathbf{X} - \mathbf{X}_{kl}^{LR} \|^2 + \|\mathbf{X}\|_{TV}, \qquad (1)$$

where the first term relates to data fidelity with k being the k-th LR series $\mathbf{X}^{LR}$ and l the l-th slice, $\|\mathbf{X}\|_{TV}$ is a TV prior introduced to regularize the solution, and λ balances the trade-off between data fidelity and regularization terms. $\mathbf{D}$ and $\mathbf{B}$ are linear downsampling and Gaussian blurring operators given by the acquisition characteristics. $\mathbf{M}$ encodes the rigid motion of slices.

Both clinical acquisitions and the corresponding simulated images are reconstructed using the SR reconstruction pipeline available in [15].

Regularization Setting. LR HASTE images of the fetal brain are simulated to mimic clinical MR acquisitions of three subjects of 26, 30 and 33 weeks of GA respectively, with particular attention to ensuring that the motion level is respected. For each subject, a SR volume of the fetal brain is reconstructed from the various orthogonal acquisitions, either real or simulated, with different values of λ (0.1, 0.3, 0.75, 1.5, 3) to study the potential of our simulation framework in optimizing the quality of the SR reconstruction in a clinical setup. A quantitative analysis is conducted on the resulting SR reconstructions to determine the value of λ that provides the sharpest reconstruction of the fetal brain with high SNR, namely the smallest normalized root mean squared error (NRMSE) with respect to a synthetic ground truth.

3 Results and Discussion

3.1 Qualitative Assessment

Fig. 2 illustrates the close resemblance between simulated HASTE images of the fetal brain and clinical MR acquisitions for two representative subjects of 26 and 30 weeks of GA respectively, in terms of MR contrast between tissues, SNR, brain anatomy and relative proportions across development, as well as typical out-of-plane motion patterns related to the interleaved slice acquisition scheme. Experts in neuroradiology and in pediatric (neuro)radiology report a good contrast between gray and white matter, which is important to investigate cortex continuity and identify the deep gray nuclei as well as any migration anomaly. They also notice good SNR in the different series and report proper visualization of the main anatomical structures: the four ventricles, the corpus callosum, the vermis, the cerebellum, even sometimes the fornix. Besides, they are able to monitor the evolution of normal gyration throughout gestation. However, they point out that small structures such as the hypophyse, the chiasma, the recesses of the third ventricle, and the vermis folds that look part of the cerebellum, are

more difficult to observe. The cortical ribbon is clearly visible but quite pixe-
lated, which is likely to complicate the diagnosis of polymicrogyria. White matter
appears too homogeneous, which makes its multilayer aspect barely distinguishable, with an MR signal that is constant across GA, thus preventing physicians
from exploring the myelination process throughout brain maturation. For these
reasons, experts feel confident in performing standard biometric measurements
on the simulated images and in evaluating the volume of white matter, but not
its fine structure.

These limitations in the resemblance of the simulated HASTE images as
compared to typical clinical acquisitions may be explained by the origin of the
simulated images and the lack of T1 and T2 ground truth measurements, both in
the multiple fetal brain tissues and throughout maturation. HASTE images are
simulated from a normative spatiotemporal MRI atlas of the fetal brain [4] where
representative images at each GA correspond to an average of fetal brain scans
across several subjects, thus resulting in smoothing of subtle inter-individual
heterogeneities, especially in the multilayer aspect of the white matter. As a
first approximation because of the lack of detailed literature on the changes that
result from maturation processes in finer structures of the brain, we consider
average T1 and T2 relaxation times of the various fetal brain tissues labeled as
gray matter, white matter or cerebrospinal fluid (see Table 1) over gestation. As
a result, our simulated images may fail to capture the fine details of the fetal
brain anatomy throughout development.

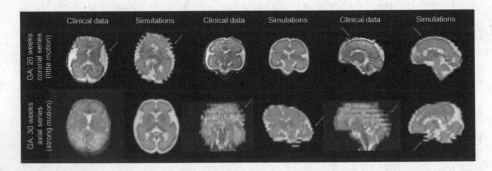

Fig. 2. Comparison between motion-corrupted clinical MR acquisitions and corresponding simulated HASTE images at two GA (26 and 30 weeks). Images are shown in
the three orthogonal planes. Red arrows point out typical out-of-plane motion patterns.

3.2 Application Example: Parameter Fine-Tuning for Optimal SR Reconstruction

In fetal MRI, the level of regularization is commonly set empirically based on
visual perception [8–10]. Thanks to its controlled environment, the presented
framework makes it possible to adjust the parameter λ for optimal SR reconstruction with respect to a synthetic 3D isotropic HR ground truth of the fetal

brain. Of note, in-plane motion artefacts like signal drops are not accounted for in the simulation pipeline at this stage, as heavily corrupted slices are commonly removed from the reconstruction.

Figure 3 explores the quality of SR fetal brain MRI depending on the weight of TV regularization. Based on the simulations, a high level of regularization ($\lambda = 0.1$) provides a blurry SR reconstruction with poor contrast between the various structures of the fetal brain, especially in the deep gray nuclei and the cortical plate. In addition, the cerebrospinal fluid appears brighter than in the reference image. A low level of regularization ($\lambda = 3$) leads to a better tissue contrast but increases the overall amount of noise in the resulting SR reconstruction. A fine-tuned regularization ($\lambda = 0.75$) provides a sharp reconstruction of the fetal brain with a high SNR and a tissue contrast close to the one displayed in the reference image. In the SR images reconstructed from clinical LR HASTE series altered by a little-to-moderate level of motion, as in the simulations, the structure of the corpus callosum and the delineation of the cortex are especially well defined for appropriate TV regularization ($\lambda = 0.75$), leading to high-SNR HR images of the fetal brain. Although the NRMSE between SR reconstructions from simulated HASTE images and the ground truth are close to each other in the three configurations studied, the error is minimal for $\lambda = 0.75$, which further supports this parameter setting for optimal SR reconstruction.

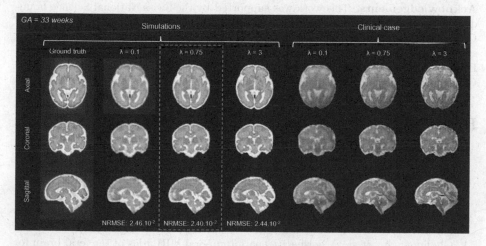

Fig. 3. Appreciation of the quality of SR reconstruction depending on the weight λ that controls the strength of the TV regularization. The potential of our framework to optimize the reconstruction quality through parameter fine-tuning in the presence of motion is illustrated for a fetus of 33 weeks of GA with three values of λ. A representative clinical case from which the synthetic HASTE images are derived is provided for comparison. The blue box highlights that the NRMSE between SR reconstructions from simulated data and a simulated 3D HR ground truth is minimal for $\lambda = 0.75$.

4   Conclusions and Perspectives

In this work, we present a novel numerical framework that simulates as closely as possible the physical principles involved in HASTE acquisitions of the fetal brain, with great flexibility in the choice of the sequence parameters and anatomical settings, resulting in highly realistic T2w images of the developing brain throughout gestation. Thanks to its controlled environment, this numerical phantom makes it possible to explore the optimal settings for SR fetal brain MRI according to the image quality of the input motion-corrupted LR HASTE series. It also enables quantitative assessment of the robustness of any SR reconstruction algorithm depending on various parameters such as the noise level, the amplitude of fetal motion in the womb and the number of series used for SR reconstruction [23]. Future work aims at investigating the ability of such synthetic images to generalize post-processing tools like fetal brain tissue segmentation to datasets acquired on other MR systems and with other parameters using domain adaptation techniques. Therefore, the developed pipeline will be extended to simulate fast spin echo sequences from other MR vendors, both at 1.5 T and 3 T. It will then be made publicly available to support reproducibility studies and provide a common framework for the evaluation and validation of post-processing strategies for fetal brain MRI.

Acknowledgements. This work was supported by the Swiss National Science Foundation (grant 182602). We acknowledge access to the facilities and expertise of the CIBM Center for Biomedical Imaging, a Swiss research center of excellence founded and supported by Lausanne University Hospital (CHUV), University of Lausanne (UNIL), Ecole polytechnique fédérale de Lausanne (EPFL), University of Geneva (UNIGE) and Geneva University Hospitals (HUG).

References

1. Kwon, E.J., Kim, Y.J.: What is fetal programming?: A lifetime health is under the control of in utero health. Obstet. Gynecol. Sci. **60**(6), 506–519 (2017)
2. O'Donnell, K.J., Meaney, M.J.: Fetal origins of mental health: the developmental origins of health and disease hypothesis. Am. J. Psychiatry **174**(4), 319–328 (2017)
3. Volpe, J.J.: Brain injury in premature infants: a complex amalgam of destructive and developmental disturbances. Lancet Neurol. **8**(1), 110–124 (2009)
4. Gholipour, A., et al.: A normative spatiotemporal MRI atlas of the fetal brain for automatic segmentation and analysis of early brain growth. Sci. Rep. **7**(1), 1–13 (2017)
5. Gholipour, A., et al.: Fetal MRI: a technical update with educational aspirations. Concepts Magn. Reson. Part A Bridging Educ. Res. **43**(6), 237–266 (2014)
6. Roy, C.W., Marini, D., Segars, W.P., Seed, M., Macgowan, C.K.: Fetal XCMR: a numerical phantom for fetal cardiovascular magnetic resonance imaging. J. Cardiovasc. Magn. Reson. **21**(1), 29 (2019)
7. Drobnjak, I., Gavaghan, D., Süli, E., Pitt-Francis, J., Jenkinson, M.: Development of a functional magnetic resonance imaging simulator for modeling realistic rigid-body motion artifacts. Magn. Reson. Med. **56**(2), 364–380 (2006)

8. Ebner, M., et al.: An automated framework for localization, segmentation and super-resolution reconstruction of fetal brain MRI. Neuroimage **206**, 116324 (2020)

9. Gholipour, A., Estroff, J.A., Warfield, S.K.: Robust super-resolution volume reconstruction from slice acquisitions: application to fetal brain MRI. IEEE Trans. Med. Imaging **29**(10), 1739–1758 (2010)

10. Kuklisova-Murgasova, M., Quaghebeur, G., Rutherford, M.A., Hajnal, J.V., Schnabel, J.A.: Reconstruction of fetal brain MRI with intensity matching and complete outlier removal. Med. Image Anal. **16**(8), 1550–1564 (2012)

11. Tourbier, S., Bresson, X., Hagmann, P., Thiran, J.P., Meuli, R., Cuadra, M.B.: An efficient total variation algorithm for super-resolution in fetal brain MRI with adaptive regularization. NeuroImage **118**, 584–597 (2015)

12. Wissmann, L., Santelli, C., Segars, W.P., Kozerke, S.: MRXCAT: realistic numerical phantoms for cardiovascular magnetic resonance. J. Cardiovasc. Magn. Reson. **16**(1), 63 (2014)

13. Malik, S.J., Teixeira, R.P.A.G., Hajnal, J.V.: Extended phase graph formalism for systems with magnetization transfer and exchange. Magn. Reson. Med. **80**(2), 767–779 (2018)

14. Weigel, M.: Extended phase graphs: dephasing, RF pulses, and echoes - pure and simple. J. Magn. Reson. Imaging **41**(2), 266–295 (2015)

15. Tourbier, S., De Dumast, P., Kebiri, H., Hagmann, P., Bach Cuadra, M.: Medical-Image-Analysis-Laboratory/mialsuperresolutiontoolkit: MIAL Super-Resolution Toolkit v2.0.1. Zenodo (2020)

16. Blazejewska, A.I., et al.: 3D in utero quantification of T2* relaxation times in human fetal brain tissues for age optimized structural and functional MRI. Magn. Reson. Med. **78**(3), 909–916 (2017)

17. Hagmann, C.F., et al.: T2 at MR imaging is an objective quantitative measure of cerebral white matter signal intensity abnormality in preterm infants at term-equivalent age. Radiology **252**(1), 209–217 (2009)

18. Nossin-Manor, R., et al.: Quantitative MRI in the very preterm brain: assessing tissue organization and myelination using magnetization transfer, diffusion tensor and T1 imaging. Neuroimage **64**, 505–516 (2013)

19. Vasylechko, S., et al.: T2* relaxometry of fetal brain at 1.5 Tesla using a motion tolerant method. Magn. Reson. Med. **73**(5), 1795–1802 (2015)

20. Yarnykh, V.L., Prihod'ko, I.Y., Savelov, A.A., Korostyshevskaya, A.M.: Quantitative assessment of normal fetal brain myelination using fast macromolecular proton fraction mapping. Am. J. Neuroradiol. **39**(7), 1341–1348 (2018)

21. BrainWeb: Simulated brain database. https://brainweb.bic.mni.mcgill.ca/brainweb/

22. Oubel, E., Koob, M., Studholme, C., Dietemann, J.-L., Rousseau, F.: Reconstruction of scattered data in fetal diffusion MRI. Med. Image Anal. **16**(1), 28–37 (2012)

23. Lajous, H., et al.: A magnetic resonance imaging simulation framework of the developing fetal brain. In: Proceedings of the 29th Annual Meeting of the International Society of Magnetic Resonance in Medicine (ISMRM), virtual. Program number 0734 (2021)

Spatio-Temporal Atlas of Normal Fetal Craniofacial Feature Development and CNN-Based Ocular Biometry for Motion-Corrected Fetal MRI

Alena Uus[1]([⊠]), Jacqueline Matthew[1], Irina Grigorescu[1], Samuel Jupp[1], Lucilio Cordero Grande[2,3], Anthony Price[2], Emer Hughes[2], Prachi Patkee[2], Vanessa Kyriakopoulou[2], Robert Wright[1], Thomas Roberts[1], Jana Hutter[1], Maximilian Pietsch[2], Joseph V. Hajnal[1,2], A. David Edwards[2], Mary Ann Rutherford[2], and Maria Deprez[1]

[1] Biomedical Engineering Department, School of Imaging Sciences and Biomedical Engineering, King's College London, St. Thomas' Hospital, London, UK
alena.uus@kcl.ac.uk
[2] Centre for the Developing Brain, School Biomedical Engineering and Imaging Sciences, King's College London, St. Thomas' Hospital, London, UK
[3] Biomedical Image Technologies, ETSI Telecomunicacion, Universidad Politécnica de Madrid and CIBER-BBN, Madrid, Spain

Abstract. Motion-corrected fetal magnetic resonance imaging (MRI) is widely employed in large-scale fetal brain studies. However, the current processing pipelines and spatio-temporal atlases tend to omit craniofacial structures, which are known to be linked to genetic syndromes. In this work, we present the first spatio-temporal atlas of the fetal head that includes craniofacial features and covers 21 to 36 weeks gestational age range. Additionally, we propose a fully automated pipeline for fetal ocular biometry based on a 3D convolutional neural network (CNN). The extracted biometric indices are used for the growth trajectory analysis of changes in ocular metrics for 253 normal fetal subjects from the developing human connectome project (dHCP).

Keywords: Motion-corrected fetal MRI · Craniofacial features · Ocular measurements · Spatio-temporal atlas · Automated biometry

1 Introduction

Arguably, an MRI scan of the fetal brain is not complete without a structural and dysmorphological assessment of the fetal craniofacial structures due to the intricate link between brain anomalies and genetic syndromes that affect the facial features [2,19]. More than 250 syndromes are associated with changes in craniofacial growth and development and can therefore result in overt anomalies or subtle changes in anatomical appearance and yet prenatal detection remains

© Springer Nature Switzerland AG 2021
C. H. Sudre et al. (Eds.): UNSURE 2021/PIPPI 2021, LNCS 12959, pp. 168–178, 2021.
https://doi.org/10.1007/978-3-030-87735-4_16

low [7,15]. Ultrasound is the primary imaging modality for fetal assessment and has well recognised limitations. High risk fetal cases are increasingly referred for MRI examinations for further characterisation and to assess for the presence of additional anomalies.

Fetal motion during MRI acquisition leads to loss of structural continuity between 2D slices and corruption of 3D information. The image degradation precludes the reliable use of MRI to assess craniofacial structures. During the past decade this has been successfully addressed by slice-to-volume registration (SVR) reconstruction tools [6,11] that can produce motion-corrected high-resolution 3D fetal brain MRI images. SVR tools also have the potential to increase the clinical reliability of extended craniofacial biometry and objective assessments of the curved structures of the fetal head and face e.g. orbits, oral hard and soft palate, and cranial shape. Formalisation of the normal trajectory of rapid development of craniofacial structures occurring during gestation that can be observed in MRI is essential for definition of the control reference. However, the existing spatio-temporal fetal atlases include only the brain region [8].

Automated segmentation and volumetry methods lower the impact of inter- and intra-observer variability and provide the means for processing large-scale studies [14]. Recently, several reported works employed semi-automated segmentation for analysis of fetal craniofacial features in MRI and ultrasound [1,12,13,21] and, more recently, an automated method was proposed for 2D slice-wise segmentation and biometry of orbits in low resolution stacks [3]. Incorporation of novel convolutional neural network (CNN) pipelines for motion-corrected fetal MRI segmentation [9,10,16] has a potential to make the application of automated biometry and volumetry of craniofacial structures feasible for large datasets and motion-corrected MRI. Quality control for automated data analysis methods also remains one of the current challenges in terms of practical application.

In this work, we propose to generate a first spatio-temporal atlas of the fetal head that includes craniofacial features. This extends the already existing brain-only fetal MRI templates for a wider application of analysis of normal craniofacial feature development. In addition, we implemented an automated pipeline for 3D CNN-based ocular biometry for motion-corrected fetal MRI with outlier detection. The biometry outputs were then used for the analysis of ocular growth trajectories for 253 normal fetal subjects with acceptable biometry results.

2   Methods

2.1   Cohort, Datasets and Preprocessing

The data used in this study included T2w MRI datasets of 291 fetuses without reported anomalies from 20 to 38 weeks gestational age (GA) acquired at St.Thomas' hospital, London as a part of the dHCP project[1] (dHCP, REC: 14/Lo/1169). The acquisition was performed on a Philips Achieva 3T system

---

[1] dHCP project: http://www.developingconnectome.org/project.

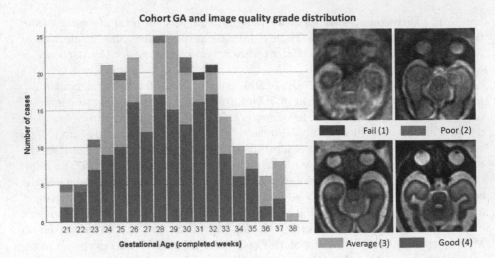

Fig. 1. Distribution of the gestational age and image quality in the investigated fetal MRI cohort. The image quality scores are: fail, poor, average and good.

with a 32-channel cardiac coil using single shot turbo spin echo (ssTSE) sequence with TE = 250 ms, TR = 2265 ms, acquisition resolution = $1.1 \times 1.1 \times 2.2$ mm (-1.1 mm gap) [18]. The datasets were reconstructed using a fully automated SVR pipeline [5] to $0.5 \times 0.5 \times 0.5$ mm resolution for the fetal head region of interest (ROI). This was followed by reorientation to the standard planes using a dedicated transformer CNN [22].

The quality of the 3D reconstructed images in terms of definition of the anatomy features, noise and contrast was assessed by an experienced researcher with the grades: good (4), average (3), poor (2) and failed (1). All available datasets were included in the biometry study irrespective of the reconstruction image quality for the purpose of testing of the proposed automated detection of outliers approach. The histograms of the cases GA and quality scores is given in Fig. 1 with the majority of scans within ≥ 3 quality window. For generation of the atlas we used only a subset of cases from 21 to 36 GA weeks.

2.2   Spatio-Temporal Atlas

For atlas generation, we selected 190 datasets with the best image quality and optimal coverage of the fetal head. The 4D spatio-temporal atlas of the fetal head was constructed using the MIRTK[2] atlas generation pipeline [20] at 16 discrete timepoints in 21 to 36 weeks GA range. We used local normalised cross-correlation similarity metric with 5 voxel window, 3 atlas generation iterations, temporal Gaussian kernel with constant 1 week sigma, and 0.7 mm output isotropic resolution settings.

---

[2] MIRTK library: https://github.com/BioMedIA/MIRTK

2.3 Automated Ocular Biometry

The proposed pipeline for fetal ocular biometry is summarised in Fig. 2. We first localise the orbit using 3D U-Net [4], then we fit a 3D line through the orbit centroids and calculate the standard ocular measurements [21]. Furthermore, the step for automated detection of outliers provides quality control of the segmentations and measurements.

For the orbit segmentation module, the 3D U-Net [4] architecture consists of 5 encoding-decoding levels with 32, 64, 128, 256 and 512 channels, respectively. Each encoder block consists of 2 repeated blocks of $3 \times 3 \times 3$ convolutions (with a stride of 1), instance normalisation and LeakyReLU activations. The first two down-sampling blocks contains a $2 \times 2 \times 2$ average pooling layers, while the others use $2 \times 2 \times 2$ max pooling layers. The decoder blocks have a similar architecture as the encoder blocks, followed by upsampling layers. The model outputs an N-channel 3D image, corresponding to our 2 classes: background and fetal orbits. The network is implementated in PyTorch[3].

The orbit masks were created manually by a trained clinician for 20 cases. The network was trained on 19 3D reconstructed images and 1 case was used for validation. The training was performed for 100 epochs with TorchIO augmentation [17] including affine transformations ($\pm 180°$ rotations and $0.9 - 1.1$ scaling), bias field and motion artifacts ($< 5°$ rotations and < 2.0 translations).

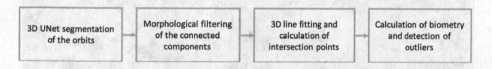

Fig. 2. Proposed pipeline for automated ocular biometry.

Next, the output 3D segmentations are post-processed using morphological filtering. The two largest components with $\pm 35\%$ difference in volume are selected as the orbits (to account for a potential intensity variability due to the presence of a bias field). The calculation of ocular biometry is performed by fitting a line through the orbit centroids followed by detection of intersection points, calculation of line segment length and extraction of the standard 2D metrics: ocular diameter (OD), binocular distance (BOD), interocular distance (IOD), see Fig. 7.f. In addition, we also calculate volumes of the orbits (OV). Similarly to [21], the mean OD and OV values are computed as an average between the left and right orbits.

Outlier detection is based on three inclusion conditions: (i) the number of detected components should not exceed 5, which is an indicator of not well defined ocular features due to low image quality; (ii) the sizes of the right and left (R/L) orbits should be comparable within $\pm 15\%$ difference in terms of both volumes and OD values; (iii) the extracted metrics should be within $\pm 30\%$ window of the GA-specific curve values [21].

---

[3] PyTorch: https://pytorch.org.

In order to confirm the correlation between the intracranial and ocular volumetry, we trained a 3D U-Net with the same architecture for brain extraction. The training was performed for 400 epochs with augmentation in 3 steps using semi-supervised approach. At the first stage we used 60 fetal brain SVR images with manual segmentations of the intracranial volume available from other research projects. Next, the results of testing on the entire cohort (291) were examined and successful brain masks that included the entire intracranial volume were used in the next stage of training. All output intracranial brain masks for good quality reconstruction cases were visually inspected by a trained researcher and manually refined in 32 cases due to the presence of errors.

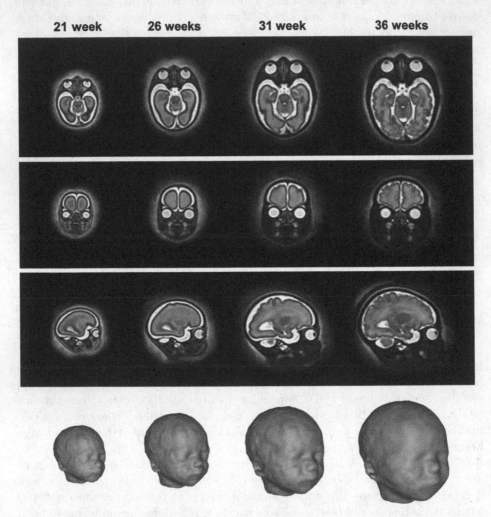

Fig. 3. The generated spatio-temporal atlas of the fetal craniofacial feature development at 21, 26, 31 and 36 weeks GA along with the corresponding face masks.

3 Results and Discussion

3.1 Spatio-Temporal Atlas of Fetal Craniofacial Feature Development

The generated spatio-temporal atlas of the head ROI at 21, 26, 31 and 36 weeks GA is shown in Fig. 3. The corresponding presented face masks were created semi-manually using combination of thresholding, manual refinement and label propagation from one to the rest of the GA timepoints. The atlas was inspected by two clinicians trained in fetal MRI who confirmed that all craniofacial features are correct, well defined and have high contrast. The atlas will be available online at the SVRTK data repository[4].

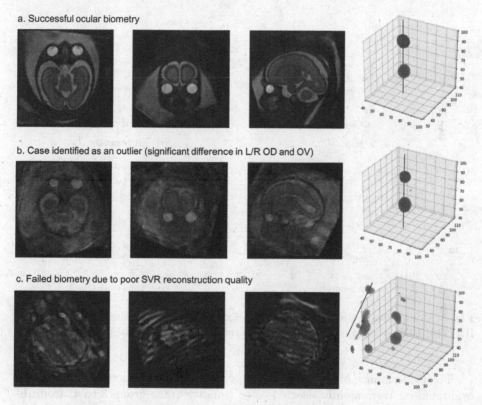

Fig. 4. Examples of a successful biometry output (a), a case detected as an outlier due to R/L OD and OV differences (b) and a completely failed case (c). The illustrations show the original 3D SVR reconstructions with orbit mask overlay and the corresponding 3D model with the fitted 3D line.

---

[4] SVRTK fetal and neonatal MRI data repository: https://gin.g-node.org/SVRTK.

3.2   Eye Biometry

Figure 4 shows an example of a successful (Fig. 4a) and failed (Fig. 4c) biometry output as well as a case that was automatically identified as an outlier (Fig. 4b) due to the difference between right and left OD and OV values. The completely failed case (Fig. 4c) was also automatically detected since there were > 5 components in the 3D U-Net output. This was caused by the low image quality due to the insufficient number of input stacks and the extreme motion that could not be resolved by SVR reconstruction [5].

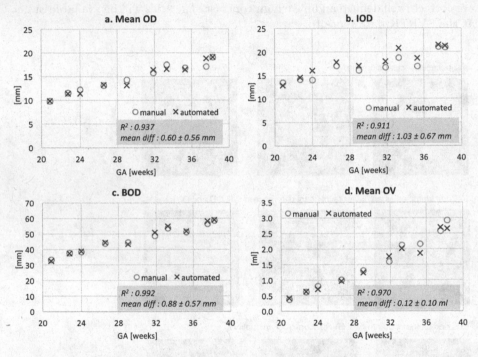

Fig. 5. Comparison between automated and manual measurements for mean OD (a), IOD (b), BOD (c) and mean OV (d) on 10 randomly selected cases.

The performance of the proposed pipeline of ocular biometry (Sect. 2.3) was evaluated on 10 randomly selected cases (quality grade group 3 to 4) from different GA groups, with automated biometry outputs compared to manual measurements. The corresponding results, presented in Fig. 5, show reasonably low absolute and relative differences (0.60 ± 0.56 mm and $3.99 \pm 3.49\%$ for mean OD, 1.03 ± 0.67 mm and $6.06 \pm 3.96\%$ for IOD and 0.88 ± 0.57 mm and $1.88 \pm 1.10\%$ for BOD, see Fig. 5) and high correlation ($R^2 > 0.91$) between the automated and manual measurements for all metrics. The slightly higher IOD and lower OV values in the automated output are primarily caused by more conservative automated segmentations that exclude the boundary around the orbits. It should be

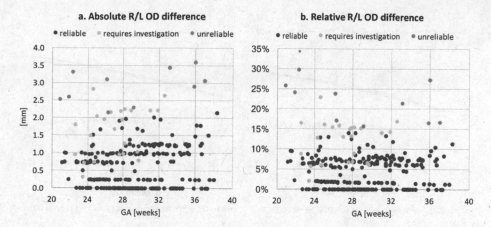

Fig. 6. Absolute (a) and relative (b) differences in right/left OD measurements. The cases in yellow and red are outliers with significant differences in either R/L OD or OV or both, respectively, while the cases in blue can be considered to be reliable for interpretation. (Color figure online)

noted that both manual and automated mean OD and OV measurements were characterised by R/L orbit differences.

The outlier detection step identified 4 cases where the reconstruction completely failed and 34 cases with high differences in right/left orbit metrics. An illustration of the absolute and relative R/L orbit OD differences for all cases is given in Fig. 6 with the outliers highlighted in yellow and red depending on whether there is a difference between either OD or OV or both these measurement. The discrete appearance of the difference values is related to the voxel size of the input images since the 2D distances are computed as voxels between intersection points along the fitted lines. The average quality scores (the manual grading in Fig. 1) in failed (4 cases), outlier (34 cases) and normal (253 cases) groups are 1.0 ± 0.0, 3.1 ± 0.9 and 3.6 ± 0.5, respectively. Notably, in addition to motion artefacts, the primary cause of the R/L differences was the presence of a strong bias field which was not taken into account during image quality grading.

3.3   Growth Charts

Prior to the analysis of growth trajectories, all automated eye segmentations and biometry results were also inspected manually which confirmed the similar size of the detected orbits in 253 cases. The automatically detected failed and outlier cases (Sect. 3.2) were excluded. The growth trajectories constructed from the selected 253 cases, shown in Fig. 7, include ocular biometry (mean OD, IOD, BOD), mean OV and total intracranial volume. The trajectories of all indices show high agreement with the existing formulas for ocular indices [21].

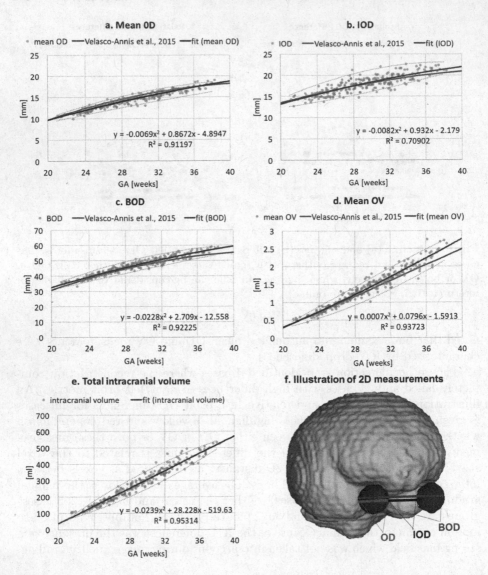

Fig. 7. Growth charts for 253 subjects from [20; 38] week GA range for mean OD (a), IOD (b), BOD (c), mean OV (d) and total intracranial volume (e) extracted from the automated measurements. The illustration of the measurements is given in (f).

4   Conclusions

In summary, we have presented the first spatio-temporal atlas of fetal cranio-facial feature development from 21 to 36 weeks GA which extends the existing brain-only fetal MRI atlases. The atlas will be available online at the SVRTK data repository[5]. We also showed that fully automated 3D CNN-based ocular biometry can be used for processing large cohort datasets as an alternative to manual measurements. In addition, the proposed solution for detection of outliers provides the means to control interpretation of the outputs of automated processing by highlighting potentially unreliable results that require manual editing. The outlier cases with significant deviations that can occur due to either low image quality, failed segmentation or anomalies should be manually inspected, if required. The growth charts from the automatically derived ocular indices showed high correlation to the previously reported trends [21]. Our future work will focus on further automation of parcellation and biometry of craniofacial structures as well as analysis of abnormal cases.

Acknowledgments. We thank everyone who was involved in acquisition and analysis of the datasets at the Department of Perinatal Imaging and Health at King's College London. We thank all participating mothers.

This work was supported by the European Research Council under the European Union's Seventh Framework Programme [FP7/ 20072013]/ERC grant agreement no. 319456 dHCP project, the Wellcome/EPSRC Centre for Medical Engineering at King's College London [WT 203148/Z/16/Z)], the NIHR Clinical Research Facility (CRF) at Guy's and St Thomas' and by the National Institute for Health Research Biomedical Research Centre based at Guy's and St Thomas' NHS Foundation Trust and King's College London.

The views expressed are those of the authors and not necessarily those of the NHS, the NIHR or the Department of Health.

References

1. Ami, O., et al.: 3D magnetic resonance imaging of fetal head molding and brain shape changes during the second stage of labor. PLoS ONE **14**(5) (2019)
2. Arangio, P., et al.: Importance of fetal MRI in evaluation of craniofacial deformities. J. Craniofac. Surg. **24**(3), 773–776 (2013)
3. Avisdris, N., et al.: Automatic fetal ocular measurements in MRI. In: ISMRM 2021, p. 1190 (2021)
4. Çiçek, Ö., Abdulkadir, A., Lienkamp, S.S., Brox, T., Ronneberger, O.: 3D U-Net: learning dense volumetric segmentation from sparse annotation. In: Ourselin, S., Joskowicz, L., Sabuncu, M.R., Unal, G., Wells, W. (eds.) MICCAI 2016. LNCS, vol. 9901, pp. 424–432. Springer, Cham (2016). https://doi.org/10.1007/978-3-319-46723-8_49
5. Cordero-Grande1, L., et al.: Automating motion compensation in 3T fetal brain imaging: localize, align and reconstruct. In: ISMRM 2019, p. 1000 (2019)
6. Ebner, M., et al.: An automated framework for localization, segmentation and super-resolution reconstruction of fetal brain MRI. Neuroimage **206**(Oct.) (2020)
7. Ettema, A., et al.: Prenatal diagnosis of craniomaxillofacial malformations: a characterization of phenotypes in trisomies 13, 18, and 21 by ultrasound and pathology. Cleft Palate-Craniofac. J. **47**(2), 189–196 (2010)

---

[5] SVRTK fetal and neonatal MRI data repository: https://gin.g-node.org/SVRTK.

8. Gholipour, A., et al.: A normative spatiotemporal MRI atlas of the fetal brain for automatic segmentation and analysis of early brain growth. Nat. Sci. Rep. **7**(476), 1–13 (2017)
9. Grigorescu, I., et al.: Harmonized segmentation of neonatal brain MRI. Front. Neurosci. **15**, 565 (2021)
10. Khalili, N., et al.: Automatic brain tissue segmentation in fetal MRI using convolutional neural networks. Magn. Reson. Imaging **64**, 77–89 (2019)
11. Kuklisova-Murgasova, M., et al.: Reconstruction of fetal brain MRI with intensity matching and complete outlier removal. MedIAn **16**(8), 1550–1564 (2012)
12. Kul, S., et al.: Contribution of MRI to ultrasound in the diagnosis of fetal anomalies. J. Magn. Reson. Imaging **35**(4), 882–890 (2012)
13. Kyriakopoulou, V., et al.: Normative biometry of the fetal brain using magnetic resonance imaging. Brain Struct. Funct. **222**(5), 2295–2307 (2016). https://doi.org/10.1007/s00429-016-1342-6
14. Makropoulos, A., et al.: The dHCP: a minimal processing pipeline for neonatal cortical surface reconstruction. Neuroimage **173**, 88–112 (2018)
15. Mossey, P., Castilla, E.E.: Global registry and database on craniofacial anomalies Report of a WHO Registry Meeting on Craniofacial Anomalies Human Genetics Programme Management of Noncommunicable Diseases (2003)
16. Payette, K., Kottke, R., Jakab, A.: Efficient multi-class fetal brain segmentation in high resolution MRI reconstructions with noisy labels. In: Hu, Y., et al. (eds.) ASMUS/PIPPI -2020. LNCS, vol. 12437, pp. 295–304. Springer, Cham (2020). https://doi.org/10.1007/978-3-030-60334-2_29
17. Pérez-García, F., et al.: TorchIO: a Python library for efficient loading, preprocessing, augmentation and patch-based sampling of medical images in deep learning. arXiv (March 2020)
18. Price, A., et al.: The developing Human Connectome Project (dHCP): fetal acquisition protocol. In: ISMRM 2019 (2019)
19. Robinson, A.J., et al.: MRI of the fetal eyes: morphologic and biometric assessment for abnormal development with ultrasonographic and clinicopathologic correlation. Pediatr. Radiol. **38**(9), 971–981 (2008)
20. Schuh, A., et al.: Unbiased construction of a temporally consistent morphological atlas of neonatal brain development. bioRxiv (2018)
21. Velasco-Annis, C., et al.: Normative biometrics for fetal ocular growth using volumetric MRI reconstruction. Prenat. Diagn. **35**(4), 400–408 (2015)
22. Wright, R., et al.: LSTM spatial co-transformer networks for registration of 3D fetal US and MR brain images. MICCAI **2018**, 107–116 (2018)

Myelination of Preterm Brain Networks at Adolescence

Beatriz Laureano[1(✉)], Hassna Irzan[1,2], Sébastien Ourselin[1,2], Neil Marlow[3], and Andrew Melbourne[1,2]

[1] School of Biomedical Engineering & Imaging Sciences, King's College London, London, UK
[2] Department of Medical Physics and Biomedical Engineering, University College London, London, UK
[3] Institute for Women's Health, University College London, London, UK

Abstract. Prematurity and preterm stressors severely affect the development of infants born before 37 weeks of gestation, with increasing effects seen at earlier gestations. Although preterm mortality rates have declined due to the advances in neonatal care, disability rates, especially in middle-income settings, continue to grow. With the advances in MRI imaging technology, there has been a focus on safely imaging the preterm brain to better understand its development and discover the brain regions and networks affected by prematurity. Such studies aim to support interventions and improve the neurodevelopment of preterm infants and deliver accurate prognoses. Few studies, however, have focused on the fully developed brain of preterm born infants, especially in extremely preterm subjects. To assess the long-term effect of prematurity on the adult brain, myelin related biomarkers such as myelin water fraction and g-ratio are measured for a cohort of 19-year-old extremely preterm subjects. Using multi-modal imaging techniques that combine T2 relaxometry and neurite density information, the results show that specific regions of the brain associated with white matter injuries due to preterm birth, such as the Posterior Limb of the Internal Capsule and Corpus Callosum, are still less myelinated in adulthood. Such findings might imply reduced connectivity in the adult preterm brain and explain the poor cognitive outcome.

1 Introduction

Prematurity is the leading cause of death in children under the age of five, with preterm birth rates continuing to increase in almost every country with reliable data [14]. Despite the medical innovations in prenatal care, extremely preterm (EPT) infants (born before 26 weeks of gestation) remain at a high risk of death (30%-50% mortality). In addition, despite improved survival, disability rates are not declining, specifically in middle-income settings[5,14]. Preterm birth and other perinatal stressors such as premature exposure to the extra-uterine environment, ischemia, hypoxia, and inflammation can lead to White Matter (WM) injuries or otherwise affect brain development; more drastically,

© Springer Nature Switzerland AG 2021
C. H. Sudre et al. (Eds.): UNSURE 2021/PIPPI 2021, LNCS 12959, pp. 179–188, 2021.
https://doi.org/10.1007/978-3-030-87735-4_17

these can subsequently result in hypomyelination and long-term alterations of the brain's connectivity and structural complexity [17]. Therefore, measurement of myelin density and the spatial variation of myelin has been used to assess brain maturation in EPT babies and as a predictor of neurodevelopmental outcome, demonstrating the link between preterm brain development and significant alterations which can relate to cognitive performance [4,17]. Few studies have focused on the fully developed brain of preterm born individuals, where myelination processes are developed [11]. This study attempts to overcome this lack of knowledge by examining a cohort of 19-year-old participants, born at both extremely preterm and term gestation, and combining the structural sensitivity, but myelin non-specificity, of DWI and the high myelin specificity, but structural insensitivity of T2 relaxometry to reveal changes in myelin related biomarkers. First, full brain analysis is performed, followed by a region-specific approach investigating areas hypothesised to be more affected by preterm stressors such as the Corpus Callosum and the Posterior Limb of the Internal Capsule (PLIC). Lastly, these biomarkers were analysed between defined cortical regions and along the pathways between regions for a more thorough look at functional connectivity, and hypomyelination [10]. Figure 1 report the main steps of the analysis.

2   Methods

2.1   Data

Imaging data were acquired for a cohort of 142 adolescents at 19 years of age. Data for 89 EPT adolescents (52 Female/37 Male) and 53 term-born socioeconomically matched peers (32 Female/22 Male) were acquired on a 3T Phillips Achieva. Diffusion weighted data (DWI) was acquired across four b-values at b $= 0, 300, 700, 2000 \mathrm{s.mm}^{-2}$ with n $= 4,8,16,32$ directions respectively at TE $= 70$ ms $(2.5 \times 2.5 \times 3.0 \mathrm{mm})$. T2 weighted data was acquired in the same space as the diffusion imaging with ten echo times at TE $= 13$, 16, 19, 25, 30, 40, 50, 85, 100, 150 ms $(2.5 \times 2.5 \times 3.0 \mathrm{mm})$. In addition we acquired 3D T1-weighted $(\mathrm{TR/TE} = 6.93/3.14 \mathrm{ms})$ volume at 1 mm isotropic resolution for segmentation and parcellation [3]. B0 field maps were acquired to correct for EPI-based distortions between the diffusion imaging and the T1-weighted volumes. All participants gave informed consent before taking part in the experiment. Ethical approval was granted by the South Central - Hampshire Research Ethics Committee.

2.2   Region of Interest Values

Based upon previous research in these areas, manual regions of interest were described in the genu and splenium of the corpus callosum in addition to the posterior limb of the internal capsule [1,13]. These regions are ordinarily highly-myelinated white matter regions providing communication between cerebral hemispheres and carrying signals relating to motor-function to the rest of the body.

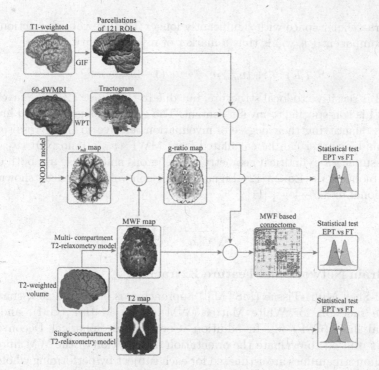

Fig. 1. Overview of methodology pipeline: T2-maps are estimated from the single compartment T2-relaxometry model, Myelin Water Fraction (MWF) maps are measured from the multi-compartment T2-relaxometry model. Neurite Density Index from the NODDI model and MWF maps are used to compute g-ratio maps (Eq. 2). A tractogram is generated by performing Whole-brain Probabilistic Tractography (WPT) for each subject from the diffusion-weighted data (DWI). Combining the streamline information with the maps of MWF, brain connectomes are derived for each subject. Statistical tests are performed to detect differences in T2, g-ratio, and MWF-derived markers between FT and EPT subjects.

2.3   Model Fitting

Myelin Water Fraction (MWF) corresponds to the volume of myelin water in a determined area of the brain and it is a surrogate measure for myelin content. Brain water T2 decreases with increasing gestational and postnatal age and is a more sensitive marker of brain maturation than visual assessment of T2-weighted MR images [9].

Ordinarily a single T2 value can be fitted to multi-echo data. This assumes that each voxel of the brain image contains a single biological compartment. Brain modelling can be enhanced by assuming that each brain voxel contains multiple compartments with different corresponding T2 values [12]. In the absence of a significant fluid component, brain tissue can be described with two compartments, one of myelin water with short T2 and one of other intra and extra-cellular space with significantly longer T2. The volume fraction of the MWF component (v_{mwf}) is thus a marker of myelin density (Eq. 1).

$$S(TE) = S_0 \left[v_{mwf} e^{-\frac{TE}{T2_1}} + (1 - v_{mwf}) e^{-\frac{TE}{T2_2}} \right] \tag{1}$$

DWI is sensitive to local structure, but due to the short T2 of the myelin signal, DWI is non-specific to myelin content. The g-ratio is a geometrical invariant of axons quantifying their degree of myelination relative to their cross-sectional size. This biomarker can be calculated from MWF (v_{mwf})and NDI (v_{ndi}) data when assuming a cylindrical geometry of the axons since these are both density related biomarkers (Eq. 2) [12]. Signal conduction velocity can be shown to be proportional to $\gamma\sqrt{-\log\gamma}$ [11]

$$\gamma = \left(\frac{v_{mwf}}{v_{ndi}} + 1 \right)^{-\frac{1}{2}} \tag{2}$$

2.4   Brain Network and Feature Extraction

A Multi-Shell Multi-Tissue (MSTMT) approach was employed to estimate the response function for White Matter (WM), Grey Matter (GM), and Cerebrospinal fluid (CSF) [8]. In addition, Constrained Spherical Deconvolution (CSD) is utilized to evaluate the orientation distribution of the WM fibres [16]. Ten million streamlines are estimated for each subject by performing whole-brain probabilistic tractography [15]. Tissue parcellations are obtained by applying the Geodesic Information Flow labelling protocol to the T1-weighted volumes [3]. Neurite Density Index (v_{ndi}) is estimated with the Neurite Orientation Dispersion and Density Imaging (NODDI) model [18]. One hundred twenty-one GM regions (ROIs) formed the nodes of the brain network. For each subject, a graph $\mathcal{G} = (V, E)$ is defined, where the nodes V are the ROI, and E corresponds to the set of edges connecting the ROIs. In $\mathcal{G}$, along each streamline connecting nodes i and j, the strength of connectivity of the edge (i, j) is defined as the mean sampled value of MWF. A more detailed description is provided by a previous study [7].

3   Results

3.1   Overall Brain

Single-compartment T2 maps of the entire cohort; as well as MWF and g-ratio maps were generated and examples are shown in Fig. 2. There was no significant difference for MWF and g-ratio values between EPT and FT participants when analysing the whole brain but there was a significant difference for T2 values which were higher in the EPT cohort (p-value 0.00137). The T2, MWF and g-ratio averages of the whole brain for EPT and FT participants were (90.85±9.00, 86.40±4.23); (0.23±0.03, 0.23±0.03) and (0.84±0.02, 0.85±0.02) respectively. Figure 3 shows the correlation between g-ratio/MWF (p-value 7.3^{-34} and $\rho -$ 0.82) and g-ratio/NDI (p-value 6.7^{-5} and ρ 0.34); with g-ratio/MWF exhibiting a more significant correlation.

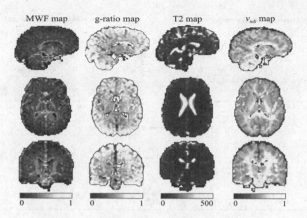

Fig. 2. Example of Myelin Water Fractio (MWF), g-ratio, T2, and Neurite Density Index (v_{ndi}) maps of an EPT participant.

3.2 Corpus Callosum and Posterior Limb of Internal Capsule

The Corpus Callosum was analysed and Fig. 4 shows an average reduction of myelin in EPT born participants (0.31 ± 0.06) when compared to FT (0.34 ± 0.05); overall (p-value 1.6^{-3}). It is worth highlighting that the Splenium of the Corpus Callosum showed the highest significant difference in the corpus callosum (higher than the genu), for EPT (0.31 ± 0.06), and FT (0.35 ± 0.050), with (p-value $1.92e^{-5}$). The PLIC region was analysed using non-parametric approaches and Fig. 5 shows a significant lower MWF value in EPT born participants (p-value 0.026). The thalamus was also analysed but showed no difference between EPT and FT born participants.

For the corpus callosum the results of g-ratio measurement show a statistically significant difference (p-value 0.0266) between EPT and FT participants; with g-ratio of EPT being higher (0.830 ± 0.031) than FT (0.819 ± 0.0254). Furthermore, g-ratio results in the Splenium of the Corpus Callosum presented a more significant difference (p-value 0.0015) between EPT (0.834 ± 0.0341) and FT (0.816 ± 0.026) participants. A difference in conduction velocity (p-value 0.022) was also found for the corpus callosum with EPT participants' values being lower (0.355 ± 0.0245) than FT (0.364 ± 0.0180); and specifically for the splenium of the corpus callosum (p-value 0.0023) where EPT (0.352 ± 0.030) and FT (0.366 ± 0.017). These results alongside the g-ratio ones are depicted in Fig. 6. No significant statistical difference was found for PLIC.

3.3 Region Connectivity and Specific Pathways

Decreased MWF was observed in the R.Thalamus-L.Calcarine Cortex (p-value = 0.03) and in the R.Thalamus- L.Precuneus (p-value = 0.018) pathways in EPT born participants. Parietal-Frontal pathways were also analysed but no significant difference was registered between EPT and FT participants. The MWF

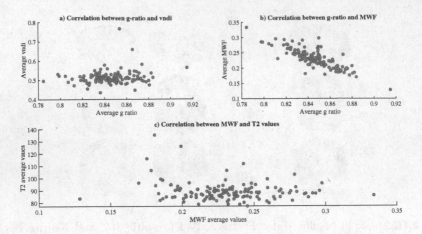

Fig. 3. Correlation plots a) v_{ndi}/g-ratio b) MWF/g-ratio c) MWF/T2.

connectomes were analysed and the average MWF of all myelin containing networks was found to be (0.240 ± 0.037) and (0.244 ± 0.036) respectively for EPT and FT born participants. No significant difference was found for these groups. Using the same approach for all networks within one region; all 121 regions were analysed and multiple regions appeared to have a decreased overall myelin density; Table 1 highlights the regions where a significant difference in MWF values was found and the corresponding p-values and CIs.

4   Discussion

The average MWF values of the whole brain showed no significant differences between EPT and FT participants. This might suggest that the effect of preterm birth is localised to specific brain regions, resulting in some brain areas being more affected. The preterm brain, having to develop *exvivo*, will reorganise its structure and develop differently in comparison to infants still in the womb, which is an indication that EPT born infants might present tracts with different characteristics than those in FT born individuals to compensate for the different developmental conditions [6,7].

Injuries in the Corpus Callosum are considerably frequent in prematurity due to being adjacent to the periventricular brain region, which is often impaired. In addition, underdevelopment can lead to disruption of intra-hemispheric communication and reduce connectivity [13]. The findings in the Corpus Callosum and specifically in the Splenium, which appears less myelinated in EPT than FT born participants, indicate the persistence of preterm birth consequences into adolescence.

The low MWF results found for the PLIC in EPT born participants might be an indicator of motor pathway underdevelopment or disruption of myelination

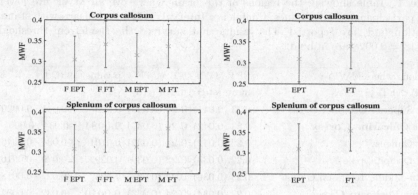

Fig. 4. Differences in the Corpus Callosum and the Splenium of the Corpus Callosum MWF information for preterm, term, female and male participants showed in 4 error bars.

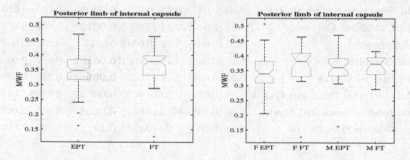

Fig. 5. Differences in the PLIC MWF information for EPT, FT, female and male participants showed in 2 boxplot graphs.

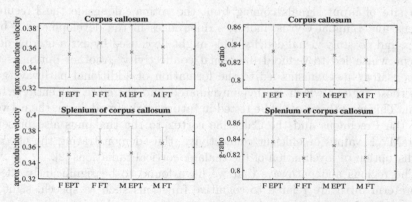

Fig. 6. g-ratio and conduction velocity error bars for the Corpus Callosum and Splenium of Corpus Callosum for all groups.

Table 1. Table showing the regions of the brain where overall MWF are lower in EPT participants. The p-values, Confidence Intervals (C.I), and means with standard deviation (std) are reported. The results that survived the Bonferroni threshold of $0.05/18 = 0.0028$ are in bold.

Regions where MWF EPT < FT	EPT MWF mean (std)	FT MWF mean (std)	p-value	95 C.I
Left Superior Occipital Gyrus	0.23 (0.041)	0.24 (0.040)	0.026	−0.031 −0.0020
Left Calcarine Cortex	0.23 (0.042)	0.25 (0.037)	**0.0084**	-0.034 −0.0051
Left Cuneus	0.22 (0.042)	0.24 (0.039)	0.0101	−0.034 −0.0046
Left Occipital pole	0.22 (0.039)	0.23 (0.039)	0.029	−0.029 −0.0016
Right Middle Temporal Gyrus	0.23 (0.036)	0.24 (0.033)	0.025	−0.027 −0.0018
Right Fusiform Gyrus	0.22 (0.034)	0.24 (0.033)	**0.0010**	−0.032 −0.0082
R Inferior Temporal Gyrus	0.22 (0.036)	0.24 (0.032)	**0.0020**	−0.032 −0.0073
R superior parietal lobe	0.22 (0.038)	0.23 (0.038)	0.035	−0.028 -0.00099
Right Angular Gyrus	0.22 (0.036)	0.23 (0.034)	0.0438	−0.026 −0.00037
Right Precuneus	0.23 (0.040)	0.24 (0.037)	0.0481	−0.028 −0.00012
Right Superior occipital Gyrus	0.20 (0.037)	0.23 (0.033)	**0.0005**	−0.035 −0.010
Right Calcarine Cortex	0.21 (0.040)	0.24 (0.035)	**0.0010**	−0.037 −0.0095
Right Cuneus	0.20 (0.038)	0.22 (0.034)	0.0043	−0.032 −0.0061
Right Occipital pole	0.19 (0.043)	0.21 (0.036)	**0.00192**	−0.037 −0.0086
Right lingual Gyrus	0.22 (0.038)	0.24 (0.038)	**0.00084**	−0.037 −0.0099
Right Occipital Fusiform Gyrus	0.19 (0.037)	0.22 (0.032)	**0.00027**	−0.0359 −0.0110
Right Inferior occipital Gyrus	0.20 (0.036)	0.22 (0.031)	**0.0005**	−0.034 −0.0097
Right Middle occipital Gyrus	0.21 (0.039)	0.22 (0.034)	0.030	−0.0279 −0.00143

processes due to preterm stressors, which have led to the hypomyelination of these pathways and possible motor deficits extended into adolescence [17].

The decrease in myelin-linked markers in the Calcarine cortex, which is the main site of input signals coming from the retina, alongside the Precuneus, Cuneus and Lingual Gyrus, indicates differences in the development of brain cortex and possibly visual deficits. This might be caused by extra-uterine development which led to reduced functional connectivity. Another option is that compensatory mechanisms led to the formation of additional pathways going from those regions which are hypomyelinated and therefore not entirely functional. This hypothesis could be tested in future work. In addition, the networks from the Precuneus and the Calcarine cortex to the thalamus also presented lower MWF values or efficient connectivity, thus demonstrating the effects of the disruption of myelination of the thalamocortical radiations [2].

The regions and pathways that we have found to be significantly affected by preterm birth are related to cognitive functions such as speech, sight and memory. Therefore, lower myelination in these areas could potentially represent the result of injuries suffered after prematurity, leading to hypomyelination and thus to functional deficits that propagated through adolescence, continuing to affect individuals born extremely prematurely.

No significant difference in results between female and male participants was registered; however, most results remained significant when distinguishing between sexes. Some regions, however, did not show a significant difference in EPT v FT individuals in the male cohort, but this might be due to the smaller sample size. In conclusion, our results support the hypothesis that preterm stressors affect brain maturation leading to changes in overall brain networks and in specific brain areas.

Acknowledgements. This work is supported by the EPSRC-funded UCL Centre for Doctoral Training in Medical Imaging (EP/L016478/1), the Department of Health NIHR-funded Biomedical Research Centre at University College London Hospitals and Medical Research Council (MR/N024869/1).

References

1. Young, J.M., et al.: Longitudinal study of white matter development and outcomes in children born very preterm. Cereb. Cortex **27**(8), 4094–4105 (2017)
2. Ball, G., et al.: The influence of preterm birth on the developing thalamocortical connectome. Cortex **49**(6), 1711–1721 (2013)
3. Cardoso, M.J., et al.: Geodesic information flows: spatially-variant graphs and their application to segmentation and fusion. IEEE Trans. Med. Imaging **34**(9), 1976–1988 (2015)
4. Dingwall, N., et al.: T2 relaxometry in the extremely-preterm brain at adolescence. Magn. Reson. Imaging **34**(4), 508–514 (2016)
5. Glass, H.C., et al.: Outcomes for extremely premature infants. Anesth. Analg. **120**(6), 1337 (2015)
6. Gozdas, E., Parikh, N.A., Merhar, S.L., Tkach, J.A., He, L., Holland, S.K.: Altered functional network connectivity in preterm infants: antecedents of cognitive and motor impairments? Brain Struct. Funct. **223**(8), 3665–3680 (2018). https://doi.org/10.1007/s00429-018-1707-0
7. Irzan, H., Molteni, E., Hütel, M., Ourselin, S., Marlow, N., Melbourne, A.: White matter analysis of the extremely preterm born adult brain. Neuroimage **237**, 118112 (2021)
8. Jeurissen, B., Tournier, J.-D., Dhollander, T., Connelly, A., Sijbers, J.: Multi-tissue constrained spherical deconvolution for improved analysis of multi-shell diffusion mri data. Neuroimage **103**, 411–426 (2014)
9. MacKay, A., et al.: Insights into brain microstructure from the T2 distribution (2006)
10. Malhotra, A., et al.: Detection and assessment of brain injury in the growth-restricted fetus and neonate. Pediatr. Res. **82**(2), 184–193 (2017)
11. Melbourne, A., et al.: Multi-modal measurement of the myelin-to-axon diameter g-ratio in preterm-born neonates and adult controls. In: Golland, P., Hata, N., Barillot, C., Hornegger, J., Howe, R. (eds.) MICCAI 2014. LNCS, vol. 8674, pp. 268–275. Springer, Cham (2014). https://doi.org/10.1007/978-3-319-10470-6_34
12. Melbourne, A., et al.: Longitudinal development in the preterm thalamus and posterior white matter: MRI correlations between diffusion weighted imaging and T2 relaxometry. Hum. Brain Mapp. **37**(7), 2479–2492 (2016)

13. Narberhaus, A., et al.: Neural substrates of visual paired associates in young adults with a history of very preterm birth: alterations in fronto-parieto-occipital networks and caudate nucleus. Neuroimage 47(4), 1884–1893 (2009)
14. World Health Organization, The Partnership for Maternal, Newborn & Child Health, and Save the Children. Born too soon: the global action report on preterm birth (2012)
15. Smith, R.E., Tournier, J.D., Calamante, F., Connelly, A.: Sift2: enabling dense quantitative assessment of brain white matter connectivity using streamlines tractography. Neuroimage 119, 338–351 (2015)
16. Tournier, J.D., Calamante, F., Connelly, A.: Robust determination of the fibre orientation distribution in diffusion MRI: non-negativity constrained super-resolved spherical deconvolution. Neuroimage 35(4), 1459–1472 (2007)
17. Wang, S., et al.: Quantitative assessment of myelination patterns in preterm neonates using T2-weighted MRI. Sci. Rep. 9(1), 1–12 (2019)
18. Zhang, H., Schneider, T., Wheeler-Kingshott, C.A., Alexander, D.C.: Noddi: practical in vivo neurite orientation dispersion and density imaging of the human brain. Neuroimage 61(4), 1000–1016 (2012)

A Bootstrap Self-training Method for Sequence Transfer: State-of-the-Art Placenta Segmentation in fetal MRI

Bella Specktor-Fadida[1(✉)], Daphna Link-Sourani[2], Shai Ferster-Kveller[1], Liat Ben-Sira[3,4], Elka Miller[5], Dafna Ben-Bashat[2,3], and Leo Joskowicz[1]

[1] School of Computer Science and Engineering, The Hebrew University of Jerusalem, Jerusalem, Israel
{bella.specktor,josko}@cs.huji.ac.il
[2] Sagol Brain Institute, Tel Aviv Sourasky Medical Center, Tel Aviv-Yafo, Israel
[3] Sackler Faculty of Medicine and Sagol School of Neuroscience, Tel Aviv University, Tel Aviv-Yafo, Israel
[4] Division of Pediatric Radiology, Tel Aviv Sourasky Medical Center, Tel Aviv-Yafo, Israel
[5] Medical Imaging, Children's Hospital of Eastern Ontario, University of Ottawa, Ottawa, Canada

Abstract. Quantitative volumetric evaluation of the placenta in fetal MRI scans is an important component of the fetal health evaluation. However, manual segmentation of the placenta is a time-consuming task that requires expertise and suffers from high observer variability. Deep learning methods for automatic segmentation are effective but require manually annotated datasets for each scanning sequence. We present a new method for bootstrapping automatic placenta segmentation by deep learning on different MRI sequences. The method consists of automatic placenta segmentation with two networks trained on labeled cases of one sequence followed by automatic adaptation using self-training of the same network to a new sequence with new unlabeled cases of this sequence. It uses a novel combined contour and soft Dice loss function for both the placenta ROI detection and segmentation networks. Our experimental studies for the FIESTA sequence yields a Dice score of 0.847 on 21 test cases with only 16 cases in the training set. Transfer to the TRUFI sequence yields a Dice score of 0.78 on 15 test cases, a significant improvement over the network results without transfer learning. The contour Dice loss and self-training approach achieve state-of-the art placenta segmentation results by sequence transfer bootstrapping.

Keywords: Deep learning segmentation · Unsupervised domain adaptation · fetal MRI

1 Introduction

The placenta plays an important role in fetal health, as it regulates the transmission of oxygen and nutrients from the mother to the fetus. Volumetric segmentation of the

C. H. Sudre et al. (Eds.): UNSURE 2021/PIPPI 2021, LNCS 12959, pp. 189–199, 2021.
https://doi.org/10.1007/978-3-030-87735-4_18

placenta in fetal MRI scans is useful for identifying pregnancy and birth complications [1, 2]. However, manual placenta delineation requires expertise, is time consuming, and has a high observer variability [3], which makes it impractical in clinical practice.

Automatic placenta segmentation in fetal MRI scans poses numerous challenges. These include MRI image related challenges, e.g., varying resolutions and contrasts, intensity inhomogeneity, image artifacts (due to the large field of view), and partial volume effect, and additional challenges specifically related to fetal MRI scanning, e.g., motion artifacts due to fetal and maternal movements, high variability in the placenta position, shape, appearance, orientation and fuzzy boundaries.

Various semi-automatic methods for volumetric placenta segmentation have been developed. Wang et al. [9] describe Slic-Seg, a Random Forest classifier followed by an iterative CRF method with multiple scans of the same patient and probability-based 4D graph cut segmentation. It yields a Dice score of 0.89 and ASSD of 1.9 mm. Wang et al. [10] describe a method that computes an initial segmentation with a CNN that is manually corrected and refined with a CRF. It achieves a Dice score of 0.893 and ASSD of 1.22 mm. [11] use a 3D U-Net whose inputs are the MRI scan, a distance map from the placenta center, and user interaction that guides the network to localize the placenta. A network trained on 80 cases and tested on 20 cases yielded a Dice score of 0.82.

Automatic methods for placenta segmentation have also been proposed, most of them using deep learning. Alansary et al. [4] describe the first automatic placenta segmentation method. It starts with a coarse segmentation with a 3D multi-scale convolution neural network (CNN) whose results are refined with a dense 3D Conditional Random Field (CRF). This method achieves a Dice score of 0.72 with 4-fold cross validation on 66 fetal MRI scans. Torrents-Barrena et al. [5] present a method based on super-resolution and motion correction followed by Gabor filters-based feature extraction and Support Vector Machine (SVM) voxel classification. This achieves a Dice score of 0.82 with 4-fold cross-validation on 44 fetal MRI scans. Han et al. [6] present a method that uses a 2D U-Net architecture trained on a small dataset 11 fetal MRI scans that achieves a mean IU score of 0.817. Quah et al. [7] compare various methods for placenta segmentation on 68 fetal MRI 3D T2* images. The best method achieves a Dice score of 0.808 using a U-Net with 3D patches. Pietsch et al. [8] describe a network that uses 2D patches and yields a Dice score of 0.76, comparable to expert variability performance.

The drawbacks of all these methods are that they apply to a single scanning sequence and thus require manually annotated data for each, which is time-consuming and impractical to acquire. Since scanning sequences used for placenta evaluation can vary both within and across clinical sites, a multi-sequence solution is needed.

To address this problem, unsupervised domain adaptation (UDA), e.g., self-training and pseudo-labeling techniques have been proposed [12, 13]. In self-training, a new network is trained with unlabeled data from a new sequence along with annotated data from the original sequence. The resulting network produces an initial segmentation that is then can be manually corrected to obtain new, high quality validated annotated datasets that are used for network training. This reduces the time and effort required to obtain annotated datasets to achieve adequate segmentation performance.

Pseudo-labeling is a method for iterative semi-supervised learning (SSL) in which a model improves the quality of pseudo annotations by learning from its own predictions on unannotated data. SSL uses a network training regime in which two or more networks are alternatively trained with an uncertainty-aware scheme [14, 15]. Nie et al. [16] describe an adversarial framework consisting of a generator segmentation network and a discriminator confidence network. Xia et al. [12] propose a UDA framework for generating pseudo-labels based on co-training multiple networks. Each network is trained on annotated and unannotated data with pseudo-annotations from the other networks using an uncertainty-aware combination. Zoph et al. [17] show the use of a simple self-training method for classification, detection and segmentation tasks for low and high annotated data regimes. The method consists of: 1) teacher training on annotated data; 2) pseudo-labels generation on unannotated data, and; 3) student training to jointly optimize the loss function on the manual and generated pseudo labels. This UDA method is effective for natural images, yet has not been applied to 3D medical images. We adopt this method for sequence transfer and combine it with active learning based on segmentation quality estimation.

2 Method

We present a new method for bootstrapping automatic placenta segmentation by deep learning on different MRI sequences with few annotated datasets. It consists of automatic placenta segmentation with two networks, each trained on labeled cases of one sequence, followed by automatic self-training of these networks to a new sequence with new unlabeled cases of this sequence. The key advantage of our method is that it streamlines the annotation of new sequences since annotating data from initial network results requires less time than annotation from scratch (<30 min vs. >60 min).

Our method consists of: 1) automatic segmentation of placenta with labeled cases from one sequence; 2) extension of automatic segmentation to a new sequence using additional unlabeled cases of that sequence. The automatic placenta segmentation method uses two cascaded networks for placenta ROI detection and segmentation and a novel combined contour and soft Dice loss function. The automatic adaptation to another sequence is performed by self-training consisting of: 1) generation of initial segmentations with the network of the original sequence on the unlabeled scans of the new sequence; 2) active learning using unsupervised segmentation quality estimation by selecting the cases with the highest estimated Dice score using Test Time Augmentations, and; 3) training of a new network with the labeled cases from the first sequence and the unlabeled cases with best estimated segmentation results of the second sequence.

2.1 Automatic Placenta Segmentation

Automatic supervised placenta segmentation is performed in two stages. First, a region of interest (ROI) is computed using segmentation network on a downscaled data by a factor of $\times 0.5$ in the in-plane axes. The placenta is then segmented within the computed ROI on the original data scale. This cascaded segmentation configuration is inspired by the nnU-Net framework [18]. We use the network architecture of [19] for the localization and the segmentation networks.

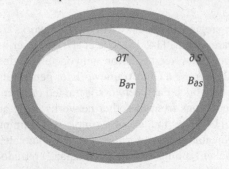 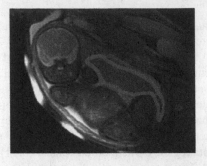

Fig. 1. Illustration of the contour Dice loss (Eq. 2). Left: diagram showing the ground truth ∂T (green line) and computed ∂S (blue line) contours and their corresponding offset bands $B_{\partial T}$ and $B_{\partial S}$. The intersection is $B_{\partial T} \cap B_{\partial S}$ (dark green). Right: offset bands of the placenta superimposed on a representative MRI slice (same colors). (Color figure online)

Contour Dice Loss. The networks are trained with a combination of the soft Dice and the contour Dice losses. We include the contour Dice score since it was shown to perform well for segmentation errors corrections [20, 21]. Since the contour Dice function is non-differentiable, Moltz et al. [22] propose to train a regression network to approximate the loss function and to combine it with a segmentation network to compute the loss during training. However, this requires training an additional regression network and is dependent on the quality of the network's output.

We propose instead to compute the contour Dice loss with an integral over an offset band around the contour. The use of a regional integral instead of a non-differentiable function was proposed in [23] to compute distances between contours. Note that this loss only considers the area of the contours and not the distances between them. It thus accounts for the desired segmentation correction of the contour Dice.

To compute the contour Dice, we first apply binary thresholding to the network output (a value in the [0, 1] interval). Then, we extract 2D offset bands around the contour for ground truth and network output masks in each slice (Fig. 1). Finally, we compute the Dice score of the offset band voxels of the ground truth and the segmentation output.

Formally, let ∂T and ∂S be the extracted surfaces of the ground truth delineation and the network results, respectively and let $B_{\partial T}$ and $B_{\partial S}$ be their respective offset bands. The contour Dice of the offset bands is:

$$Contour\,Dice(T, S) = \frac{|\partial T \cap B_{\partial S}| + |\partial S \cap B_{\partial T}|}{|\partial T| + |\partial S|} \tag{1}$$

To make it a loss function, we take the negative value and enlarge the contour to an offset band around the contour with boundaries $B_{\partial T}$ and $B_{\partial S}$:

$$ContourDiceLoss\,(L_{CD}) = -\frac{2|B_{\partial T} \cap B_{\partial S}|}{|B_{\partial T}| + |B_{\partial S}|} \tag{2}$$

Since contour-based methods require an additional regional term to avoid empty-region solutions, we combine the contour Dice score with a regional loss [23]. For this, we use the soft Dice score, as it was shown to perform well for fetal structures

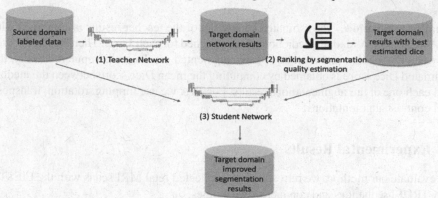

Fig. 2. Flow diagram of the self-training method for two MRI sequences: 1) teacher network training on annotated FIESTA scans and inference on unlabeled TRUFI scans; 2) ranking by segmentation quality estimation and selection of TRUFI cases with best estimated Dice scores; 3) student network training with labeled FIESTA and selected TRUFI cases with pseudo-labels.

segmentation [19]. Indeed, Isensee et al. [18] report that a combination of soft Dice and Cross Entropy loss yields good results in 53 diverse segmentation tasks.

Formally, let L_{DSC} be the soft Dice loss, let L_{CE} be the Cross Entropy loss, let L_{CD} be the contour Dice loss and let β and γ balancing p arameters. The total loss is:

$$Combined\ Loss\ (L) = L_{DSC} + \beta L_{CE} + \gamma L_{CD} \tag{3}$$

2.2 Self-training for Sequence Transfer

We now describe how to use semi-supervised self-training to extend the above mentioned method to additional scanning sequence without requiring new annotated labels of the new sequence. Our method is a variant of the teacher-student framework, where the original networks serve as teachers for the new student networks [17]. We illustrate it with the FIESTA (original) and TRUFI (new) sequences.

Our self-training method consists of three steps (Fig. 2): 1) automatic placenta segmentation using detection and segmentation networks from the FIESTA sequence on unlabeled TRUFI sequence scans; 2) segmentation quality estimation using estimated Dice score. We use Test Time Augmentations (TTA) to estimate the Dice score, as this does not require training of an additional regression network [19, 24] and; 3) training of new detection and segmentation networks using pseudo-labeled TRUFI sequence scans using the best estimated Dice score dataset obtained with TTA, combined with the labeled FIESTA data.

Both the teacher and the student networks are trained in a cascaded framework with a combination of the Dice and contour Dice losses. The training of student detection and segmentation networks starts with the weights computed for the original sequence. The cases with best estimated Dice score are used for both networks. TTA is used during inference of the teacher and student segmentation networks.

Quality Estimation. The estimated Dice score of the teacher results is computed with TTA as follows. A median prediction is first assigned for each voxel. Then the Dice scores between the median and each one of the augmented results are computed. Finally, the estimated Dice score is obtained by computing the mean Dice score between the median and each one of the augmentations results. For TTA we use flipping, rotation, transpose and contrast augmentations.

3   Experimental Results

To evaluate our method, we retrospectively collected fetal MRI scans with the FIESTA and TRUFI sequences and conducted two studies.

Datasets and Annotations: We collected fetal MRI placenta scans of patients acquired with the FIESTA and the TRUFI sequences as part of routine fetal assessment from the Sourasky Medical Center (Tel Aviv, Israel) with gestational ages (GA) 28–39 weeks. The FIESTA dataset consists of 40 labeled cases acquired on a 1.5T GE Signa Horizon Echo speed LX MRI scanner using a torso coil. Each scan has 50–100 slices, 256×256 pixels/slice, resolution $1.56 \times 1.56 \times 3.0 \, mm^3$. The TRUFI dataset has 15 labeled and 59 unlabeled cases acquired on the Siemens Skyra 3T, Prisma 3T, Auera 1.5T scanners. Each scan has 50–120 slices, $320–512 \times 320–512$ pixels/slice. The labeled and unlabeled cases resolutions are $0.78 \times 0.78 \times 2.0$ and $0.78–1.34 \times 0.78–1.34 \times 2–4.8 \, mm^3$.

Ground truth segmentations were created as follows. All 31 FIESTA and 15 TRUFI cases were annotated from scratch; 9 additional FIESTA cases were manually corrected from network results. Both the annotations and the corrections were performed by a clinical trainee. All test cases segmentations were validated by two expert radiologists.

Studies: We conducted two studies. Study 1 evaluates the accuracy of the supervised placenta segmentation for the FIESTA sequence. Study 2 evaluates the efficacy of sequence transfer using semi-supervised placenta segmentation on the TRUFI sequence.

In both studies, the segmentation quality is evaluated with the Dice, Hausdorff and 2D ASSD (slice wise Average Symmetric Surface Difference) metrics. The 3D ASSD was not evaluated, as it is highly dependent on the surface extraction method. The Volume Difference ratio (VD, the absolute volume difference divided by ground truth volume) was also computed, since placenta volume is a clinically relevant measure [2].

A network architecture similar that of Dudovitch et al. [19] was utilized with a patch size of $128 \times 128 \times 48$ to capture a large field of view and comply with GPU limitations. For the ROI detection network, the scans were downscaled by $\times 0.5$ in the in-plane axes. The segmentation results were refined by standard post-processing techniques.

Study 1: Supervised Placenta Segmentation. The method was evaluated on training, validation and test sets of 16, 3 and 21 annotated FIESTA scans respectively. Ablation experiments were performed to evaluate the contour Dice loss and the cascaded framework. The contour Dice loss was set to $\gamma = 0.5$ and the Cross Entropy loss to $\beta = 1$. For contour dice surface extraction we binarize segmentation results with a threshold of 1.

Table 1. Study 1: results of the various segmentation network loss functions with and without Test Time Augmentations (TTA): Dice, Cross Entropy (CE), contour Dice (CD). The measures are Dice, Hausdorff distance, volume difference (VD) ratio, and 2D ASSD.

	Dice	Hausdorff	VD ratio	ASSD (2D)
Dice	0.773 ± 0.117	57.73 ± 44.24	0.27 ± 0.24	8.35 ± 7.43
Dice with TTA	0.772 ± 0.126	56.22 ± 46.06	0.26 ± 0.21	7.79 ± 7.92
Dice + CE	0.807 ± 0.098	50.48 ± 40.15	0.21 ± 0.21	5.83 ± 3.34
Dice + CE with TTA	0.817 ± 0.096	44.33 ± 43.08	0.18 ± 0.18	5.18 ± 3.17
Dice + CD	**0.847 ± 0.058**	44.60 ± 42.31	0.18 ± 0.17	4.46 ± 2.45
Dice + CD with TTA	0.847 ± 0.061	**42.29 ± 42.44**	**0.16 ± 0.14**	**4.43 ± 2.50**
Dice + CE + CD	0.789 ± 0.127	47.75 ± 41.58	0.25 ± 0.22	5.87 ± 3.67
Dice + CE + CD with TTA	0.792 ± 0.128	50.11 ± 43.03	0.20 ± 0.21	6.39 ± 5.17

Fig. 3. Four examples of placenta segmentation results (red) on FIESTA sequence scan slices (rows). Left to right (columns): 1) original slice; 2) Dice loss; 3) Dice loss and Cross Entropy loss; 4) Dice loss and contour Dice loss; 5) ground truth. Arrows point to relevant differences. (Color figure online)

Table 1 shows the results of the different loss functions and Test Time Augmentations (TTA) on the segmentation accuracy. The contour Dice loss in combination with Dice

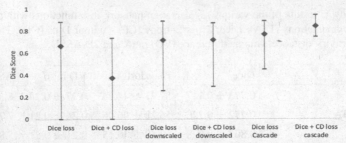

Fig. 4. Placenta segmentation Dice scores of (left to right): single network with Dice loss, single network with Dice and CD (contour Dice) loss, single downscaled network with Dice loss, single downscaled network with Dice and CD loss, cascaded networks with Dice loss, and cascaded networks with Dice and CD loss (mean and minimum maximum intervals).

loss yields the best placenta segmentation results, with Dice score of 0.847, with and without TTA. Note that TTA slightly improved the Volume Difference ratio.

The second-best performance achieved dice score of 0.817 with a combination of Dice and Cross Entropy losses and TTA Fig. 3 shows illustrative placenta segmentation results. The network with Dice loss and contour Dice yielded better contours segmentations and decreased the over/under segmentation produced by the other networks.

An ablation study was performed to quantify the effectiveness of the cascaded two networks framework on the placenta segmentation. Dice scores for a single network, for a single network with downscaled datasets and for the cascaded framework of a detection network with downscaled datasets followed by the segmentation network were computed using either the Dice loss by itself or combined with contour Dice loss.

The cascade framework significantly improved the segmentation Dice score, while the contour Dice loss was effective only when used for segmentation network inside an ROI in the cascaded framework. Figure 4 illustrates this with four cases.

Study 2: Semi-supervised (SSL) Placenta Segmentation for Sequence Transfer. The semi-supervised self-training method on the TRUFI sequence used annotated data from the FIESTA sequence and unannotated data from the TRUFI sequence. For semi-supervised method we applied the FIESTA network on 59 TRUFI cases and selected 16 cases with estimated Dice scores ≥ 0.96. These cases were then used in combination with the FIESTA annotated cases to train a new network. Prior to training, all cases were matched to the FIESTA cases resolution of $1.56 \times 1.56 \times 3.0\,\mathrm{mm}^3$.

Four scenarios were tested: 1) a network trained on annotated FIESTA cases without TTA; 2) a network trained on annotated FIESTA cases with TTA; 3) SSL using a network trained on annotated FIESTA cases and unannotated TRUFI cases with best estimated dice without TTA, and; 4) SSL using a network trained on annotated FIESTA cases and unannotated TRUFI cases with best estimated dice with TTA.

Table 2 shows the results. The self-training semi-supervised method improves by a significant margin the placenta segmentation Dice score from 0.495 to 0.78. Out of the 15 TRUFI test cases, the segmentation completely failed in 5 cases when only

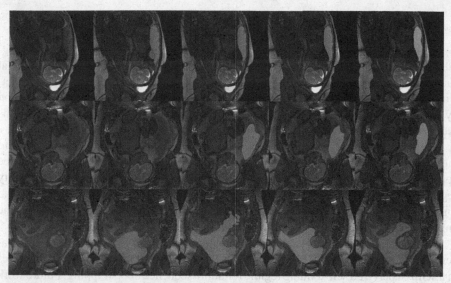

Fig. 5. Three examples of placenta segmentation results (red) on TRUFI sequence scan slices (rows). Left to right (columns): 1) original slice; 2) Teacher networks; 3) Student networks; 4) Student networks with TTA; 5) ground truth. (Color figure online)

Table 2. Study 2: results of the self-training semi-supervised and supervised only approaches with and without TTA for the FIESTA (original) and TRUFI (new) sequences.

	Dice ↑	Hausdorff (mm) ↓	2D ASSD (mm) ↓
Supervised FIESTA	0.495 ± 0.355	107.97 ± 72.27	41.22 ± 64.83
Supervised FIESTA with TTA	0.471 ± 0.352	108.61 ± 72.17	43.08 ± 64.23
Supervised SSL FIESTA and unsupervised TRUFI	0.776 ± 0.067	52.30 ± 24.42	9.49 ± 4.91
Supervised SSL FIESTA and unsupervised TRUFI with TTA	**0.780** ± **0.071**	**51.91** ± **24.88**	**8.64** ± **4.23**

annotated FIESTA cases were used for training. Adding TRUFI unsupervised cases in a self-training regime resulted in a TRUFI network that yields useful segmentations that can be further refined manually, thereby yielding annotation bootstrapping. Using TTA slightly improved the SSL results: the 2D ASSD decreased from 9.49 mm to 8.64 mm. Figure 5 shows illustrative placenta segmentation results for the TRUFI sequence. The use of unsupervised TRUFI cases in the training set improved placenta segmentation quality; TTA further decreased over segmentation.

4 Conclusion

We present a new, fully automatic method for bootstrapping automatic placenta segmentation by deep learning on different MRI sequences without requiring additional supervision. The method consists of automatic placenta segmentation with two networks trained on labeled cases of one sequence followed by the automatic adaptation by self-training of the same network to a new sequence with new unlabeled cases of this sequence. It uses a novel combined contour and soft Dice loss function.

The key advantages of this method are that it does not require modifications to the network architecture, that it is self-training, that it is automatic, and that it can be applied to a variety of fetal MRI scanning sequences. Correcting reasonable quality segmentations automatically generated by a self-trained network is less tedious and time-consuming than generating annotation from scratch. This helps to speed up the annotation process required to develop a multi-sequence placenta segmentation.

Our experimental results show that the contour Dice loss yields state-of-the-art placenta segmentation results. They also show the value of using additional unlabeled cases in a semi-supervised self-training regime for multi-sequence bootstrapping. Future work includes exploring other self-training unsupervised domain adaptation techniques, e.g., adversarial loss functions [15], to improve the initial teacher segmentation.

Our method was tested on two sequences and 36 placenta segmentation test cases in total. Future work may include a further large-scale validation study on FIESTA and TRUFI sequences and method extensions to additional sequences and acquisition types.

Acknowledgments. This research was supported in part by Kamin Grants 72061 and 72126 from the Israel Innovation Authority.

References

1. Leyendecker, J.R., et al.: MRI of pregnancy-related issues: abnormal placentation. Am. J. Roentgenol. **198**(2), 311–320 (2012)
2. Dahdouh, S., et al.: In vivo placental MRI shape and textural features predict fetal growth restriction and postnatal outcome. J. Magn. Resonan. Imaging **47**(2), 449–458 (2018)
3. Kveller-Fenster, S.: Placenta segmentation in fetal MRI scans by deep learning: a bootstrapping approach. MSc. thesis, The Hebrew University of Jerusalem, March 2021
4. Alansary, A., et al.: Fast fully automatic segmentation of the human placenta from motion corrupted MRI. In: Ourselin, S., Joskowicz, L., Sabuncu, M.R., Unal, G., Wells, W. (eds.) MICCAI 2016. LNCS, vol. 9901, pp. 589–597. Springer, Cham (2016). https://doi.org/10.1007/978-3-319-46723-8_68
5. Torrents-Barrena, J., et al.: Fully automatic 3D reconstruction of the placenta and its peripheral vasculature in intrauterine fetal MRI. Med. Image Anal. **54**, 263–279 (2019)
6. Han, M., et al.: Automatic segmentation of human placenta images with U-Net (2019). https://doi.org/10.1109/ACCESS.2019.2958133
7. Quah, B., et al.: Comparison of pure deep Learning approaches for placental extraction from dynamic functional MRI sequences between 19 and 37 gestational weeks. In: Proceedings of International Society for Magnetic Resonance in Medicine (2021)

8. Pietsch, M., et al.:. APPLAUSE: automatic prediction of PLAcental health via U-net segmentation and statistical evaluation. In: Proceedings of Conference on International Society for Magnetic Resonance in Medicine (2021)
9. Wang, G., et al.: Slic-Seg: a minimally interactive segmentation of the placenta from sparse and motion-corrupted fetal MRI in multiple views. Med. Image Anal. **34**, 137–147 (2016)
10. Wang, G., et al.: DeepIGeoS: a deep interactive geodesic framework for medical image segmentation. IEEE Trans. Pattern Anal. Mach. Intell. **41**(7), 1559–1572 (2019)
11. Shahedi M, et al.: Segmentation of uterus and placenta in MR images using a fully convolutional neural network. In: Proceedings of SPIE Conference on Computer-Aided Diagnosis (2020)
12. Xia, Y., et al.: Uncertainty-aware multi-view co-training for semi-supervised medical image segmentation and domain adaptation. Med. Image Anal. **65**, 101766 (2020)
13. Zou Y, Yu Z, Kumar BV, Wang J.: Unsupervised domain adaptation for semantic segmentation via class-balanced self-training. In: Proceedings of European Conference on Computer Vision, pp. 289–305 (2018)
14. Cheplygina, V., de Bruijne, M., Pluim, J.P.: Not-so-supervised: a survey of self-supervised, multi-instance, and transfer learning in medical image analysis. Med. Image Anal. **54**, 280–296 (2019)
15. Tajbakhsh, N., Jeyaseelan, L., Li, Q., Chiang, J.N., Wu, Z., Ding, X.: Embracing imperfect datasets: a review of deep learning solutions for medical image segmentation. Med. Image Anal. **63**, 101693 (2020)
16. Nie, D., Gao, Y., Wang, L., Shen, D.: ASDNet: attention based semi-supervised deep networks for medical image segmentation. In: Frangi, A., Schnabel, J., Davatzikos, C., Alberola-López, C., Fichtinger, G. (eds.) Medical Image Computing and Computer Assisted Intervention – MICCAI 2018. MICCAI 2018. Lecture Notes in Computer Science, vol. 11073, pp. 370–378. Springer, Caam (2018). https://doi.org/10.1007/978-3-030-00937-3_43
17. Zoph, B., et al.: Rethinking pre-training and self-training. Adv. Neural. Inf. Process. Syst. **33**, 3833–3845 (2020)
18. Isensee, F., Jaeger, P.F., Kohl, S.A., Petersen, J., Maier-Hein, K.H.: nnU-net: a self-configuring method for deep learning-based biomedical image segmentation. Nat. Methods **18**(2), 203–211 (2021)
19. Dudovitch, G., Link-Sourani, D., Ben Sira, L., Miller, E., Ben Bashat, D., Joskowicz, L.: Deep learning automatic fetal structures segmentation in mri scans with few annotated datasets. In: Martel, A.L., et al. (eds.) MICCAI 2020. LNCS, vol. 12266, pp. 365–374. Springer, Cham (2020). https://doi.org/10.1007/978-3-030-59725-2_35
20. Nikolov S, et al.: Deep learning to achieve clinically applicable segmentation of head and neck anatomy for radiotherapy. arXiv preprint arXiv:1809.04430, 12 September 2018
21. Kiser, K., Barman, A., Stieb, S., Fuller, C.D., Giancardo, L.: Novel autosegmentation spatial similarity metrics capture the time required to correct segmentations better than traditional metrics in a thoracic cavity segmentation workflow. medRxiv preprint, January 2020
22. Moltz, J.H., et al.: Learning a loss function for segmentation: a feasibility study. In: Proceedings of IEEE 17th International Symposium on Biomedical Imaging, pp. 357–360 (2020)
23. Kervadec, H., Bouchtiba, J., Desrosiers, C., Granger, E., Dolz, J., Ben Ayed, I.: Boundary loss for highly unbalanced segmentation. Med. Image Anal. **67**, 101851 (2021)
24. Wang, G., Li, W., Aertsen, M., Deprest, J., Ourselin, S., Vercauteren, T.: Aleatoric uncertainty estimation with test-time augmentation for medical image segmentation with convolutional neural networks. Neurocomputing **338**, 34–45 (2019)

Segmentation of the Cortical Plate in Fetal Brain MRI with a Topological Loss

Priscille de Dumast[1,2]([✉]), Hamza Kebiri[1,2], Chirine Atat[1], Vincent Dunet[1], Mériam Koob[1], and Meritxell Bach Cuadra[1,2]

[1] Department of Radiology, Lausanne University Hospital (CHUV) and University of Lausanne (UNIL), Lausanne, Switzerland
priscille.guerrierdedumast@unil.ch
[2] CIBM Center for Biomedical Imaging, Lausanne, Switzerland

Abstract. The fetal cortical plate undergoes drastic morphological changes throughout early *in utero* development that can be observed using magnetic resonance (MR) imaging. An accurate MR image segmentation, and more importantly a topologically correct delineation of the cortical gray matter, is a key baseline to perform further quantitative analysis of brain development. In this paper, we propose for the first time the integration of a topological constraint, as an additional loss function, to enhance the morphological consistency of a deep learning-based segmentation of the fetal cortical plate. We quantitatively evaluate our method on 18 fetal brain atlases ranging from 21 to 38 weeks of gestation, showing the significant benefits of our method through all gestational ages as compared to a baseline method. Furthermore, qualitative evaluation by three different experts on 26 clinical MRIs evidences the out-performance of our method independently of the MR reconstruction quality. Finally, as a proof of concept, 3 fetal brains with abnormal cortical development were assessed. The proposed topologically-constrained framework outperforms the baseline, thus, suggesting its additional value to also depict pathology.

Keywords: Fetal brain · Cortical plate · Deep learning · Topology · Magnetic resonance imaging

1 Introduction

The early *in utero* brain development involves complex intertwined processes, reflected in both physiological and structural changes [23]. The developing cortical plate specifically undergoes drastic morphological transformations throughout gestation. Nearly all gyri are in place at birth, even though the complexification of their patterns carries on after birth [18]. T2-weighted (T2w) magnetic resonance imaging (MRI) offers a good contrast between brain tissues, hence allowing to assess the brain growth and detect abnormalities *in utero*. In the

© Springer Nature Switzerland AG 2021
C. H. Sudre et al. (Eds.): UNSURE 2021/PIPPI 2021, LNCS 12959, pp. 200–209, 2021.
https://doi.org/10.1007/978-3-030-87735-4_19

clinical context, fetal MRI is performed with fast, 2D orthogonal series in order to minimize the effect of unpredictable fetal motion but results in low out-of-plane spatial resolution and significant partial volume effect. In order to combine these multiple series, advanced imaging techniques based on super-resolution (SR) algorithms [10,24] allow the reconstruction of 3D high-resolution motion-free isotropic volumes. Together with improved visualization, these SR volumes open up to more accurate quantitative analysis of the growing brain anatomy. Consequently, based on 3D reconstructed volumes, multiple studies explored semi-automated fetal brain tissue segmentation [19] and cortical folding patterns *in-utero* [6,25]. Cortical plate is crucial in early brain development as pathological conditions, e.g. ventriculomegaly, are proved to manifest along with altered foldings [2]. However, cortical plate segmentation remains challenging as it undergoes significant changes due to the brain growth and maturation, respectively modifying the morphology and the image contrast [19]. Furthermore, being a thin layer easily altered by partial volume effect in MRI, anatomical topology is prone to be incorrectly represented by automatic segmentation methods.

In this respect, we present a fully automated and topologically correct age-invariant segmentation method of the cortical plate. In [4,5], the first topological-based segmentation of the fetal cortex was introduced, based on geometrical constraints that integrated anatomical and topological priors. Regrettably, their topological correctness was not further evaluated and qualitative results on only 6 fetuses were presented. More recently, deep learning (DL) methods have also focused on fetal brain MRI cortical gray matter segmentation. Using a neonatal segmentation framework as initialization, [12] proposes a multi-scale approach for the segmentation of the developing cortex, while [9] implements a two-stage segmentation framework with an attention refinement module. Nevertheless, while the segmentation accuracy of these recent DL methods is promising, none of these works assess the topological correctness of their results. In fact, these works report high overlap metrics but illustrated results show lack of topological consistency with notably discontinuous/broken cortical ribbons.

To our knowledge, only two works explore topological fidelity of the segmentation in different applications. In [15], they proposed a topological loss for neuronal membrane segmentation. More recently, topological constraints for MR cardiac image segmentation have been presented [3], although prior topological knowledge is required. In this paper we integrate for the first time a topological constraint, from [15], in a deep image segmentation framework to overcome the limitation of disjoint cortical plate segmentation in fetal MRI and further improve DL architectures (Fig. 1).

2 Methodology

2.1 Topological Loss

Our approach, is based on the topological loss function proposed in [15]. The topology-preserving loss compares the predicted likelihood to the ground truth segmentation using the concept of persistent homology [11]. In a nutshell, homology structures are obtained by filtration to all possible threshold values of the

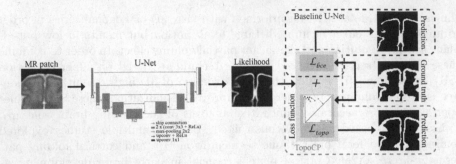

Fig. 1. Figure adapted from [15]. TopoCP, integrates a topological loss based on persistent homology to a 2D U-Net segmentation of cortical plate fetal MRI.

predicted likelihood and reported in a persistence diagram (Fig. 1). Both 0-dimensional and 1-dimensional Betti numbers [13], corresponding respectively to the number of connected components and the number of holes, are tracked. The persistence diagrams of the likelihood and the ground truth are matched, finding the best one-to-one structure correspondence, and the topological loss is computed as the distance between the matched pairs. We refer the reader to the original paper for advanced technical details [15].

2.2   Network Architecture

The topological loss introduced above is indeed compatible with any deep neural network providing a pixel-wise prediction. We chose as baseline the well-established U-Net [22] image segmentation method, as it recently proved its ability to deal with 2D fetal brain MRI tissue segmentation [17]. The baseline 2D U-Net uses a binary cross-entropy loss function $\mathcal{L}_{bce}$. The proposed framework *TopoCP* is based on a 2D U-Net trained using

$$\mathcal{L} = \mathcal{L}_{bce} + \lambda_{topo}\mathcal{L}_{topo}, \tag{1}$$

where $\mathcal{L}_{topo}$ is the topological term in [15] and λ_{topo} the weight of the contribution of $\mathcal{L}_{topo}$ in the final loss.

The 2D U-Net architecture is composed of encoding and decoding paths. The encoding path in our study is composed of 5 repetitions of the followings: two 3×3 convolutional layers, followed by a rectified linear unit (ReLu) activation function and a 2×2 max-pooling downsampling layer. Feature maps are hence doubled from 32 to 512. In the expanding path, 2×2 upsampled encoded features concatenated with the corresponding encoding path are 3×3 convolved and passed through ReLu. The network prediction is computed with final 1×1 convolution. Both Baseline and TopoCP are implemented in Tensorflow. In TopoCP, the topological loss is implemented in C++ and built as a Python library.

2.3 Training Strategy

The publicly available dataset Fetal Tissue Annotation and Segmentation Dataset (FeTA) is used in the training phase [20,21]. Discarding pathological and non-annotated brains, our training dataset results in 15 healthy fetal brains (see details summarized in Table 1). Both networks are fed with 64×64 patches of axial orientation (see Fig. 1), containing cortical gray matter. Intensities of all image patches are standardized and data augmentation is performed by randomly flipping and rotating patches (by $n \times 90°$, $n \in [\![0; 3]\!]$). As in [15], to overcome the high computational cost of persistent homology, we adopted the following optimization strategy: 1) our baseline model was trained over 23 epochs with a learning rate decay scheduled at epochs 11, 16, 17, 22 and a decay factor of 0.5, initialized at 0.0001; 2) from the pretrained model in the first step, both networks were fine-tuned over 35 epochs, with a learning rate decay scheduled at epochs 14, 23 for Baseline U-Net and none for TopoCP. TopoCP was trained with $\lambda_{topo} = 1$. A 7-fold cross-validation approach was used to determine the epochs for learning rate decay.

3 Evaluation

3.1 Quantitative Evaluation

Data. In the training dataset (FeTA), label maps were sparse (annotations were performed on every 2nd to 3rd slice) and their interpolation resulted in *noisy* labels with topological inconsistencies. Therefore, we rather evaluate our method on an independent pure testing dataset, presenting a topologically accurate segmentation. The normative spatiotemporal MRI atlas of the fetal brain [14] provides 3D high-quality isotropic smooth volumes along with tissue label maps, including more than fifty anatomical regions, for all gestational age between 21 and 38 weeks (see Table 1 for details). Atlas labels were merged to match the tissue classes represented in our training dataset.

Analysis. Though inferred segmentation rely on 2D patches, performance of the methods is evaluated on the whole 3D segmentation. Three complementary types of evaluation metrics are used: 1) the overlap between the ground truth and the predicted segmentation is quantified with the Dice similarity coefficient (DSC) [8]; 2) a boundary-distance-based metric is measured to evaluate the contours: the 95th percentile of the Hausdorff distance (HD95) [16]; 3) finally, the topological correctness is quantified with the error of a topological invariant: the Euler characteristic (EC), defined as a function of the k-dimensional (k-dim) Betti numbers (B_k), topologically invariant themselves. The 3D Euler characteristic is defined as:

$$EC = B_0 - B_1 + B_2, \tag{2}$$

where B_0 counts the number of connected components, B_1 the number of holes (tunnels) and B_2 counts number of void/cavities encapsulated in the binary objects. Topology errors are defined as the absolute difference of the ground truth and the prediction measures. For completeness, k-dim Betti errors (BE) are also reported. To assess the significance of the observed differences between the two methods, we perform a Wilcoxon rank sum test for each metrics. p-values were adjusted for multiple comparisons using Bonferroni correction and statistical significance level was set to 0.05.

3.2 Qualitative Evaluation

Data. In order to better represent the diversity of the cortical variability and to prove the generalization of our approach to SR reconstructions of clinical acquisitions, we introduce a second pure testing set of T2w SR images of 26 healthy fetuses. Two subsets were created, from a consensus of three experts evaluation, based on the quality of the reconstructed 3D volumes: 1) excellent (N = 16) and 2) acceptable (N = 10) - with remaining motion artifacts or partial volume effects. Additionally, as a proof of concept, three subjects with cortical plate pathologies were segmented (schizencephaly (1); polymicrogyria (1); corpus callosum agenesis (CCA) and schizencephaly (1)). MR image patches were preprocessed for intensity standardization with no further intensity-based domain adaptation performed. Nevertheless, prior to the segmentation inference, clinical images were resampled to match the resolution of the training data using ANTs [1] in order to present a similar field of view (see Table 1).

Analysis. Three experienced raters (two radiologists and one engineer) performed independently a qualitative analysis of the baseline and TopoCP segmentations. For healthy subjects, randomly-ordered segmentation of axial slices from healthy subjects were presented. The experts were asked to indicate if they preferred either the segmentation A or B or if they were of equivalent quality. The inter-rater reliability was assessed with their percentage agreement before considering a consensus evaluation resulting from the majority voting of the experts' evaluations. For the pathological cases, three radiologists, blindly assessed the whole 3D volume to ensure that the pathological area was included.

Table 1. Summary of the data used for training and quantitative and qualitative evaluation.

Dataset	Field strength	Vendor	Num. of subjects	Gestational age (weeks)	Reconstruction method	Resolution (mm^3)
Training	1.5T; 3T	General electric	15	[22.6–33.4] (28.7 ± 3.5)	mialSRTK [7, 24]	0.5 × 0.5 × 0.5
Evaluation quantitative	1.5T; 3T	Siemens; Philips	18	21–38	Gholipour et al., 2017 [14]	0.8 × 0.8 × 0.8
Evaluation qualitative	1.5T	Siemens	29	[18–25] (27.8 ± 4.1)	mialSRTK [7, 24]	1.12 × 1.12 × 1.12

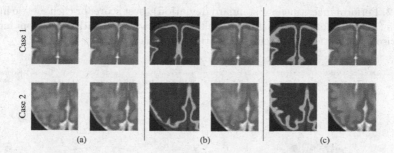

Fig. 2. Segmentation results on 35 weeks of gestation atlas. (a) T2w (left) and ground truth segmentation overlaid (right). (b) Baseline U-Net and (c) TopoCP: predicted likelihood (left) and estimated segmentation (right). Likelihood probabilities: 0 ▰▰▰▭ 1. Case 1 illustrates a net improvement in the segmentation of the midsagittal area and frontal cortical foldings. Case 2 shows a more accurate detection of the deep sulci with TopoCP.

4   Results

Figure 2 shows the ground truth of two representative patches with their predicted likelihood and segmentation overlaid on the T2w SR image. These results illustrate the benefits of TopoCP on the estimated probability maps, detecting more subtle variation of the cortex. The improved likelihood echoes with a better segmentation. A summary of the 3D performance (Sect. 3.1) metrics on the fetal brain atlas is presented in Table 2. TopoCP outperformed the Baseline U-Net in both similarity- and distance-based evaluation metrics. Corrected p-values between both methods (shown in italics) indicate that our method significantly improves the baseline segmentation. Regarding the topological correctness, the holistic EC error shows significant improvement with TopoCP. The 1-dim BE is the most improved Betti Error and with the highest impact on the global topological assessment. We recall that it represents the error of bored cortical ribbon compared to the ground truth, which is the initial problem addressed. Besides, it should be noted that the 0-dim BE is deteriorated with TopoCP. Visual inspection shows the presence of small isolated false positives in the deep gray matter area. Although, these false positives do not echo with impaired similarity and distance-based metrics. We hypothesise that this behaviour would be due to the fact that training was done on positive (cortex-aware) patches only. We believe these false positive can be reduced with the integration of negative patches in the training phase. Nonetheless, the 3D topology of the cortical plate with TopoCP is much closer to the reality than with Baseline U-Net (see Fig. 3a). Moreover, we observe large standard deviations in the topology-based metrics, although they are slightly reduced with TopoCP (Table 2). Figure 3b shows that the performance metrics varies over the gestational age. For both methods, we observe better performances in the middle of the gestational age range, which we explain as this corresponds to the age range present in the training set (see Table 1). Furthermore, third trimester fetuses benefits more from TopoCP than

Table 2. Performances (mean ± standard deviation), best score for each metric in bold. *p*-values (in italics) of paired Wilcoxon rank sum test adjusted with Bonferroni multiple comparisons correction, between both methods for each metric.

	DSC↑		HD95 (mm)↓		0-d BE↓		1-d BE↓		2-d BE↓		EC Error↓	
	Baseline U-Net	TopoCP	Baseline U-Net	TopoCP	Baseline U-Net	TopoCP	Baseline U-Net	TopoCP	Baseline U-Net	TopoCP	Baseline U-Net	TopoCP
3D	0.57 ± 0.07	**0.72 ± 0.05**	3.5 ± 0.87	**2.58 ± 0.96**	**10.1 ± 10.8**	13.3 ± 9.6	61 ± 30.3	**35.4 ± 23.7**	8.5 ± 13.4	**8.0 ± 12.2**	60.1 ± 33.4	**30.0 ± 25.3**
	2e-07		*0.0053*		*0.076*		*0.00099*		*0.96*		*0.00075*	

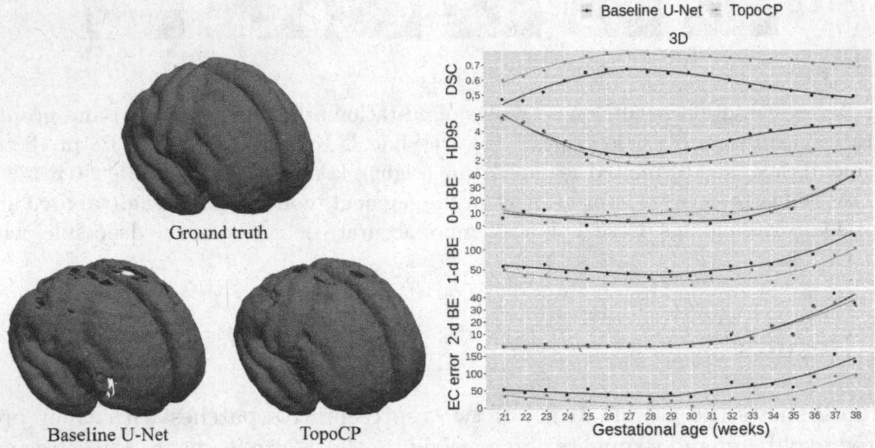

Fig. 3. (a) 3D rendering of 28 weeks-old atlas cortical plate segmentation from both automatic methods compared to the ground truth. (b) Performance metrics at the subject-level computed on the whole 3D volume for all atlas images.

others. TopoCP is more valuable to older fetuses, as they undergo the more complex cortical gyrification patterns. While the overlap metric constantly improves throughout gestation, distance error is mainly enhanced from the third trimester. The topological loss has a stronger positive effect on the topological errors for old subjects, although the whole range of gestational age presented benefits from it.

Qualitative assessment of healthy fetuses indicates a good inter-rater agreement of 74%. Figure 4a shows the consensus of the experts' blind evaluation of the cortical plate segmentation on SR volumes based on T2w clinical acquisitions. For both excellent and acceptable sets, TopoCP was selected as giving the best segmentation (overall on 81% of the slices), showing the robustness of our method to the SR quality. Figure 4b illustrates a representative slice segmented with both methods. Similarly, all raters preferred TopoCP segmentation in the three pathological cases (CCA and schizencephaly shown in Fig. 5).

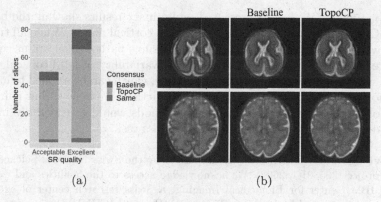

(a)   (b)

Fig. 4. (a) Experts' qualitative evaluation results in the comparison of Baseline U-Net and TopoCP automatic segmentations. (b) Segmentation results on 23 (top) and 32 (bottom) gestational weeks fetuses.

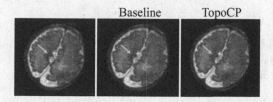

Fig. 5. Segmentation results of a 33 weeks old subject with corpus callosum agenesis and schizencephaly. Yellow arrows indicate the pathological area, where TopoCP is better performing. (Color figure online)

5   Discussion and Conclusion

This work assesses for the first time the integration of a topological constraint in DL-based segmentation of the fetal cortical plate on MRI. Our results on a wide range of gestational ages (21 to 38 weeks) (measured with 3D topology error) and qualitative assessment on 29 clinical subjects (including 3 with cortical pathologies) demonstrate the resulting improved topological correctness of the fetal cortex, despite noisy training labels and 2D inference. Our approach can possibly be extended to 3D, although, one should note that an increase in the input dimension will echo to an increase of the computational cost. In this study, we arbitrarily set to 1 the weight of the topological loss, as done in [15]. We acknowledge the loss contribution has its influence in the training phase and should be fine tuned for improved performance. By testing our method on different acquisitions than those of the training phase, we observe that the segmentation quality of our method seems robust to different scanners and reconstruction methods. Nevertheless, the main drawback of our work is its sensitivity to the resolution of the input image. Resampling of both the input image and result segmentation introduces interpolation that might embed the final results. We hypothesize that training on images of various resolutions would make our method more robust to this parameter. We briefly presented preliminary results showing the benefits of TopoCP in the segmentation of pathological cortical plates. While all training images were of neurotypical fetal brains, we assume pathological brains could be added to training set to better represent the variability of fetal cortical plates. Finally, we emphasize the genericity of this loss, which can be applied to any segmentation network providing a pixel-wise prediction. We believe that pairing up the topological loss with state-of-the-art methods would considerably improve the resulting segmentation, even in a multi-class task.

Acknowledgments. This work was supported by the Swiss National Science Foundation (project 205321-182602). We acknowledge access to the facilities and expertise of the CIBM Center for Biomedical Imaging, a Swiss research center of excellence founded and supported by Lausanne University Hospital (CHUV), University of Lausanne (UNIL), Ecole polytechnique fédérale de Lausanne (EPFL), University of Geneva (UNIGE) and Geneva University Hospitals (HUG).

References

1. Avants, B., et al.: A reproducible evaluation of ANTs similarity metric performance in brain image registration. NeuroImage **54**(3), 2033–2044 (2011). https://doi.org/10.1016/j.neuroimage.2010.09.025

2. Benkarim, O.M., et al.: Cortical folding alterations in fetuses with isolated nonsevere ventriculomegaly. NeuroImage Clin. **18**, 103–114 (2018). https://doi.org/10.1016/j.nicl.2018.01.006

3. Byrne, N., Clough, J.R., Montana, G., King, A.P.: A persistent homology-based topological loss function for multi-class CNN segmentation of cardiac MRI. In: Puyol Anton, E., et al. (eds.) STACOM 2020. LNCS, vol. 12592, pp. 3–13. Springer, Cham (2021). https://doi.org/10.1007/978-3-030-68107-4_1

4. Caldairou, B., et al.: Segmentation of the cortex in fetal MRI using a topological model. In: 2011 IEEE International Symposium on Biomedical Imaging: From Nano to Macro, Chicago, IL, USA, pp. 2045–2048. IEEE, March 2011. https://doi.org/10.1109/ISBI.2011.5872814

5. Caldairou, B., et al.: Data-driven cortex segmentation in reconstructed fetal MRI by using structural constraints. In: Real, P., Diaz-Pernil, D., Molina-Abril, H., Berciano, A., Kropatsch, W. (eds.) CAIP 2011. LNCS, vol. 6854, pp. 503–511. Springer, Heidelberg (2011). https://doi.org/10.1007/978-3-642-23672-3_61

6. Clouchoux, C., et al.: Quantitative in vivo MRI measurement of cortical development in the fetus. Brain Struct. Funct. **217**(1), 127–139 (2012). https://doi.org/10.1007/s00429-011-0325-x

7. Deman, P., et al.: meribach/mevislabFetalMRI: MEVISLAB MIAL superresolution reconstruction of fetal brain MRI v1.0 (2020). https://doi.org/10.5281/zenodo.3878564

8. Dice, L.R.: Measures of the amount of ecologic association between species. Ecology **26**(3), 297–302 (1945)

9. Dou, H., et al.: A deep attentive convolutional neural network for automatic cortical plate segmentation in fetal MRI (2020). arXiv: 2004.12847

10. Ebner, M., et al.: An automated framework for localization, segmentation and super-resolution reconstruction of fetal brain MRI. NeuroImage **206** (2020). https://doi.org/10.1016/j.neuroimage.2019.116324

11. Edelsbrunner, H., et al.: Topological persistence and simplification. Discret. Comput. Geom. **28**(4), 511–533 (2002). https://doi.org/10.1007/s00454-002-2885-2
12. Fetit, A.E., et al.: A deep learning approach to segmentation of the developing cortex in fetal brain MRI with minimal manual labeling. In: Medical Imaging with Deep Learning (MIDL) (2020). https://openreview.net/forum?id=SgZo6XA-l
13. Gardner, M.: The Sixth Book of Mathematical Games from Scientific American. WH Freeman, New York (1984)
14. Gholipour, A., et al.: A normative spatiotemporal MRI atlas of the fetal brain for automatic segmentation and analysis of early brain growth. Sci. Rep. **7**(1), 476 (2017). https://doi.org/10.1038/s41598-017-00525-w
15. Hu, X., et al.: Topology-preserving deep image segmentation. In: Advances in Neural Information Processing Systems, vol. 32, pp. 5657–5668. Curran Associates, Inc. (2019)
16. Huttenlocher, D.P., et al.: Comparing images using the Hausdorff distance. IEEE Trans. Pattern Anal. Mach. Intell. **15**(9), 850–863 (1993). https://doi.org/10.1109/34.232073
17. Khalili, N., et al.: Automatic brain tissue segmentation in fetal MRI using convolutional neural networks. Magn. Reson. Imaging **64**, 77–89 (2019). https://doi.org/10.1016/j.mri.2019.05.020
18. Lenroot, R.K., Giedd, J.N.: Brain development in children and adolescents: insights from anatomical magnetic resonance imaging. Neurosci. Biobehav. Rev. **30**(6), 718–729 (2006). https://doi.org/10.1016/j.neubiorev.2006.06.001
19. Makropoulos, A., et al.: A review on automatic fetal and neonatal brain MRI segmentation. NeuroImage **170**, 231–248 (2018). https://doi.org/10.1016/j.neuroimage.2017.06.074
20. Payette, K., Jakab, A.: Fetal tissue annotation dataset feta, February 2021. https://doi.org/10.5281/zenodo.4541606. https://doi.org/10.5281/zenodo.4541606
21. Payette, K., et al.: A comparison of automatic multi-tissue segmentation methods of the human fetal brain using the feta dataset (2020). arXiv: 2010.15526
22. Ronneberger, O., Fischer, P., Brox, T.: U-Net: convolutional networks for biomedical image segmentation. In: Navab, N., Hornegger, J., Wells, W.M., Frangi, A.F. (eds.) MICCAI 2015. LNCS, vol. 9351, pp. 234–241. Springer, Cham (2015). https://doi.org/10.1007/978-3-319-24574-4_28
23. Tierney, A.L., Nelson, C.A.: Brain development and the role of experience in the early years. Zero Three **30**(2), 9–13 (2009)
24. Tourbier, S., et al.: An efficient total variation algorithm for super-resolution in fetal brain MRI with adaptive regularization. NeuroImage **118**, 584–597 (2015). https://doi.org/10.1016/j.neuroimage.2015.06.018
25. Wright, R., et al.: Automatic quantification of normal cortical folding patterns from fetal brain MRI. NeuroImage **91**, 21–32 (2014). https://doi.org/10.1016/j.neuroimage.2014.01.034

Fetal Brain MRI Measurements Using a Deep Learning Landmark Network with Reliability Estimation

Netanell Avisdris[1,2(✉)] , Dafna Ben Bashat[2,3,4], Liat Ben-Sira[3,4,5],
and Leo Joskowicz[1]

[1] School of Computer Science and Engineering,
The Hebrew U. of Jerusalem, Jerusalem, Israel
{netana03,josko}@cs.huji.ac.il
[2] Sagol Brain Institute, Tel Aviv Sourasky Medical Center, Tel Aviv, Israel
[3] Sackler Faculty of Medicine, Tel Aviv University, Tel Aviv, Israel
[4] Sagol School of Neuroscience, Tel Aviv University, Tel Aviv, Israel
[5] Division of Pediatric Radiology, Tel Aviv Sourasky Medical Center, Tel Aviv, Israel

Abstract. We present a new deep learning method, FML, that automatically computes linear measurements in a fetal brain MRI volume. The method is based on landmark detection and estimates their location reliability. It consists of four steps: 1) fetal brain region of interest detection with a two-stage anisotropic U-Net; 2) reference slice selection with a convolutional neural network (CNN); 3) linear measurement computation based on landmarks detection using a novel CNN, FMLNet; 4) measurement reliability estimation using a Gaussian Mixture Model. The advantages of our method are that it does not rely on heuristics to identify the landmarks, that it does not require fetal brain structures segmentation, and that it is robust since it incorporates reliability estimation. We demonstrate our method on three key fetal biometric measurements from fetal brain MRI volumes: Cerebral Biparietal Diameter (CBD), Bone Biparietal Diameter (BBD), and Trans Cerebellum Diameter (TCD). Experimental results on training ($N = 164$) and test ($N = 46$) datasets of fetal MRI volumes yield a 95% confidence interval agreement of 3.70 mm, 2.20 mm and 2.40 mm for CBD, BBD and TCD, in comparison to measurements performed by an expert fetal radiologist. All results were below the interobserver variability, and surpass previously published results. Our method is generic, as it can be directly applied to other linear measurements in volumetric scans and can be used in a clinical setup.

Keywords: fetal MRI · Linear measurements · Reliability estimation

1 Introduction

Magnetic resonance imaging (MRI) is increasingly used to assess fetal brain development. Clinical assessment of fetal brain development based on MRI is

© Springer Nature Switzerland AG 2021
C. H. Sudre et al. (Eds.): UNSURE 2021/PIPPI 2021, LNCS 12959, pp. 210–220, 2021.
https://doi.org/10.1007/978-3-030-87735-4_20

mainly subjective and is complemented with a few biometric linear measurements [17]. Three key biometric linear measurements currently performed on fetal brain MRI are the Cerebral Biparietal Diameter (CBD), the Bone Biparietal Diameter (BBD), and the Trans Cerebellum Diameter (TCD) [18]. These measurements are used to assess fetal development according to the gestational age. They are manually acquired on individual MR reference slices by a fetal radiologist following guidelines that indicate how to establish the scanning imaging plane, how to select the reference slice in the MR volume for each measurement, and how to identify the two endpoint landmarks of the linear measurement [9].

Various methods have been developed for computing biometric linear measurements in 2D ultrasound images, e.g., the biparietal diameter [13], the fetal head circumference [12], and the fetus femur length [13]. Recently, Avisdris et al. [3] describe an automatic method for computing fetal brain linear measurements in MRI scans. The method mimics the radiologist manual annotation workflow, relies on a fetal brain segmentation and is based on measurement specific geometric heuristics for identifying the anatomical landmarks of each linear measurement. While it yields acceptable measurements, its reliance on accurate fetal brain segmentation and ad hoc heuristics may not always be robust.

Methods for the automatic computation of linear measurements of a structure in volumetric scans have been proposed in the past. For example, Yan et al. [21] describe a deep learning method for the computation of the length and width of a lesion following the RECIST guidelines. The method uses the Mask-RCNN network [10] to detect and segment each lesion from which the linear measurements are computed. The training segmentation masks are obtained from the ground truth measurements by fitting an ellipse bounded by the long and short axes measurement endpoints. This method is specific to lesions and RECIST measurements and is not applicable to fetal brain measurements.

Automatic landmark detection in images is a common task in a variety of computer vision applications, e.g., face alignment [20], pose estimation and in medical image analysis [16,22]. Two popular CNN-based methods consist of computing the spatial coordinates of each landmark by direct regression [22] or by heat map regression [16,20]. In the latter, the network computes a heat map defined by a Gaussian function centered at the landmark coordinates whose covariance describes the landmark location uncertainty. HRNet [20], a heat map regression network, achieves state of the art results in face landmark detection, human pose estimation, object classification and semantic segmentation.

Uncertainty estimation is an essential aspect of a variety of related tasks, e.g., classification, regression and segmentation with deep neural networks in computer vision [8] and medical image analysis [4]. Wang et al. [19] describes a test time augmentation (TTA) based uncertainty estimation method. TTA consists of generating similar new cases, computing the voxel predictions for each, and then obtaining the final voxel prediction on the original image by taking the mean or median value or voxel uncertainty by computing the entropy. This approach is not directly applicable to landmark detection. Payer et al. [15] describes a Gaussian-based uncertainty estimation method for landmark localization in hand X-ray images and in lateral cephalograms datasets. The method fits a Gaussian for each landmark from its predicted heat map. Their results show that the predicted uncertainties correlate with the landmark spatial error and the interobserver variability.

2   Method

We present a new deep learning method, called FML (Fetal Measurement by Landmarks), to automatically compute landmark-based linear measurements in a fetal brain MRI volume and to estimate their reliability. We demonstrate FML on 3 key fetal biometric measurements: CBD, BBD, TCD.

The method consists of four steps: 1) fetal brain region of interest (ROI) detection with a two-stage anisotropic U-Net; 2) reference slice selection with a 2D CNN; 3) linear measurement computation based on landmarks detection using a novel CNN, FMLNet; 4) measurement reliability estimation with a Bayesian score using a Gaussian Mixture Model (GMM).

2.1   Fetal Brain ROI Detection

The first step computes the fetal brain ROI in the fetal MRI volume with the method described in [6]. The ROI is a 3D axis-aligned bounding box that contains the fetal brain. The method uses a custom anisotropic 3D U-Net network trained with a Dice loss on a ×4 downsized fetal MRI volume. It outputs a coarse fetal brain segmentation from which a tight fitting ROI is computed.

2.2   Reference Slice Selection

The second step computes for each measurement, the reference slice of the fetal MRI volume on which the linear measurements will be performed from input fetal MRI volume with the method described in [3]. The method uses a slice-based 2D CNN that predicts for each slice in the ROI its probability to be a reference slice. It then selects the one with the highest probability. One CNN is trained for each measurement reference slice to detect the slice that was manually selected by the radiologist.

2.3   Linear Measurement Computation

The third step computes for each measurement the two anatomical landmarks of the measurement endpoints on the selected reference slice. The landmarks are computed using FMLNet, a variant of the HRNet network [20].

HRNet is a CNN type network whose key characteristic is that it maintains a high resolution representation of the input throughout network layers. This is achieved by connecting the high-to-low resolution convolution streams in parallel and by repeatedly spreading the intermediate results of each layer across the other layers at different resolutions.

We describe next the FMLNet architecture and its training and inference pipelines.

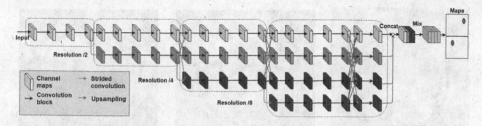

Fig. 1. Diagram of the FMLNet architecture. FMLNet is CNN network that consists of four streams (rows) at subsequently lower resolutions: full, 1/2, 1/4, 1/8 (each in a different color). Each stream consists four convolutional blocks (dotted boxes); in each block, boxes represent feature maps and arrows correspond to layers. After each block, the feature maps are combined across streams (red and green arrows). At the end of the blocks (two upper right 4-box clusters), the features maps of all four resolutions are concatenated and combined (pink box). The outputs are the two landmark Gaussian heat maps, one for each measurement endpoint (top rightmost box with red ovals). (Color figure online)

FMLNet Architecture: Figure 1 shows the architecture of FMLNet. It is a CNN that combines the representations from four high-to-low resolution parallel streams into a single stream. The representations are then input to a two-layer convolution classifier. The first layer combines the feature maps of all four resolutions; the second layer computes a Gaussian heat map for each of the two landmark endpoint. One network is trained for each measurement with the Mean Squared Error (MSE) loss between the Gaussian maps created from the ground truth measurement landmarks and the predicted heat maps. At inference time, the two measurement landmark locations are defined by the coordinates of the pixel with the maximal value on each heat map.

FMLNet Training: Three FMLNet networks are trained, one for each of the linear measurements, CBD, BBD and TCD. The input is the reference slice image; the outputs are the two measurement endpoint locations on the image.

Two training time augmentations are used: 1) rotations around the image center at angles randomly sampled in the $[-180, 180]^o$ range; 2) image scaling at scales randomly sampled in the $[-5, +5]\%$ range. In addition, landmark class (left/right) reassignment (LCR) is performed on the resulting landmarks.

Landmark class reassignment is necessary because rotations may cause the left/right labeling of the two measurement landmarks to be inconsistent with the image coordinates, i.e. the left and right points may be switched, which will hamper the network training. This inconsistency is corrected by performing landmark class reassignment (Fig. 2). During each training epoch, the left/right assignment for each rotated image is verified with respect to the image coordinate system and if needed, is corrected by switching the left/right labels.

Note that these augmentations are different than the ones used in the original HRNet. Unlike faces, which are almost always vertical, the fetal brain can be in any orientation, so the full range of rotations (beyond flipping) should be accounted for. The network is trained for 200 epochs on a batch size of 16

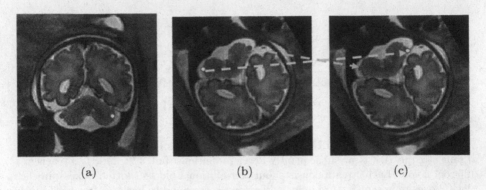

(a)   (b)   (c)

Fig. 2. Illustration of the landmark class reassignment: (a) reference slice image with ground-truth left (blue) and right (green) landmarks; (b) image after rotation with inconsistent landmark labeling (left/right switched); (c) reassignment of labels. (Color figure online)

images with the ADAM optimizer [14] with an initial learning rate of 10^{-4} and a dropping factor of 0.2 in epochs 10, 40, 90, and 150.

FMLNet Inference: the inference pipeline consists of three steps (Fig. 3): 1) test time augmentation (TTA) of the reference slice image; 2) landmarks location prediction with FMLNet; and 3) robust landmarks fusion (RLF).

1. **Test time augmentation:** new reference slice images are generated with a set of reversible spatial transformations $T = \{t_i\}$ applied to the original image I. Rotation transformations are applied at a equally spaced angles in the $[0, 360]^o$ range (in our case, $a = 12$). The result is a set of transformed images $I' = \{I'_i = t_i(I)\}$.

2. **Landmarks prediction with FMLNet:** two measurement landmarks, $L'_i = \{l_i'^{(k)} | k \in K\}, K = \{left, right\}$ are computed for each image I'_i with the trained FMLNet. The resulting landmark predictions, L'_i, are then mapped back to their original location on the reference slice image by applying the reverse spatial transformation t_i^{-1} to the corresponding image I'_i. The result is a set of landmarks $L = \{l_i^{(k)} | i \in [1..a], k \in K\}$ in the original image coordinates.

3. **Robust landmarks fusion:** the final landmark predictions, $l^{(right)}, l^{(left)}$, are computed from the landmarks prediction set L with the Density Based Spatial Clustering of Applications with Noise (DBSCAN) [7] algorithm. First, DBSCAN clusters together a minimum of q points that are within a pre-defined distance of d between them (in our case, $q = 4$ and $d = 2$ pixels). Next, outlier points outside clusters are discarded, and two point clusters corresponding to the left and right measurement landmark endpoints are computed with the K-means algorithm [2]. Finally, the left and right landmark coordinates are obtained by computing the centroid of the points in each cluster.

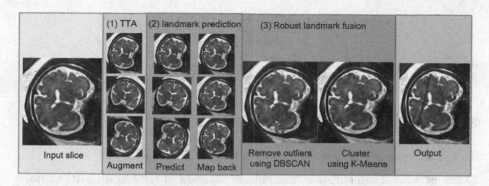

Fig. 3. FMLNet inference pipeline. The input is the reference slice of the measurement; the outputs are the measurement endpoints and the measurement value (blue line). The pipeline consists of: 1) test time augmentation of the reference slice image (TTA, illustrated with three augmentations); 2) landmarks prediction with FMLNet, and; 3) robust landmarks fusion. The dots on the images corresponds to the left (blue), right (green), unassigned (yellow) and outlier (red cross) landmark predictions. (Color figure online)

2.4   Measurement Reliability Estimation

The fourth step estimates the reliability of the landmarks predictions using the set of landmark predictions L generated in the landmark prediction step of the FMLNet inference. It models the landmarks location distribution, irrespective of their landmark class (left/right), with a bi-modal Gaussian Mixture Model (GMM). The GMM Bayesian likelihood is computed to obtain an estimate of the prediction reliability. When likelihood value is low, the landmark locations are spatially dispersed, so their distribution is not bi-modal (two clusters). In this case, the measurement is labeled as unreliable and should be performed manually by an expert radiologist.

Formally, the GMM is defined as $X \sim \sum_{k \in K} \pi_k N(\mu_k, \Sigma_k)$, where for each of the two landmark clusters k (left/right), π_k is the cluster probability, $N(\mu_k, \Sigma_k)$ is a multivariate Gaussian distribution, $\mu_k \in \mathbb{R}^2$ is the cluster mean location and $\Sigma_k \in \mathbb{R}^{2 \times 2}$ is the cluster location covariance.

We estimate the GMM parameters, $\Phi = (\pi_k, \mu_k, \Sigma_k)$, from the set of landmark predictions L by Expectation-Maximization (EM) [5]. The mean values of each cluster location are initialized with the final landmark locations, e.g., $\mu_k = l^{(k)}$, and are as described in the robust landmark fusion inference step. To estimate the reliability of the predicted landmark points in L, we compute the GMM log-likelihood score:

$$LLscore(\Phi|L) = \sum_{l \in L} \sum_{k \in K} \log(\pi_k N(l|\mu_k, \Sigma_k))$$

where l is a point in L (the labels left/right are ignored), and $N(l|\mu_k, \Sigma_k)$ is the Gaussian probability of cluster k for l.

3   Experimental Results

To evaluate our method, we conducted three studies on fetal MRI dataset.

Dataset and Annotations: The dataset consists of fetal brain MRI volumes acquired with the FRFSE protocol at the Sourasky Medical Center (Tel Aviv, Israel) as part of routine fetal assessment. The dataset includes 210 MRI volumes of 154 singleton pregnancies (cases) with mean gestational age of 32 weeks (std $= 2.8$, range 22–38). Of these, 113 volumes (87 cases) were diagnosed as normal, and 107 volumes (67 cases) as abnormal. To allow direct comparison, we use the same train/test splits of 164/46 volumes (121/33 disjoint cases) as in [3]. CBD, BBD and TCD measurements for all volumes were manually performed by a senior pediatric neuro-radiologist.

Studies: We conducted three studies. Study 1 evaluates the accuracy of the FMLNet method and the contribution of its various components. Study 2 analyzes the impact of selected reference slices. Study 3 evaluates the measurement reliability estimation.

In all studies, we use the following metrics: L_1 difference, bias and agreement. For two sets of n measurements, $M_1 = \{m_i^1\}$, $M_2 = \{m_i^2\}$, m_i^1 and m_i^2 ($1 \leq i \leq n$) are two values of the measurement, e.g., ground-truth and computed. The difference between two linear measurement sets M_1, M_2 is defined as $L_1(M_1, M_2) = 1/n \sum_{i=1}^{n} |d_i|$, where the difference between each measurement is $d_i = m_i^1 - m_i^2$. For repeatability estimation, we use the Bland-Altman method [1] to estimate the bias and agreement between two observers. Agreement is defined by the 95% confidence interval $CI_{95}(M_1, M_2) = 1.96 \times \sqrt{1/n \sum_{i=1}^{n} (L_1(M_1, M_2) - d_i)^2}$. The measurements bias is defined as $Bias(M_1, M_2) = 1/n \sum_{i=1}^{n} d_i$. These three metrics represent different aspects of algorithm performance.

Study 1: Accuracy Analysis and Ablation Study. We evaluate the accuracy of the FMLNet method and the contribution of its three main components: test time augmentation (TTA), robust landmark fusion (RLF) and landmark class reassignment (LCR). We compare its performance to that of HRNet, to the geometric method [3] and to the interobserver variability on the test dataset and its ground-truth values.

Table 1 shows the results. The original HRNet (row 2) performs poorly, as it yields high measurement agreement CI_{95} and difference L_1 values. Removing each one of algorithmic components from FMLNet - TTA (row 3), RLF (row 4) or LCR (row 5), yields better results than standalone HRNet, but still not acceptable. FMLNet with all its components (row 6) yields the best results, which are always reliable (46 out of 46 for CBD, BBD, TCD). Using a two-sided

Table 1. Study 1 and 2 results for the CBD, BBD, TCD fetal brain measurements. For each, the number of MR volumes N, Bias, agreement CI_{95}, and difference L_1 with respect to the manual annotation are listed. Row 1 lists the variability of manual measurements of two radiologists. Rows 2–5 list the results of the ablation study including HRNet (standalone), FMLNet without test time augmentation (-TTA), without robust landmark fusion (-RLF) and without landmark class reassignment (-LCR). Row 6–8 list FMLNet results, while row 6 is for all test set and 7, 8 is for normal and abnormal cases, respectively. Row 9 list the results of FMLNet on ground truth reference slice (FMLNet+). Rows 10–11 list geometric method results on the predicted reference slice, and on ground truth reference slice (Geometric+). Bold face results indicates the best results for each metric.

Measurement	CBD				BBD				TCD			
Metrics	N	Bias [mm]	CI_{95} [mm]	L_1 [mm]	N	Bias [mm]	CI_{95} [mm]	L_1 [mm]	N	Bias [mm]	CI_{95} [mm]	L_1 [mm]
Interobserver	45	0.03	4.12	1.60	45	−0.09	3.18	1.27	45	0.26	2.39	0.97
HRNet	**46**	−10.40	54.36	13.37	**46**	−8.42	45.82	9.51	**46**	−3.96	28.97	6.21
FMLNet												
-TTA	**46**	−0.70	20.06	3.38	**46**	−1.17	23.90	3.20	**46**	−3.59	24.83	5.83
-RLF	**46**	−1.45	8.58	3.11	**46**	−0.90	6.28	2.00	46	0.72	4.60	1.78
-LCR	39	0.72	5.35	1.85	44	−2.48	22.14	3.43	45	2.28	9.45	2.49
FMLNet	**46**	0.57	3.70	1.60	**46**	0.21	2.20	0.90	**46**	0.88	2.40	1.16
*Normal	25	0.27	3.62	1.56	25	−0.05	2.00	0.85	25	0.69	1.94	0.97
*Abnormal	21	0.92	3.52	1.66	21	0.52	2.35	1.05	21	1.12	2.79	1.39
FMLNet+	**46**	0.20	**2.32**	**0.96**	45	**0.00**	**2.08**	**0.89**	**46**	0.70	**2.15**	1.06
Geometric	40	**0.00**	3.63	1.28	40	−0.80	3.58	1.28	37	−0.51	2.45	**0.89**
Geometric+	**46**	−0.02	3.99	1.47	**46**	−0.50	3.07	1.17	**46**	−0.56	2.58	1.09

t-test, no statistically significant difference was found in the performance between Normal (row 7) and Abnormal cases (row 8) in all measurements ($p > 0.05$).

Comparison of our method results with those of the previously published geometric method [3] with reliability estimation (row 10) shows that for all three measurements, FMLNet (row 6) performs better in terms of reliable measurements on the number of MR volumes N, agreement CI_{95} and difference L_1. Specifically, for TCD, our method yields reliable measurements in all cases ($N = 46$) with comparable agreement CI_{95}, while the geometric method fails in 20% of the cases (9 out of 46). Using a two-sided t-test, in BBD and TCD our method performs significantly better ($p < 0.01$). These results shows superiority of our new method.

Study 2: Impact of Selected Reference Slices. Differences were detected between the reference slices selected by the radiologist and the slice selected by the algorithm on the test set of: 20/46 cases in CBD/BBD slices and 12/46 cases in TCD slices. Table 1 shows the measurements accuracy results. When analyzing the impact of selected reference slices on the accuracy of measurements (FMLNet - uses the slices selected by the algorithm vs FMLNet+ uses the slices selected by radiologist) - the main impact was on CBD measurements (almost doubles the variability), while the TCD and BBD accuracy remains similar. A

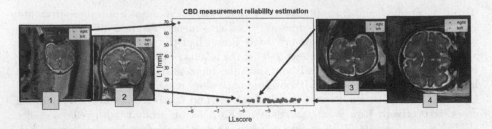

Fig. 4. CBD Measurement reliability estimation results. The plot graph (center) shows the log-likelihood score *LLscore* (horizontal axis) with respect to the measurement difference L_1(vertical axis) for the test set cases (blue dots). Representative images of unreliable (left) and reliable (right) measurements: (1) failure of brain ROI detection; (2) blurry image; (3) reliable predictions with outliers; and (4) reliable predictions. (Color figure online)

possible explanation is that the TCD and BBD are measured on smooth contours (cerebellum and skull, respectively) while CBD is measured on a more complex contour, the brain sulcation and gyri. Furthermore, when normalizing the interobserver variability in the absolute value of measurements, the relative error variability results in 6% for CBD and 4% for BBD and TCD. For the geometric algorithm (Geometric vs Geometric+), no significant differences were observed when working on slices selected by a radiologist. These results show that improving the reference slice selection algorithm may improve the accuracy of the automatic measurements.

Study 3: Measurement Reliability Estimation. Table 1 shows the number of cases (column N) for each measurement for which the computed values are reliable.

We evaluate the reliability estimation by computing the difference between the measurements computed and ground truth values, L_1, and by establishing their correlation with the *LLscore*, on the test set cases. Figure 4 illustrates this correlation for the CBD measurement. For all three measurements, we observe that *LLscore* correlates well with L_1. We establish from the plot graphs the threshold value of *LLscore* that adequately discriminates between reliable and unreliable computed measurements – set to −5.75 for all three measurements.

4   Conclusion

We have presented a new fully automatic method to compute landmark-based linear measurements in a fetal brain MRI volume, to estimate their reliability, and to identify the unreliable measurements that should be obtained manually for the case at hand. The computed reliability estimation value can be used to rank the measurement predictions for inspection and validation by a radiologist.

We demonstrate our method on three key fetal biometric measurements from fetal brain MRI scans, CBD, BBD, TCD, and show that it yields state-of-the-art results for all three measurements, within the interobserver variability. These results also comparable to the best reported results for fetal head circumference (HC) measurements in ultrasound images of the HC18 challenge [11] (mean L_1 of 1.72 mm, CI_{95} of 3.16 mm).

The main novelties of our method are three-fold. First, it robustly handles the wide span of fetal brain orientations and correctly identifies left and right measurement landmark endpoints by test time augmentation, robust landmark fusion, and landmark class reassignment. Second, it directly computes landmarks by heat map regression, obviating the need for structure segmentation and its data annotation effort [3,21]. Third, it computes landmarks location uncertainty estimation with a new method that combines test time augmentation [19] and landmark Gaussian-based uncertainty estimation [15] and that simultaneously computes the estimates on multiple landmarks with a GMM instead of using a single Gaussian for each landmark, thereby yielding a single reliability score.

The advantages of our method are that it only requires a small number ($\sim$150) of manual linear measurements for the training dataset, that it does not rely on heuristics to identify the landmarks, that it does not require fetal brain structures segmentation, and that it is robust since it incorporates reliability estimation. Note that our method is generic, i.e., it is not tailored to specific linear measurements, so it can be applied directly to other measurements.

Acknowledgments. This research was supported in part by Kamin Grants 72061 and 72126 from the Israel Innovation Authority.

References

1. Altman, D.G., Bland, J.M.: Measurement in medicine: the analysis of method comparison studies. J. R. Stat. Soc. Ser. D (Stat.) **32**(3), 307–317 (1983)
2. Arthur, D., Vassilvitskii, S.: K-means++: the advantages of careful seeding. In: Proceedings of ACM-SIAM Symposium on Discrete Algorithms (2007)
3. Avisdris, N., et al.: Automatic linear measurements of the fetal brain with deep neural networks. Int. J. Comput. Assist. Radiol. Surg. (2021). https://doi.org/10.1007/s11548-021-02436-8
4. Ayhan, M.S., Berens, P.: Test-time data augmentation for estimation of heteroscedastic aleatoric uncertainty in deep neural networks. In: International Conference on Medical Imaging with Deep Learning (2018)
5. Dempster, A.P., Laird, N.M., Rubin, D.B.: Maximum likelihood from incomplete data via the EM algorithm. J. R. Stat. Soc. Ser. B **39**(1), 1–22 (1977)
6. Dudovitch, G., Link-Sourani, D., Sira, L.B., Miller, E., Bashat, D.B., Joskowicz, L.: Deep learning automatic fetal structures segmentation in MRI scans with few annotated datasets. In: Proceedings of the International Conference on Medical Image Computing and Computer Assisted Intervention (2020)
7. Ester, M., Kriegel, H.P., Sander, J., Xu, X.: A density-based algorithm for discovering clusters in large spatial databases with noise. In: Proceedings of the International Conference on Knowledge Discovery and Data Mining (1996)
8. Gal, Y., Ghahramani, Z.: Dropout as a Bayesian approximation: representing model uncertainty in deep learning. In: Proceedings of International Conference on Machine Learning (2016)

9. Garel, C.: MRI of the Fetal Brain. Springer, Heidelberg (2004). https://doi.org/10.1007/978-3-642-18747-6

10. He, K., Gkioxari, G., Dollár, P., Girshick, R.: Mask R-CNN. In: Proceedings of International Conference on Computer Vision (2017)

11. van den Heuvel, T.L.: HC18 challange leaderboard (2021). https://hc18.grand-challenge.org/evaluation/challenge/leaderboard/. Accessed 29 June 2021

12. van den Heuvel, T.L., de Bruijn, D., de Korte, C.L., van Ginneken, B.: Automated measurement of fetal head circumference using 2d ultrasound images. PloS ONE **13**(8), e0200412 (2018)

13. Khan, N.H., Tegnander, E., Dreier, J.M., Eik-Nes, S., Torp, H., Kiss, G.: Automatic detection and measurement of fetal biparietal diameter and femur length-feasibility on a portable ultrasound device. Open J. Obstetr. Gynecol. **7**(3), 334–350 (2017)

14. Kingma, D.P., Ba, J.: Adam: a method for stochastic optimization. In: Proceedings of International Conference Learning Representations (2015)

15. Payer, C., Urschler, M., Bischof, H., Štern, D.: Uncertainty estimation in landmark localization based on Gaussian heatmaps. In: Sudre, C.H., et al. (eds.) UNSURE/GRAIL -2020. LNCS, vol. 12443, pp. 42–51. Springer, Cham (2020). https://doi.org/10.1007/978-3-030-60365-6_5

16. Payer, C., štern, D., Bischof, H., Urschler, M.: Integrating spatial configuration into heatmap regression based CNNs for landmark localization. Med. Image Anal. **54**, 207–219 (2019)

17. Prayer, D., et al.: ISUOG practice guidelines: performance of fetal magnetic resonance imaging. Ultrasound Obstetr. Gynecol. **49**(5), 671–680 (2017)

18. Salomon, L., et al.: ISUOG practice guidelines: ultrasound assessment of fetal biometry and growth. Ultrasound Obstetr. Gynecol. **53**(6), 715–723 (2019)

19. Wang, G., Li, W., Aertsen, M., Deprest, J., Ourselin, S., Vercauteren, T.: Aleatoric uncertainty estimation with test-time augmentation for medical image segmentation with convolutional neural networks. Neurocomputing **338**, 34–45 (2019)

20. Wang, J., et al.: Deep high-resolution representation learning for visual recognition. IEEE Trans. Pattern Anal. Mach. Intell. **43**, 3349–3364 (2019)

21. Yan, K., et al.: MULAN: multitask universal lesion analysis network for joint lesion detection, tagging, and segmentation. In: Shen, D., et al. (eds.) MICCAI 2019. LNCS, vol. 11769, pp. 194–202. Springer, Cham (2019). https://doi.org/10.1007/978-3-030-32226-7_22

22. Zhang, J., Liu, M., Shen, D.: Detecting anatomical landmarks from limited medical imaging data using two-stage task-oriented deep neural networks. IEEE Trans. Pattern Anal. Mach. Intell. **26**(10), 4753–4764 (2017)

CAS-Net: Conditional Atlas Generation and Brain Segmentation for Fetal MRI

Liu Li[1]([✉]), Matthew Sinclair[1], Antonios Makropoulos[1], Joseph V. Hajnal[2], A. David Edwards[2], Bernhard Kainz[1,2,3], Daniel Rueckert[1,4], and Amir Alansary[1]

[1] BioMedIA Group, Department of Computing, Imperial College London, London, UK
liu.li20@imperial.ac.uk
[2] King's College London, London, UK
[3] FAU Erlangen–Nürnberg, Erlangen, Germany
[4] Technical University of Munich, Munich, Germany

Abstract. Fetal Magnetic Resonance Imaging (MRI) is used in prenatal diagnosis and to assess early brain development. Accurate segmentation of the different brain tissues is a vital step in several brain analysis tasks, such as cortical surface reconstruction and tissue thickness measurements. Fetal MRI scans, however, are prone to motion artifacts that can affect the correctness of both manual and automatic segmentation techniques. In this paper, we propose a novel network structure that can simultaneously generate conditional atlases and predict brain tissue segmentation, called CAS-Net. The conditional atlases provide anatomical priors that can constrain the segmentation connectivity, despite the heterogeneity of intensity values caused by motion or partial volume effects. The proposed method is trained and evaluated on 253 subjects from the developing Human Connectome Project (dHCP). The results demonstrate that the proposed method can generate conditional age-specific atlas with sharp boundary and shape variance. It also segment multi-category brain tissues for fetal MRI with a high overall Dice similarity coefficient (DSC) of 85.2% for the selected 9 tissue labels.

1 Introduction

The perinatal period is an important time for the study of human brain development. Cellular connections start to form across the brain, and the cerebral cortex becomes more complex. These brain structure developments are closely related to the formation of human cognitive functions [5]. Magnetic Resonance Imaging (MRI) plays an essential role in fetal diagnosis and for studying neurodevelopment, as it can capture different brain tissues in detail compared to conventional fetal Ultrasound [11].

Brain tissue segmentation is important for the quantitative evaluation of the cortical development, and is a vital step for standard surface reconstruction pipelines [19]. However, during the scanning process, the fetus is moving and not

C. H. Sudre et al. (Eds.): UNSURE 2021/PIPPI 2021, LNCS 12959, pp. 221–230, 2021.
https://doi.org/10.1007/978-3-030-87735-4_21

sedated and the mother breathes normally, which can produce motion artifacts. Recent works for super-resolution and motion correction in fetal MRI [1,10,16] can reconstruct the scanned image with less motion artifacts between the slices; nevertheless, in-plane motion can remain.

Other limitations are partial volume effects and lower signal-to-noise-ratio caused by the small size of the brain. Furthermore, the structure of the fetal brain has a large shape variance because of the rapid development during the perinatal period. This can hinder the learning of automatic segmentation models, especially using a limited number of training subjects from a wide age range.

In order to address problems caused by poor image quality and inaccurate training labels, we propose a novel architecture that learns to predict the segmentation maps and a conditional atlas simultaneously in an end-to-end pipeline. The atlas enables the model to learn anatomical priors without depending solely on the intensity values of the input image. This can improve the segmentation performance especially if there is no gold standard label for training due to the poor image quality.

1.1 Related Work

Traditionally, the two tasks of atlas generation and tissue segmentation are learned separately. For tissue segmentation, atlas-based segmentation results are obtained by warping the atlas label maps to the target image with a restrained deformation field [4,18]. Generating an atlas with higher similarity to the target image will ease the calculation for the deformation field thus improving the segmentation performance. For atlas generation, the atlas is usually generated by registering and aggregating all the images to the same atlas coordinate space. The performance of these methods largely relies on the initial similarity of the atlas and the testing samples [18]. Since atlas-based approaches require expensive registration optimization, the processing time may take hours to days for a single subject.

With the evolution of deep learning methods, UNet-based [21] architectures have defined state-of-the-art performance for segmentation [2,14,24], including fetal brain scans [9,15,20]. Such methods can be sensitive to the consistency of the intensity distribution of the input images and the accuracy of the segmentation label maps used for training. Fetit et al. [13] proposed a segmentation pipeline for neonatal CGM that utilized the silver labels from the DrawEM method [18], and further used a small part of the manually labeled slices to refine the segmentation results.

Dalca et al. [7] trained an additional auto-encoding variational network to encode the anatomical prior in a decoder. However, the anatomical priors are implicitly encoded in these methods, which can be difficult to interpret manually. Recently, Sinclair et al. [23] proposed a network that jointly learns segmentation and registration, and atlas construction for cardiac and brain scans. It explicitly learns the shape prior based on a structure-guided image registration [17].

Contributions: Inspired by previous works [7,13,23], we propose a novel segmentation network that utilizes anatomical priors conditioned on an age-specific

4D atlas to address the challenges caused by noisy labels and bad image quality. The main contributions of this work include:

- Novel multi-task network for learning end-to-end tissue segmentation and conditional atlas generation simultaneously, called CAS-Net.
- Age-conditioned 4D atlas construction that is constrained by the smoothness and continuity of the diffeomorphic transformation field.
- Detailed evaluation of the quality of the spatio-temporal atlas and the performance of the segmentation, whereas the proposed CAS-Net outperforms other baseline methods with an overall 85.2% dice score.

2 Method

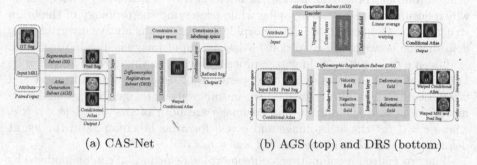

(a) CAS-Net   (b) AGS (top) and DRS (bottom)

Fig. 1. The main architecture of the proposed CAS-Net for brain tissue segmentation and conditional atlas generation (a). The detailed architecture of the atlas generation (AGS) and diffeomorphic registration (DRS) subnets (b).

CAS-Net learns end-to-end the conditional atlas generation and tissue segmentation, whereas the age-specific atlas can be considered as an anatomical prior for the segmentation task. It consists of three main sub-networks: segmentation subnet (SS), atlas generation subnet (AGS), and diffeomorphic registration subnet (DRS), see Fig. 1.

The input of the CAS-Net, Fig. 1a, is the paired MR image, segmentation labelmap and the conditional attribute (*e.g.*, gestational age) from the training set $\mathcal{D} = \{I_i, S_i, a_i\}$, where i is the index of training samples. Note that the MRI and its labelmap are in the size of $l \times w \times h \times 1$ and $l \times w \times h \times c$, where l, w and h are the length, width and height of the 3D volume, and c is the number of anatomical labels in the segmentation maps. In our setting, the conditional attribute is the gestational age (GA), which is used to generate an age-specific atlas for studying the development of brain structure. However, other attributes such as sex or pathologies camayn can also be used for conditional atlas generation.

The input of the AGS, Fig. 1b, are the GA attribute a_i, the global averaging atlas image A_g and its labelmap A_g^s, and the output are the conditional atlas image A and its labelmap A^s. At the same time, the segmentation subnet takes the MR image I_i as an input, and predicts the multi-label segmentation map $\hat{S}_i$.

As such, both I_i and $\hat{S}_i$ are in the image space, and conditional atlas space (A and A^s). Then, the four images (I_i, $\hat{S}_i$, A and A^s) are concatenated along the channel axis as the input for the diffeomorphic registration subnet. Based on the concatenated input, a deformation field Φ_i is predicted by the DRS to warp the conditional atlas to the image space, and output the atlas-based segmentation result $\hat{S}_i^a$. Finally, $\hat{S}_i$ and $\hat{S}_i^a$ are merged by a convolutional layer, i.e., a 3D convolutional layer with the kernel size of $1 \times 1 \times 1$, to output the final refined segmentation results $\hat{S}_i^r$.

Diffeomorphic Registration Subnet (DRS): The purpose of the DRS is to learn a deformation field that registers the conditional atlas to the target input image, so that the atlas label maps can also be warped and propagated to segment the target image. Inspired by [8,23], we model the registration step using a diffeomorphic deformation parameterized by a stationary velocity field, which ensures a one-to-one mapping while preserving the topology of the brain structure [3]. Given that the diffeomorphic deformation is always invertible, the deformation between the conditional atlas and input image is also bidirectional.

Given a deformation field Φ, the process of warping the moving atlas can be formulated as $A \circ \Phi$, where $\circ$ is the operation of using the values from A and voxel locations from Φ to calculate the output warped image. Note that the 3D moving atlas and the output warped image are both of size of $l \times w \times h \times c$, where $c = 1$ for the atlas image and $c = k$ for atlas labelmap, and the learnt deformation field Φ is in the size of $l \times w \times h \times 3$.

The procedure for computing a diffeomorphic deformation can be modeled by an ordinary differential equation (ODE) parameterized with a stationary velocity field V [3]:

$$\frac{\mathrm{d}\Phi}{\mathrm{d}t} = V(\Phi^{(t)}). \tag{1}$$

As shown in Fig. 1b (bottom), V is simulated by an encoder-decoder structure with the concatenated input. Given the velocity field, the final deformation field $\Phi^{(t=1)}$ is obtained by integrating the stationary velocity field V over unit time:

$$\Phi^{(t=1)} = \Phi^{(t=0)} + \int_{t=0}^{t=1} V(\Phi^{(t)})\mathrm{d}t. \tag{2}$$

Here, $\Phi^{(t=0)}$ is the initial condition of identity deformation, i.e., $X \circ \Phi_{t=0} = X$, where X is a map with the size of $l \times w \times h \times c$.

In practice, the computation of the integration is approximated by the Euler methods with small time steps h, and is interpolated by the scaling and squaring layer in CAS-Net, similar to [6]. Specifically, the integration is recursively calculated with small time steps h as:

$$\Phi^{(t+h)} = \Phi^{(t)} + \int_{t}^{t+h} V(\Phi^{(t)})\mathrm{d}t \approx \Phi^{(t)} + hV(\Phi^{(t)}) = (x + hV) \circ \Phi^{(t)}, \tag{3}$$

with $x = \Phi^{(0)}$. When the integration steps in the scaling and squaring layer is the power of 2, given the predicted velocity field V, the final deformation field $\Phi^{(1)}$ can be calculated iteratively through Euler integration.

Since the diffeomorphic registration is invertible, the image can be warped to the atlas space using the inverse deformation field $\boldsymbol{\Phi}_i^{-1}$, which is calculated by integrating the negative velocity field $-\boldsymbol{V}_i$, as shown in Fig. 1b (bottom).

Atlas Generation Subnet (AGS): The main task of the AGS is to generate an age specific deformation field that can warp the global average atlas to be age-specific. Different to [12], which only constructs the conditional atlas image by adding a displacement field, AGS predicts the deformation field, atlas image and labelmap simultaneously. The age-specific atlas labelmap is then used by the DRS to predict the atlas-based segmentation result.

The structure of this subnet is shown in Fig. 1b (top). The input attribute is first decoded from the low-dimensional attribute space $\mathbf{a}$ to a high-dimensional feature space $\boldsymbol{Q}_i$. Specifically, the decoder consists of a fully connected layer, an upsampling layer and several convolutional layers. In order to deal with the problem of the imbalanced distribution of GA, we divide the training samples into 4 age groups, and encode this attribute in a one-hot label (instead of directly encoding the GA as a continuous scalar). Based on the decoder feature map $\boldsymbol{Q}_i$, we also model the deformation with a diffeomorphism parameterized by a velocity field, where $\boldsymbol{Q}_i$ is treated as the velocity field. Then, the age-specific $\boldsymbol{Q}_i$ is integrated by the scaling and squaring layer $\boldsymbol{\Psi}_i$. Consequently, the conditional atlas image and labelmap are constructed as: $A = A_{\mathrm{g}} \circ \boldsymbol{\Psi}_i$, and $A^{\mathrm{s}} = A_{\mathrm{g}}^{\mathrm{s}} \circ \boldsymbol{\Psi}_i$. Here, A_{g} and $A_{\mathrm{g}}^{\mathrm{s}}$ are the global atlas image and labelmap, respectively, which are initialized by averaging all the input images I_i and their corresponding labelmaps S_i in the training set.

In addition, since the deformation between global atlas space to conditional atlas space ($\boldsymbol{\Phi}_i$ and $\boldsymbol{\Phi}_i^{-1}$) and the deformation between conditional atlas space and input MRI space ($\boldsymbol{\Psi}_i$ and $\boldsymbol{\Psi}_i^{-1}$) are diffeomorphic, following [23], the global atlas and its labelmap in our CAS-Net are also updated at the end of each epoch to improve the segmentation performance using: $A_{\mathrm{g}}^{(j)} = \frac{1}{N} \sum_{i=1}^{N} I_i \circ \boldsymbol{\Phi}_i^{-1(j)} \circ \boldsymbol{\Psi}_i^{-1(j)}$, and $A_{\mathrm{g}}^{\mathrm{s}(j)} = \frac{1}{N} \sum_{i=1}^{N} S_i \circ \boldsymbol{\Phi}_i^{-1(j)} \circ \boldsymbol{\Psi}_i^{-1(j)}$. Here, j is the index of epoch, and N is the number of the training samples.

Loss Function: In order to achieve end-to-end training, the training process is optimized using four loss terms, namely, segmentation loss $\mathcal{L}_{\mathrm{S}}$, registration loss $\mathcal{L}_{\mathrm{R}}$, combination loss $\mathcal{L}_{\mathrm{C}}$ and the regularization term $\mathcal{L}_{\mathrm{Reg}}$.

The segmentation loss is a standard L^2-norm between the predicted $\hat{S}_i$ and groundtruth labels S_i, and used to update the parameters of the segmentation subnet.

The parameters of the AGS and DRS are supervised by L_{R} in both image and labelmap space as follows:

$$
\begin{aligned}
\mathcal{L}_{\mathrm{R}} &= \lambda_{\mathrm{l}} \| A^{\mathrm{s}} \circ \boldsymbol{\Phi}_i - S_i \|_2 + \lambda_{\mathrm{i}} \| A \circ \boldsymbol{\Phi}_i - I_i \|_2 \\
&= \lambda_{\mathrm{l}} \| A_{\mathrm{g}}^{\mathrm{s}} \circ \boldsymbol{\Psi}_i \circ \boldsymbol{\Phi}_i - S_i \|_2 + \lambda_{\mathrm{i}} \| A_{\mathrm{g}} \circ \boldsymbol{\Psi}_i \circ \boldsymbol{\Phi}_i - I_i \|_2.
\end{aligned}
\tag{4}
$$

Here, λ_{i} and λ_{l} are the weights for image and labelmap space loss. The first term is beneficial for the generated atlas quality, while the second term contributes to the accuracy of the segmentation. At the beginning of the training, λ_{i} has higher values in order to learn an accurate conditional atlas, and later λ_l is increased for a better segmentation performance.

The combination loss ($\mathcal{L}_C$) is defined as the L^2-norm between the refined $\hat{S}_i^r$ and ground truth S_i segmentation.

In order to preserve the topology of the warped images, L_{Reg} is defined as: $\lambda_g \|\nabla U_i\|_2 + \lambda_d \|U_i\|_2 + \lambda_m \|\bar{U}_i\|_2$, which regularizes the continuity and smoothness of the predicted deformation field [12]. U_i represents the displacement field ($\Phi^{(t=1)} - \Phi^{(t=0)}$), and λ_g, λ_d and λ_m are the hyper-parameters for tuning the weight of the regularization loss. Finally, the overall loss $\mathcal{L}$ is the linear combination of all four losses: $\mathcal{L} = \mathcal{L}_S + \mathcal{L}_R + \mathcal{L}_C + \mathcal{L}_{Reg}$.

3   Evaluation and Results

Data: We train and validate our model on 274 T2 fetal MRI scans (253 patients) from the Developing Human Connectome Project (dHCP)[1]. These images are randomly split into 202 training images (from 186 subjects), 18 validation images (from 18 subjects), and 54 testing images (from 49 subjects). Note that there is no overlap of any single subject in different subsets. The GA attribute in this dataset ranges from 20.6 to 38.2 weeks. In order to improve the performance of learning-based deformation, all the input images are affinely aligned to a coarse fetal atlas [22].

We use the revised segmentation results provided by the dHCP pipeline using DrawEM [18] as the ground truth for training and evaluation, from which the atlas used in DrawEM is advanced to a fetal atlas [22] to improve the segmentation performance. The ground truth label maps can be imperfect because of the fetal image artifacts, which is the main motivation of using a conditional atlas as an anatomical prior. The segmentation labels consist of nine classes, namely, cerebrospinal fluid (CSF), cortical gray matter (CGM), white matter (WM), outliers, ventricles, cerebellum, deep grey matter (DGM), brainstem, and hippocampus.

Table 1 shows the selected hyper-parameters used in the CAS-Net. All the experiments are conducted on a PC with an NVIDIA GTX 3080 GPU. Our model took 2 hours for 500 training epochs (generating the conditional atlas), and took around 2 seconds during inference per MRI (outputting the segmentation results), which is much faster comparing to traditional atlas generation and brain tissue segmentation.

Evaluation: A 3D-UNet is used as a baseline for evaluating the segmentation performance compared to different variants of the CAS-Net. In order to compensate for the class imbalance, the loss term for the 3D-UNet is re-scaled for the different tissues according to the average volume of each tissue. Table 2 demonstrates the Dice Similarity Coefficient (DSC) scores used to evaluate the accuracy of the results. It also shows that the proposed CAS-Net achieves a higher overall average accuracy of 85.2%, compared to 70.1% using a 3D-UNet. Furthermore,

---

[1] http://www.developingconnectome.org/.

Table 1. Hyper-parameters in CAS-Net.

Hyperparameters	l, w, h	c	T	λ_i	λ_l	λ_g	λ_d	λ_m
Value	64	10	6	2 <200 epoch 1 ≥200 epoch	1 <200 epoch 2 ≥ 200 epoch	200	500	200

Table 2. The segmentation performance of the different variants of CAS-Net compared to a 3D-UNet baseline in terms of Dice similarity coefficient (%). The metrics are presented in the format of mean and standard deviation (sd).

Methods	CSF	CGM	WM	Outlier	Ventricles	Cerebellum	DGM	Brainstem	Hippocampus	Overall
3D-UNet	18.0	80.3	90.1	66.0	**85.9**	86.5	80.0	54.1	70.0	70.1
(sd)	2.3	2.8	3.2	7.4 ·	3.5	3.9	2.2	5.0	6.5	4.1
SS	86.7	83.2	91.9	**71.6**	0	0	0	0	0	37.0
(sd)	7.2	4.7	3.8	13.8	0	0	0	0	0	3.3
DRS	84.7	83.4	91.8	66.0	76.0	90.2	90.5	89.6	77.0	83.2
(sd)	7.0	3.9	3.1	19.6	6.0	4.5	1.8	3.7	4.7	6.0
CAS-Net	**87.7**	**85.1**	**92.9**	71.5	78.3	**91.7**	**91.1**	**90.1**	**78.5**	**85.2**
(sd)	7.2	4.1	2.8	13.5	5.7	4.0	2.0	2.8	4.6	5.2

CAS-Net significantly improves the accuracy of the small or complicated structures that are more likely to be affected by motion or partial volume artifacts. Whereas intensity-based methods, e.g. 3D-UNet, may fail to segment such labels accurately from bad quality images.

Table 2 also demonstrates that the segmentation results from using different variants of the proposed CAS-Net based on the selected sub-networks, namely, segmentation subnet (SS) and diffeomorphic registration subnet (DRS). Note that the SS is not supervised by the re-scaled loss, which results in a better segmentation performance for the salient tissue (larger volume) but fails to segment smaller tissues. Although the overall segmentation performance of the SS is inferior to the 3D UNet with a re-scaled loss, the main contribution of the SS within the CAS-Net architecture is to provide valuable information for the DRS to learn a good deformation field. The extension of the SS with the DRS significantly improves the overall accuracy to 83.2% for all tissues, which can indicate that the predictions from the SS can provide valuable information for the DRS to learn the deformation field.

At last, the final CAS-Net output is the combination of the SS and DRS, while SS learns to segment the input images based on solely the intensity values, and DRS learns to preserve the structural topology. The quantitative results from Table 2 show that the performance from the final CAS-Net is better than both SS and DRS by 1.9% and 1.7%, respectively, in terms of DSC for the CGM segmentation. Similar results can be found for the other tissues.

Discussion: Figure 2a shows that the complexity of the cortex of the generated conditional atlases increases with the GA. Furthermore, based on the conditional atlas, the segmentation maps from our model, including the output from the intermediate SS and DRS, as well as the final CAS-Net output, are shown in

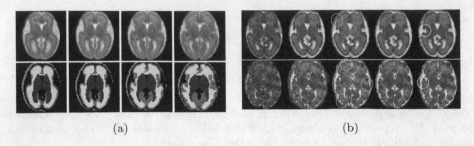

(a)   (b)

Fig. 2. Generated conditional atlas image and labelmap (a): From left to right: lower than 25, 26–28, 29–32 and over than 33 weeks GA. Samples in the test set with imaging artifacts and their segmentation results from CAS-Net. (b): From left to right: original MRI in axial axis, groundtruth labelmap, segmentation results from SS, DRS, and combined layer, and their Dice score are 81.3, 82.5, 83.4 (upper for 27.6 GA subject) and 68.5, 71.2, 72.4 (lower for 37.0 GA subject), respectively. Note that the MRIs are shown in original resolution, and the segmentation maps are produced at lower resolution.

Fig. 2b. Compared with the SS output (the 3rd column), the segmentation result from the DRS (the 4th column), i.e., the atlas-based method, can preserve the connectivity for the CGM structure, while the discontinuous part from SS is highlighted by green circles. Comparing with the segmentation result between DRS and final CAS-Net, the final CAS-Net output can tightly follow the intensity changes of the tissues boundaries benefiting from the output from SS, which is highlighted by yellow circles. Consequently, our model combines the advantage from UNet-based (SS) and atlas-based (DRS) methods, which achieves better segmentation performance in terms of both connectivity and tissue accuracy. As shown in the second column, the ground truth we used to train the CAS-Net is not the golden labelmap that the segmentation errors are highlighted in red circle, but the output from our method can also correct this mislabeling, which indicates the potential for our pipeline to the noisy label problem.

4   Conclusion

Fetal MRI is an important tool to monitor the development of brain structure. In this paper, we first generate a model that learns a conditional atlas based on fetal MRI which shows the averaged morphological change across different age groups in the dHCP dataset. Given the conditional atlas with an anatomical prior, tissue segmentation performance improved with better mesh connectivity for the tissues when facing imaging artifacts. In future work, quantitative evaluation of the generated conditional atlas will be extended, and ablation studies for different subnets will be performed.

Acknowledgements. Data in this work were provided by ERC Grant Agreement no. [319456]. We are grateful to the families who generously supported this trial.

References

1. Alansary, A., et al.: PVR: patch-to-volume reconstruction for large area motion correction of fetal MRI. IEEE Trans. Med. Imaging **36**(10), 2031–2044 (2017)
2. Alom, M.Z., Hasan, M., Yakopcic, C., Taha, T.M., Asari, V.K.: Recurrent residual convolutional neural network based on U-Net (R2U-Net) for medical image segmentation. arXiv preprint arXiv:1802.06955 (2018)
3. Ashburner, J.: A fast diffeomorphic image registration algorithm. NeuroImage **38**(1), 95–113 (2007)
4. Cabezas, M., Oliver, A., Lladó, X., Freixenet, J., Cuadra, M.B.: A review of atlas-based segmentation for magnetic resonance brain images. Comput. Methods Programs Biomed. **104**(3), e158–e177 (2011)
5. Casey, B., Giedd, J.N., Thomas, K.M.: Structural and functional brain development and its relation to cognitive development. Biol. Psychol. **54**(1–3), 241–257 (2000)
6. Dalca, A.V., Balakrishnan, G., Guttag, J., Sabuncu, M.R.: Unsupervised learning for fast probabilistic diffeomorphic registration. In: Frangi, A.F., Schnabel, J.A., Davatzikos, C., Alberola-López, C., Fichtinger, G. (eds.) MICCAI 2018. LNCS, vol. 11070, pp. 729–738. Springer, Cham (2018). https://doi.org/10.1007/978-3-030-00928-1_82
7. Dalca, A.V., Guttag, J., Sabuncu, M.R.: Anatomical priors in convolutional networks for unsupervised biomedical segmentation. In: Proceedings of the IEEE Conference on Computer Vision and Pattern Recognition, pp. 9290–9299 (2018)
8. Dalca, A.V., Rakic, M., Guttag, J., Sabuncu, M.R.: Learning conditional deformable templates with convolutional networks. arXiv preprint arXiv:1908.02738 (2019)
9. Dou, H., et al.: A deep attentive convolutional neural network for automatic cortical plate segmentation in fetal MRI. IEEE Trans. Med. Imaging **40**(4), 1123–1133 (2020)
10. Ebner, M., et al.: An automated framework for localization, segmentation and super-resolution reconstruction of fetal brain MRI. NeuroImage **206**, 116324 (2020)
11. Ertl-Wagner, B., Lienemann, A., Strauss, A., Reiser, M.F.: Fetal magnetic resonance imaging: indications, technique, anatomical considerations and a review of fetal abnormalities. Eur. Radiol. **12**(8), 1931–1940 (2002). https://doi.org/10.1007/s00330-002-1383-5
12. Yu, E.M., Dalca, A.V., Sabuncu, M.R.: Learning conditional deformable shape templates for brain anatomy. In: Liu, M., Yan, P., Lian, C., Cao, X. (eds.) MLMI 2020. LNCS, vol. 12436, pp. 353–362. Springer, Cham (2020). https://doi.org/10.1007/978-3-030-59861-7_36
13. Fetit, A.E., et al.: A deep learning approach to segmentation of the developing cortex in fetal brain MRI with minimal manual labeling. In: Medical Imaging with Deep Learning, pp. 241–261. PMLR (2020)
14. Isensee, F., et al.: nnU-Net: self-adapting framework for U-Net-based medical image segmentation. arXiv preprint arXiv:1809.10486 (2018)
15. Khalili, N., et al.: Automatic brain tissue segmentation in fetal MRI using convolutional neural networks. Magn. Reson. Imaging **64**, 77–89 (2019)
16. Kuklisova-Murgasova, M., Quaghebeur, G., Rutherford, M.A., Hajnal, J.V., Schnabel, J.A.: Reconstruction of fetal brain MRI with intensity matching and complete outlier removal. Med. Image Anal. **16**(8), 1550–1564 (2012)

17. Lee, M.C.H., Oktay, O., Schuh, A., Schaap, M., Glocker, B.: Image-and-spatial transformer networks for structure-guided image registration. In: Shen, D., et al. (eds.) MICCAI 2019. LNCS, vol. 11765, pp. 337–345. Springer, Cham (2019). https://doi.org/10.1007/978-3-030-32245-8_38

18. Makropoulos, A., et al.: Automatic whole brain MRI segmentation of the developing neonatal brain. IEEE Trans. Med. Imaging **33**(9), 1818–1831 (2014)

19. Makropoulos, A., et al.: The developing human connectome project: a minimal processing pipeline for neonatal cortical surface reconstruction. NeuroImage **173**, 88–112 (2018)

20. Payette, K., et al.: An automatic multi-tissue human fetal brain segmentation benchmark using the fetal tissue annotation dataset. arXiv preprint arXiv:2010.15526 (2020)

21. Ronneberger, O., Fischer, P., Brox, T.: U-Net: convolutional networks for biomedical image segmentation. In: Navab, N., Hornegger, J., Wells, W.M., Frangi, A.F. (eds.) MICCAI 2015. LNCS, vol. 9351, pp. 234–241. Springer, Cham (2015). https://doi.org/10.1007/978-3-319-24574-4_28

22. Serag, A., et al.: Construction of a consistent high-definition spatio-temporal atlas of the developing brain using adaptive kernel regression. NeuroImage **59**(3), 2255–2265 (2012)

23. Sinclair, M., et al.: Atlas-ISTN: joint segmentation, registration and atlas construction with image-and-spatial transformer networks. arXiv preprint arXiv:2012.10533 (2020)

24. Zhou, Z., Rahman Siddiquee, M.M., Tajbakhsh, N., Liang, J.: UNet++: a nested U-Net architecture for medical image segmentation. In: Stoyanov, D., et al. (eds.) DLMIA/ML-CDS -2018. LNCS, vol. 11045, pp. 3–11. Springer, Cham (2018). https://doi.org/10.1007/978-3-030-00889-5_1

Detection of Injury and Automated Triage of Preterm Neonatal MRI Using Patch-Based Gaussian Processes

Russell Macleod[1(✉)], Serena Counsell[1], David Carmichael[2], Ralica Dimitrova[1,3], Maximilian Pietsch[1,2], A. David Edwards[1], Mary Ann Rutherford[1], and Jonathan O'Muircheartaigh[1,3,4]

[1] Centre for the Developing Brain, School of Biomedical Engineering and Imaging Sciences, King's College London, London, UK
russell.macleod@kcl.ac.uk

[2] Department of Bioengineering, School of Biomedical Engineering and Imaging Sciences, King's College London, London, UK

[3] Department of Forensic and Neurodevelopmental Sciences, King's College London, London, UK

[4] MRC Centre for Neurodevelopmental Disorders, King's College London, London, UK

Abstract. Automatic detection or highlighting of neonatal brain injury could be a valuable adjunct to radiological interpretation. Here we propose a normative modeling-based detection method for preterm neonatal neuroimaging using gaussian processes (GPs). These GPs incorporates local image intensity information from image patches and demographics such as age. Z-score images can then be created from the scaled difference between the model predictions and a neonate's T1 and T2 weighted MRI. To test the use of these GP Z-scores as a form of automated triage, we trained a logistic regression classifier to separate normal and abnormal images. We used 133 preterm neonatal images with normal-reported MRI to train a GP model and optimized lesion detection performance on 36 preterm neonatal images with manually annotated lesion masks. The automated triage model was trained on 100 preterm neonates with normal reported MRI and 109 preterm neonates with MRI detectable lesions. It was tested on the same 36 manually annotated abnormal MRI preterm neonates and 33 normal-reported preterm neonates. Using a patch diameter of 7 voxels and integrating both T1w and T2w Z-score images provided our highest performing GP model for within image lesion detection, achieving an AUC of 0.961. By combining the output probabilities of a T1w and a T2w Z-score histogram classifiers allows for the correctly identification of 32/36 abnormal and 28/33 normal images. These results indicate patch-based normative model can accurately detect lesions in a highly interpretable fashion in preterm neonates with abnormal MRI. Using outputs from these predictions, the classifier is effective at separating abnormal and normal images.

Keywords: Abnormality detection · Neonatal imaging · Gaussian processes · Normative modeling

© Springer Nature Switzerland AG 2021
C. H. Sudre et al. (Eds.): UNSURE 2021/PIPPI 2021, LNCS 12959, pp. 231–241, 2021.
https://doi.org/10.1007/978-3-030-87735-4_22

1   Introduction

The developing human brain is a complex structure, with large variability in shape and volume between individuals and across age [1]. Developmental processes such as gyrification, myelination and volumetric growth undergo rapid progress over the first few months of life [2]. This makes identification of abnormal tissue challenging, requiring expert knowledge of typical developmental brain anatomy and image contrast changes, as well as the imaging appearances of injuries themselves.

The expertise needed for abnormality identification makes an automated method that can assist detection highly desirable [3]. Although several methods have successfully focused on detection in specific pathological domains in adults (e.g. tumours [4] or multiple sclerosis lesions [5]), these uses have focused only on the information relevant to the specific illness or abnormality. Neonatal brain injury covers a wide range and comprehensive detection of the various pathologies (e.g. periventricular leukomalacia or interventricular haemorrhage) could require multiple specialised classifiers.

Normative modelling-based approaches [6] model what is "normal" for a given context (e.g., age, clinical background) and detect values that deviate from this norm, falling into the family of anomaly detection methods. Recent neurological applications include dementia [7], white matter lesions in adults [8] and punctate white matter lesions (PWML) in neonates [9]. The latter [9] is particularly relevant as they modelled normal neonatal development as a function of age and prematurity to model the fast changes in structure (gyrification, volume) and image intensity (myelination).

However, voxel-wise models such as these provide predictions that are spatially smooth, as models are specific to the point being modelled and do not take into account neighbouring tissue. This makes them sensitive to local shifts in intensity, typical brain shape variability, and misregistration errors. In predicted images, highly variable areas (especially at the cortical surface) and structural boundaries can appear blurry, reducing interpretability. This is due to tissue type variability across the sample (WM, grey matter, GM, and cerebrospinal fluid, CSF) and differences in exact position (registration performance). This can result in areas of rare, but still typical, anatomical variants being classified as abnormal (false positives). Ideally, a useful model should incorporate normal local tissue heterogeneity into the prediction while inferring normal tissue intensities from the training set in abnormal areas.

Here, we incorporate subject-specific local structural information in the form of patches of voxels into the model. The aim is to increase spatial accuracy of models by incorporating subject specific local anatomical and image heterogeneity into the prediction, increasing interpretability. An additional advantage could be to reduce false positives in highly variable areas and improve anomaly detection specificity due to better intensity value inference. Using the resulting outlier maps, we also implement a classifier that triages abnormal MRI in neonates for radiological follow-up.

2   Methods

2.1   Sample and Dataset

Ethical approval was granted by a research ethics committee and written informed parental consent was obtained prior to scanning. An MRI dataset of 423 preterm neonates were acquired on a Philips 3T scanner. Modeling was completed on 3 types of acquisitions (Table 1). After quality control, 145 were discarded due to not having both a T1w and T2w modality images of sufficiently high quality, leaving 278 complete datasets. The age distribution of the preterm neonatal cohort is shown in Fig. 1A.

Table 1. Acquisition parameters.

Scan	TR	TE	Voxel Size	Subjects
T2w	7000–8000 ms	160 ms	$0.859 \times 0.859 \times 1$ mm	278
T1w	17 ms	4.6 ms	$0.8 \times 0.82 \times 0.82$ mm	241
High Res T1w	17 ms	4.6 ms	$0.5 \times 0.82 \times 0.82$ mm	37

Of the 278 images, 133 were considered "normal" in a preterm context and used as training data for the GP models as they contained no obvious lesions. Of the 145 considered abnormal, 36 were selected as held-out testing set and the tissue injury manually labelled. They included the following pathologies: Germinal matrix hemorrhage (n = 11), intra-ventricular hemorrhage (n = 6), cerebellar hemorrhage (n = 11), hemorrhagic parenchymal infarct (n = 4), temporal horn cysts (n = 1), sub-arachnoid cysts (n = 2), pseudocysts (n = 1), cystic periventricular leukomalacia (CPVL) (n = 3), large/many PWML (n = 7) and few/single PWML (n = 9). 13 neonates had multiple pathologies. Subject demographics are displayed in Table 2. Manual segmentations were created by 2 raters and inter-rater reliability assessed (opinion on lesion extent can vary [10]).

Table 2. Cohort information.

Parameter	Value
Total Subjects (Female)	278 (125)
Mean Post Menstrual Age at scan (PMA) (Weeks+Days) [Range]	37+0 [25+1 to 55+0]
Mean Gestational Age at birth (GA) [Range]	29+3 [23+2 to 36+1]
Preterm without Lesions (Training) (Female)	133 (66)
PMA Mean [Range]	39+6 [27+5 to 51+0]
GA Mean [Range]	29+5 [23+2 to 35+4]
Preterm with Lesions (Female)	145 (59)
PMA Mean [Range]	34+2 [25+1 to 55+0]
GA Mean [Range]	29+3 [24+3 to 36+1]
Testing Abnormalities with Ground Truth (Female)	36 (18)
PMA Mean [Range]	32+2 [25+1 to 37+5]
GA Mean [Range]	29+4 [24+3 to 35+2]

2.2   Pre-processing

Individual subject T1w and T2w images were rigidly co-registered [11, 12], B0 inhomogeneity corrected using N4BiasCorrection [13] and FSLBET [14] removed non-brain tissue. Affine registrations between the T2w images and weekly developing human connectome project templates [15] were calculated using FSL [11], followed by a non-linear registration to the same template using ANTS [13]. These transformations and warps were applied to the T1w images. Datasets below 28 weeks PMA were registered to week 28, the youngest template available. Each template had an affine registration to the week 36 template, also applied to the images. Finally, all voxels within an image were standardized to image intensity mean and standard deviation.

2.3   Base Gaussian Process Tissue Intensity Model

An independent GP [16] model was fit to each voxel using GPyTorch [17]. GPs model a distribution of possible functions that fit some dependent variable given an input dataset. A prior ε (an initial estimation of the intensity distribution of voxel n) is specified based on a normal distribution N with a mean (usually 0) and variance σ^2.

$$\varepsilon \sim N\left(0, \sigma_n^2\right) \tag{1}$$

The covariance between datapoints is quantified by the widely used radial basis function (RBF) as a covariance matrix Σ.

$$K(x, x) = exp\left(-\frac{\|x - x'\|}{2\sigma^2}\right) \tag{2}$$

$$\Sigma = K(x, x) \tag{3}$$

The RBF kernel hyperparameters (length-scale and variance) were optimized using the Adam optimizer [18] which minimizes the log marginal likelihood. Predictions were made based on the joint conditional probability of the training and testing samples.

$$\begin{bmatrix} f \\ f^* \end{bmatrix} \sim N\left(0, \begin{bmatrix} k\ (x, x)\ k(x, x^*) \\ k\ (x^*, x)\ k(x^*, x) \end{bmatrix}\right) \tag{4}$$

Here, each voxel's GP incorporated GA at birth, PMA at scan and sex in the design matrix to model image intensity. This is described as the base model from here on out.

2.3.1   Patch Extension to the Gaussian Process Tissue Intensity Model

Patch-based models used the same structure as the base model but incorporated additional spatial information at training and inference time. Intensity values from a patch of voxels surrounding the target voxel were added into the design matrix. This provides a prior on target voxel intensity from surrounding patch voxels making the model less dependent on global image standardisation. We investigated patch neighbourhoods of $3 \times 3 \times 3, 5 \times 5 \times 5, 7 \times 7 \times 7$ and $9 \times 9 \times 9$ using two approaches of voxel contribution. A dense patch used all voxels *except* the target voxel for prediction, while a surface patch only those voxels on the outermost edge of the patch (Fig. 1B).

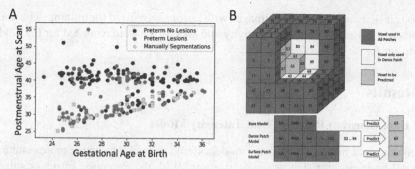

Fig. 1. A: Age distribution of preterm neonatal dataset. B: 5 × 5 × 5 patch model. A model predicts the intensity of the green voxel based on the GA, PMA, sex, and any patch voxel intensities included in the design matrix. For this 5 × 5 × 5 patch, the dense patch included white and blue voxel intensities while the surface patch included only blue voxel intensities. (Color figure online)

2.3.2 Iterative Patch Extension Gaussian Process Tissue Intensity Model

Lesions, when contained within a patch, will have values far outside those in the training range, leading to high predictive variance. To address this, we investigated the effect of replacing the initial input image with the model predicted image. This acts as a form of lesion infilling by iteratively replacing outlier tissue intensities. All patch models were iterated in this way up to 5 times with no updates to model parameters.

2.4 Detecting Pathology

Z-score images were created by calculating the difference between the observed and predicted images and scaling by the predicted standard deviation. In the 36 held-out images with labelled pathology, receiver operating characteristic's (ROC) area under the curve (AUC) were calculated for all neonate's absolute Z-score image against their corresponding manual segmentations and averaged across subjects for each model. The sensitivity and specificity for fixed Z thresholds of 3–5 standard deviations were calculated for all models at the iteration with the best results from the testing AUC.

2.5 Automated Clinical Image Triage

A logistic regression classifier was used to separate images into those with (0) or without (1) lesions using Z-score histograms as inputs. We used T1w Z-score, T2w Z-score and joint Z-score histograms, each with 99 bins between −100 and 100. The logistic regression classifier outputs the probability of a sample belonging to a class based on a threshold ($\geq 0.5 = 1$, $< 0.5 = 0$). This allows for an additional classification based on the combined probability of the T1w and T2w classifier predictions. We simply taking the mean of a sample's probability prediction from each classifier and apply the threshold. 5-fold cross application of the model with the best AUC from the previous analysis was applied to the 133 preterm neonates without obvious lesions for three iterations. 33 of these, stratified by age, were added to the 36 manually segmented images with abnormalities to act as a classifier test set. The remaining 100 preterm images with no obvious lesions and 109 preterm images with lesions were used for training. The number of correctly classified images, sensitivity and specificity were recorded for the held-out testing set for each classifier.

3   Results

3.1   Base Gaussian Process Tissue Intensity Model

The synthesized images of the base model offered a good qualitative representation of neonatal brain structure but were very smooth. With the exception of major sulci and gyri, heterogeneous cortical folding patterns (as in frontal and parietal areas) were not modelled well, appearing blurry. Nonetheless, for detecting lesions, the base model had a high joint AUC of 0.948 and a T2w AUC of 0.951, Table 3.

Table 3. AUC for all models on the 1st and 3rd iteration.

Model AUC (Iteration)	Base	Dense 3 $\times$ 3 $\times$ 3	Dense 5 $\times$ 5 $\times$ 5	Dense 7 $\times$ 7 $\times$ 7	Dense 9 $\times$ 9 $\times$ 9	Surface 5 $\times$ 5 $\times$ 5	Surface 7 $\times$ 7 $\times$ 7	Surface 9 $\times$ 9 $\times$ 9
T1w (1)	0.918	0.901	0.919	0.919	0.929	0.926	0.937	**0.938**
T2w (1)	**0.951**	0.904	0.92	0.92	0.928	0.927	0.937	0.938
Joint (1)	0.948	0.915	0.924	0.924	0.942	0.935	0.951	**0.953**
T1w (3)		0.908	0.932	0.941	0.941	0.941	**0.947**	0.946
T2w (3)		0.911	0.934	0.943	0.942	0.943	0.948	0.946
Joint (3)		0.925	0.94	0.953	0.952	0.953	*0.961*	0.959

3.2   Patch Gaussian Process Tissue Intensity Model

Dense patch model synthesized images incorporated more subject specific structural heterogeneity but also pathological hypo/hyperintensity into target voxels (the latter being undesirable in this application), Fig. 2. Patch size affected generalisation, smaller patches closely matched heterogeneous brain structure, but included more local pathology in the prediction. Larger patches were smoother and less sensitive to predicting lesioned tissue. In most of the brain, target voxel patches were consistent with those in the training set and reduced variance to extremely low values. In lesioned areas, patches could contain intensity values drastically different from those in the training set, dramatically increasing model variance compared to areas of uninjured tissue. This led to Z-score values in abnormal areas within normal range (Fig. 2).

The surface patch model increased generalisation and reduced variance inflation. Fewer voxels, farther from a target voxel, reduced the model's ability to accurately infer intensity and ranges for a given input. Synthesis of "normal" brain tissue remained similar to the dense patch model, although the 9 × 9 × 9 surface patch model demonstrated an obvious blur in the cortical surface. The 9 × 9 × 9 surface patch joint Z-score achieved an AUC of 0.953 compared to 0.948 joint AUC and 0.951 T2w AUC of the base model, Table 3. The T1w AUC scores were consistently lower than the T2w AUC score but when combined gave the highest AUC scores for all except the base model.

All patch models had higher AUC scores after patch-based prediction iteration. However, after 3 iterations most models demonstrated no further improvement. The 7 × 7 × 7 surface patch model had the best performance with an AUC of 0.961, Table 3. We examined the trade of between sensitivity and specificity for the models with AUC above 0.95 after 3 iterations for a range of thresholds, Table 4. For Z thresholds of 2 to 4 the 7 × 7 × 7 surface patch model gave the best sensitivity, specificity trade-off but the 7 × 7 × 7 dense patch model had higher sensitivity for higher thresholds.

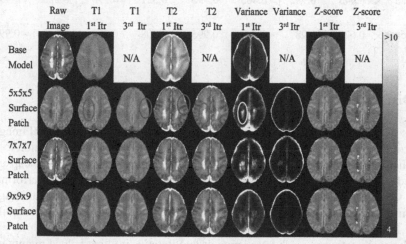

Fig. 2. T1w/T2w predicted images, variance maps and absolute Z-scores (thresholded at 4 standard deviations) for the 1st and 3rd model iteration in example surface patch models. Blue circles show heterogenous cortical structure patch models were able to duplicate. Red circles show lesion areas retained by the patch model but with high associated variance (yellow circles). (Color figure online)

3.3 Automated Triage of Abnormal Images

The classifier used 7 × 7 × 7 surface patch model output data as it had the best lesion detection AUC. The joint Z-scores classifier had a sensitivity of 0.767, a specificity of 0.744 and correctly identified 29/36 abnormal and 23/33 normal images. The T1w classifier a sensitivity of 0.778, specificity of 0.848 and identified 28/36 abnormal and 28/33 normal images. The T2w classifier a sensitivity of 0.818, specificity of 0.833 and identified 30/36 abnormal and 27/33 normal images. Combining the T1w and T2w classifiers probability outputs achieved 0.875 for sensitivity 0.865 for specificity and correctly identifying 32/36 abnormal images and 28/33 normal images.

Table 4. Sensitivity and Specificity using the joint Z-scores for all models that achieved an AUC higher than 0.95 with thresholds ranging from 2 to 5 standard deviations.

Z Thresh (Sens/Spec)	2 (Sens)	2 (Spec)	3 (Sens)	3 (Spec)	4 (Sens)	4 (Spec)	5 (Sens)	5 (Spec)
Base	0.358	0.993	0.198	0.998	0.108	0.999	0.056	0.999
Dense 7 × 7 × 7	0.656	0.972	0.467	0.993	0.319	0.998	**0.232**	**0.999**
Dense 9 × 9 × 9	0.632	0.971	0.436	0.992	0.308	0.997	0.222	0.999
Surface 5 × 5 × 5	0.641	0.98	0.451	0.995	0.306	0.998	0.211	0.999
Surface 7 × 7 × 7	**0.645**	**0.982**	**0.463**	**0.995**	**0.321**	**0.998**	0.223	0.999
Surface 9 × 9 × 9	0.629	0.98	0.445	0.995	0.308	0.998	0.213	0.999

4   Discussion

In this work, we developed a patch-based model to synthesise "normal" MRI images of preterm neonates. We tested different variations of patch size and contribution to evaluate their predictive ability. We showed that the patch model qualitatively encapsulated individual anatomy accurately while demonstrating good lesion detection performance. Next, we constructed a classifier based on the GP outputs to detect abnormal images as a form of triage. Combined, this strategy could allow a 2-stage process of automated triage followed by lesion identification in preterm neonates MRI.

The combination of patch-based models and iterative infilling helped maintain normal subject-specific expected variance when synthesizing a "typical" image while preserving abnormal tissue intensity detection. The $7 \times 7 \times 7$ surface patch iterated 3 times was our best performing model. It produced more accurate predictions, attained a higher AUC (0.961) and higher sensitivity for a given specificity. Combining T1w and T2w absolute Z-scores further increased AUC of patch models, likely through a higher specificity due to noise removal but relies on good quality images. Figure 3 shows a single preterm neonate with germinal matrix haemorrhage whose AUC improved with use of a patch-based models (A), multiple iterations (B) and combining Z-scores from multiple modalities. The corresponding absolute Z-score images for the base (T2w) and $7 \times 7 \times 7$ surface patch model (joint) are shown in C and D.

Patch model outputs were more interpretable than the base model as the predicted image was anatomically closer to the subjects observed MRI image. Small haemorrhages, groups of PWML and other small/subtle pathologies had more complete highlighting of lesioned tissue at the same Z-score threshold compared to the base model. However, patch models had lower performance when looking at some larger pathologies as abnormal tissue was present in the prediction. This can be attributable to two factors. One, the model infers target intensity from the voxels within the patch, even if those voxels are abnormal. Two, if there is a training image patch with intensities close to the abnormal intensities in the target patch it will interpolate from that image. Due to this, large lesions like CPVL often miss the lesion core but detected the edges. Some pathologies (e.g., a single PWML, tiny cerebellar haemorrhage) could not be easily detected due to partial volume or intensity blurring during registration.

In classification, combining the single modality probabilities achieved the best results (32/36 abnormal and 28/33 normal images). This supports the argument that different modalities make pathologies more detectable [19] and output combination is advantageous. The joint Z-score histogram classifier gave the worst performance being poor at identifying normal images. The missed abnormal images contained single/very small PWML and small cerebellar hemorrhages with small intensity differences.

Inclusion of lesioned tissue in the predicted image could be corrected by better standardization of the tissue intensities between images. Alternatively, average intensity values could be created from tissue segmentations and either incorporated into the model or used to group similar images together. The classification could be improved by using a probabilistic method such as GPs [20] or a non-linear method that can better separate the input data. A final note is that both the GP model and the classifier could be improved with access to larger datasets of both lesion free (more robust normative model) and lesioned (type and variational coverage) neonates.

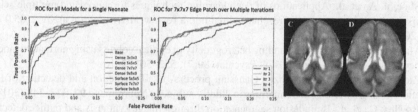

Fig. 3. A: ROCs for all models for a single neonate. B: ROCs for the same neonate for multiple iterations of the $7 \times 7 \times 7$ Surface Patch model. C: The base model absolute T2 Z-score. D: The $7 \times 7 \times 7$ surface patch model absolute joint Z-score. C and D both have a threshold of 4.

5 Conclusion

The inclusion of patch information into MRI normative models results in synthesized "normal" images that more closely match subject specific structural heterogeneity while maintaining excellent lesion detection. Optimal performance was attained by selection of patch type, size and by iteratively applying the once trained model. The $7 \times 7 \times 7$ surface patch with prediction iteration improved abnormality detection giving higher AUCs, a better sensitivity specificity trade-off and better coverage of smaller lesions. A classifier built using the combined probability output from T2w and T1w Z-score histograms was able to correctly identify 32/36 abnormal images only missing images with small lesions that were almost imperceptible after non-linear registration.

Acknowledgements. This work was supported by the Wellcome/EPSRC Centre for Medical Engineering at Kings College London (WT 203148/Z/16/Z), the NIHR Clinical Research Facility (CRF) at Guy's and St Thomas' and by the National Institute for Health Research Biomedical Research Centres based at Guy's and St Thomas' NHS Foundation Trust, and South London, Maudsley NHS Foundation Trust. J.O. is supported by a Sir Henry Dale Fellowship jointly funded by the Wellcome Trust and the Royal Society (grant 206675/Z/17/Z). J.O. received support from the Medical Research Council Centre for Neurodevelopmental Disorders, King's College London (grant MR/N026063/1). The project includes data from a programme of research funded by the NIHR Programme Grants for Applied Research Programme (RP-PG-0707-10154.)

References

1. Groeschel, S., et al.: Developmental changes in cerebral grey and white matter volume from infancy to adulthood. J. Dev. Neurosci. **28**(6), 481–489 (2010)
2. Miller, J.H., Bardo, D.M.E., Cornejo, P.: Neonatal neuroimaging. In: Seminars in Pediatric Neurology, vol. 33. WB Saunders (2020)
3. Makropoulos, A., et al.: A review on automatic fetal and neonatal brain MRI segmentation. Neuroimage **170**, 231–248 (2018)
4. Bahadure, N.B., et al.: Image analysis for MRI based brain tumor detection and feature extraction using biologically inspired BWT and SVM. Int. J. Biomed. Imaging **2017** (2017)
5. Shoeibi, A., et al.: Applications of deep learning techniques for automated multiple sclerosis detection using magnetic resonance imaging: a review. arXiv preprint arXiv:2105.04881 (2021)
6. Marquand, A., et al.: Understanding heterogeneity in clinical cohorts using normative models: beyond case-control studies. Biol. Psychiat. **80**(7), 552–561 (2016)
7. Ziegler, G., et al.: Individualized Gaussian process-based prediction and detection of local and global gray matter abnormalities in elderly subjects. Neuroimage **97**, 333–348 (2014)
8. Bowles, C., et al.: Brain lesion segmentation through image synthesis and outlier detection. NeuroImage Clin. **16**, 643–658 (2017)
9. O'Muircheartaigh, J., et al.: Modelling brain development to detect white matter injury in term and preterm born neonates. Brain (2020)
10. Bogner, M.S.: Human Error in Medicine. CRC Press, Boca Raton (2018)
11. Jenkinson, M.: A global optimisation method for robust affine registration of brain images. Med. Image Anal. **5**(2), 143–156 (2001)
12. Jenkinson, M., et al.: Improved optimisation for the robust and accurate linear registration and motion correction of brain images. Neuroimage **17**(2), 825–841 (2002)
13. Avants, B.B., et al.: Advanced normalization tools (ANTS). Insight J. **2**, 1–35 (2009)
14. Smith, S.M.: Fast robust automated brain extraction. Hum. Brain Mapp. **17**(3), 143–155 (2002)
15. Makropoulos, A., et al.: The developing human connectome project: a minimal processing pipeline for neonatal cortical surface reconstruction. Neuroimage **173**, 88–112 (2018)
16. Rasmussen, C.E.: Gaussian processes in machine learning. In: Bousquet, O., von Luxburg, U., Rätsch, G. (eds.) ML 2003. LNCS (LNAI), vol. 3176, pp. 63–71. Springer, Heidelberg (2004). https://doi.org/10.1007/978-3-540-28650-9_4
17. Gardner, J., et al.: Gpytorch: blackbox matrix-matrix gaussian process inference with gpu acceleration. In: Advances in Neural Information Processing Systems (2018)
18. Kingma, D.P., et al.: Adam: a method for stochastic optimization. arXiv preprint arXiv:1412.6980 (2014)

19. Tusor, N., et al.: Punctate white matter lesions associated with altered brain development and adverse motor outcome in preterm infants. Sci. Rep. **7**(1), 1–9 (2017)
20. Salvador, R., et al.: Evaluation of machine learning algorithms and structural features for optimal MRI-based diagnostic prediction in psychosis. PLoS One **12**(4), e0175683 (2017)

Assessment of Regional Cortical Development Through Fissure Based Gestational Age Estimation in 3D Fetal Ultrasound

Madeleine K. Wyburd[1] , Linde S. Hesse[1]([✉]) , Moska Aliasi[5],
Mark Jenkinson[2,3,4], Aris T. Papageorghiou[6], Monique C. Haak[5],
and Ana I. L. Namburete[1]

[1] Institute of Biomedical Engineering, University of Oxford, Oxford, UK
linde.hesse@seh.ox.ac.uk
[2] Wellcome Center for Integrative NeuroImaging, FMRIB, University of Oxford,
Oxford, UK
[3] Australian Institute for Machine Learning (AIML), Adelaide, Australia
[4] South Australian Health and Medical Research Institute (SAHMRI),
Adelaide, Australia
[5] Department of Obstetrics and Fetal Medicine, Leiden University Medical Center,
Leiden, The Netherlands
[6] Nuffield Department of Woman's and Reproductive Health,
University of Oxford, Oxford, UK

Abstract. The relationship between fetal cortical development and gestational age has been commonly studied, with cortical folding events found to be temporally consistent across the healthy population. In order to utilise this relationship in clinical practice, manual fissure grading charts have been proposed to compare fissure appearance or measurements to the known fetal gestational age. However, these techniques are found to be extremely user-dependent, time-consuming and error-prone. In this study, we propose a deep learning-based automated method to assess the development of three fissures: the Sylvian fissure (SF), Parieto-occipital fissure (POF) and Calcarine sulcus (CLC), by predicting fetal gestational age based on their respective morphology. This fissure-specific age prediction can then be compared to the true gestational age to determine if regional cortical development is healthy, delayed, or advanced. Our best-performing CNN estimated the gestational age with an error of 3.4, 5.0, 4.9 and 4.1 days, for the SF, POF, CLC and whole-brain, respectively, outperforming previously reported ultrasound whole-brain age prediction techniques.

Keywords: Age prediction · Fissures · 3D ultrasound · Gyrification · Deep learning · Fetal brain

---

M. K. Wyburd and L. S. Hesse—Equal contribution.

1   Introduction

As a fetus develops, the once smooth cortical plate progressively becomes heavily folded in a process known as *gyrification*. This event creates peaks (*gyri*) and troughs (*sulci*) within the cortex, with larger sulci referred to as fissures. In healthy development, the emergence and growth of fissures is well timetabled and found to be temporally consistent across a large population [17]. Therefore, the development of the fissures can be utilised as landmarks to monitor fetal development. Abnormal cortical folding has been commonly linked to disease, with the development of specific fissures or regions significantly affected [2]. Therefore, individual fissures have the potential to be used as biomarkers for disease, however, a fast and accurate method of assessing and grading the fissure development is required.

During routine ultrasound (US) scans, the appearance of fissures is usually not assessed [12]. However, if any fetal abnormalities are identified, detailed neurosonography is performed, which usually includes the assessment of the Sylvian fissure (SF) [12]. The SF is one of the earliest and most prominent features seen in the fetal brain, emerging around the 14th gestational week (GW) and developing into a deep fissure in each hemisphere. Several methods to grade SF development have been proposed in the past. Initial studies primarily focused on manual measurements of the fissure depth and angle in 2D US planes [15], but such methods are extremely time-consuming and require consistent localisation of the correct US plane. Alternatively, several grading charts have been proposed to quantify the fissure development in more detail [14,16]. However, manually grading fissures based on such charts is still very user-dependent, resulting in high inter-observer variability [3]. For these reasons, there is a clear need for reproducible and non-user dependent techniques to quantify fissure development.

An alternative method to using grading charts to quantify fissure development is to predict the age of the fetus based on fissure appearance and compare this to the known true gestational age (GA). Whole-brain age prediction has been commonly performed for the adult brain, to quantify healthy development and detect any deviation related to disease [4]. These methods have since been applied in-utero using magnetic resonance imaging (MRI) [9] and US [10]. However, as these methods quantify the GA of the whole brain in one measure, specific regions of advanced or delayed gyrification can be easily missed. Convolutional neural networks (CNNs), such as the VGG-Net [18] and Residual network (ResNet) [8], are often used to perform automated age estimation tasks.

In this work, we propose a CNN-based method to estimate the development of 3 key fissures; the SF, Parieto-occipital (POF) and Calcarine (CLC) from 3D US scans between 19 and 30 GWs, by predicting the fetal GA based on their respective morphology. These fissures were selected as they undergo rapid development across this GA range and are well visualised in US scans. Furthermore, these fissures were found to show significant differences in diseased groups [2] and were previously identified as age-discriminating regions [10].

2   Methods

2.1   Age Prediction Pipeline

The processing pipeline is shown in Fig. 1. For each volume, a 3D region of fixed size is extracted around each of the fissures of interest $\mathbf{X}$. Subsequently, a separate CNN, f, is trained for each of these 3D boxes, predicting the GA of the region $\hat{\mathbf{y}} = f(\mathbf{X})$.

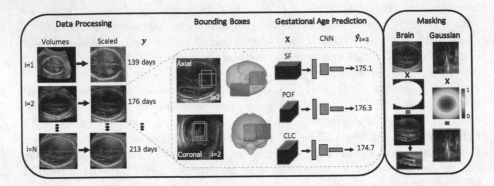

Fig. 1. Left panel: Schematic of the age prediction pipeline. Bounding boxes were extracted from each scaled volume and used as inputs to individual CNNs trained for each fissure. Right panel: Masking control experiments (described in Sect. 3.4).

In this study, the aim of predicting age based on individual image patches is not to predict the most accurate overall GA (i.e. by averaging the predictions of the individual fissures), but to determine the degree of maturation of a specific cortical region. For this reason, all image volumes are scaled to the same brain volume, ensuring that our CNN is not predicting GA only based on size, but focuses on the shape, and thus degree of maturation, of the fissures.

2.2   Network Architectures

We compared 3D VGG-Net [18] and 3D ResNet [8] architectures with each other to determine the optimal network architecture for our task. For the VGG-Net, we adapted the batch-normalised VGG-Net 11 [18] to make it suitable for our 3D regression task. Due to our relatively small input volumes, we removed the last two convolutional layers as well as the last max-pool operation, resulting in a VGG-Net of depth 9. Furthermore, the number of feature maps in the convolutional layers was set to 24, 48, 96, 96, 192, 192, and the dimensions of the two linear layers preceding the final layer were reduced to 64. The 3D ResNet architecture used in this work is an adapted version of a previously published architecture for video data classification [7]. The deviating stride for the temporal axis was set equal to the stride in the spatial dimensions (=2) and, as for the VGG-Net, the last fully connected layer was replaced by a single neuron without an activation function. For the ResNet, we explored depths of 10 and 18 layers.

2.3   Attention Maps

To improve our understanding of the CNN GA estimation, we created attention maps for our input volumes using Guided Backpropagation from each network [19]. This saliency method determines the pixels in the image that most strongly activate the neurons in a certain layer, and, as such, are most discriminative for the output of this layer. We computed attention maps with respect to the last layer of our network, producing maps of the same dimensions as the input X, showing the importance of pixels with respect to the predicted GA.

2.4   Dataset

A total of 811 3D US volumes from the INTERGROWTH-21th dataset between the GA of 19^{+0} and 30^{+6} weeks were used throughout. As only the brain hemisphere proximal to the US probe is visible on volumes acquired in the axial plane, it was ensured that volumes from both the left and right hemisphere were included. The GA of each fetus was determined by the last menstrual period (LMP) and confirmed with an US crown-rump length measurement between 9^{+0} and 13^{+6} GWs after the LMP, agreeing within 7 days.

All volumes were resampled to an isotropic voxel size of 0.6 mm (using trilinear interpolation), cropped around the centre to a total size of $160 \times 160 \times 160$ voxels and manually aligned to a common coordinate system [11]. To ensure the network was not predicting based on size differences, each volume was upscaled to the maximum brain volume present in our dataset, based on the size of a whole-brain mask. The whole-brain masks were derived from MRI fetal atlas masks which were rigidly aligned to the individual US volumes [5]. Next, a bounding box was extracted around each fissure. As all volumes were aligned and scaled to the same size, the same bounding box size and location was selected across all volumes. However, as the three fissures vary considerably in size, the bounding box dimensions differed for each fissure (SF: $66 \times 44 \times 46$ voxels, CLC: $33 \times 40 \times 34$ voxels, POF: $46 \times 37 \times 36$ voxels). To perform baseline experiments, a bounding box in the less visible hemisphere was also selected (referred to as *random* crop, 38,49,25 voxels), as well as a crop within the US beam but outside the fetal skull (referred to as *background* crop, $38 \times 35 \times 41$ voxels).

2.5   Experimental Setup

The dataset was split into a training (85%) and test set (15%). Five-fold cross-validation was performed across the training set, with an even distribution of each GW in each fold. All five models were applied to the test set and, throughout this study, the mean performance across these folds is reported.

We trained our CNNs with a standard mean squared error loss $L_{mse} = \frac{1}{N}\sum_{i=1}^{N}(y_i - \hat{y}_i)$ between the predicted, $\hat{y}$, and true GA, y. As augmentation, we applied scaling (0.8–1.2), rotation around the centre of the bounding box ($\pm10°$) and translation (±3 voxels) during training. Furthermore, to ensure that the network could not predict GA based on interpolation patterns introduced during up-scaling of the volumes, we also applied a random level of down-sampling, followed by up-sampling to the original resolution. The sampling was performed using trilinear interpolation, with the maximum down-sampling factor specified per GW, given by the factor required to down-sample the average volume at the respective GW to the average brain volume at 19 GWs (computed from the whole brain masks).

Implementation. All experiments were implemented in Python 3.8 with Pytorch (version 1.8.1) and trained on an NVIDIA GeForce RTX 2080 Ti with 12 GB of memory. We trained all models for 200 epochs using the Adam optimizer with an initial learning rate of 0.001, and a batch size of 10. Attention maps were generated with the Pytorch library M3d-CAM [6] and augmentation was implemented with TorchIO (version 0.18) [13].

3   Results and Discussion

3.1   Network Architecture

The age prediction performance for each of the individual fissures is reported in Table 1. Across the three architectures, the performance was relatively consistent. As the ResNet-10 performed marginally better, it was used for the rest of this work. The ResNet-10 estimated the GA of the SF, POF, CLC and scaled whole-brain with a mean error of 4.1, 5.1, 4.9 and 4.1 days, respectively, outperforming previously reported US whole-brain age-prediction, which achieved an accuracy of ±6.10 days [10]. Moreover, as the true GA (estimated from the LMP) can contain inaccuracies up to 7 days, some variation is expected.

Both control areas (*random* and *background*) obtained lower performance than each of the fissures. The performance of the *random* crop, which was only significantly worse than the SF and POF, indicates that the network was still able to learn age-discriminative patterns in an uninformative area for a human observer. However, as the whole brain develops and changes in appearance during gestation, all areas within the fetal skull will contain changing patterns that the network can learn. Although obtaining large prediction errors, the performance of the *background* area was higher than randomly predicted ages. Most likely, this can be attributed to the fact that the network could learn some GA-specific features from both the US cone shape, which may vary due to up-scaling of the volumes, as well as from the tissue surrounding the fetal head, such as the amniotic fluid composition.

3.2   Age Prediction

In Fig. 2 the predicted GA versus the true GA is shown for each of the fissures. It can be observed that for all fissures, good performance is obtained across the GA range. Although there is a small tendency towards the mean (i.e. over-predicting at the younger ages and under-predicting later in gestation), which

Table 1. MAE for the different network configurations. Values between brackets are standard deviations. For each fissure, significance with respect to the *random* area, calculated using paired t-tests with Bonferonni correction for multiple comparisons, is noted with a *($p < 0.05$) or **($p < 0.005$). Masking experiments described in Sect. 3.4.

	SF	CLC	POF	Random	Background
VGG-Net	5.4 (4.3)**	5.7 (4.5)*	6.1 (4.9)	7.1 (5.2)	10.2 (7.6)
ResNet-18	4.1 (3.7)**	**5.0** (4.5)	5.2 (4.6)	6.3 (5.4)	11.1 (8.7)
ResNet-10	4.1 (3.7)**	5.1 (4.2)	**4.9** (4.2)*	6.1 (5.1)	11.2 (8.5)
+ brain mask	**3.4** (3.1)	–	–	–	–
+ Gaussian mask	–	5.1 (4.1)	5.2 (4.4)	–	–

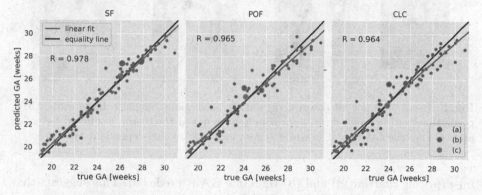

Fig. 2. True versus predicted GA for the SF, CLC and POF, shown with a linear fit. The separately colored samples match the examples in Fig. 3.

can be observed by the linear fit slightly deviation from the equality line, this deviation is small and is to be expected for a regression task.

In Fig. 3, qualitative examples of each fissure and their corresponding labels are shown. Examples were included of outliers (Fig. 3a), which were age-matched to more accurate predictions by both the predicted age and the true age (Fig. 3b and c, respectively). The outlier shown for the SF has a true GA of 183 and a predicted GA of 192 days. However, it can be observed from the age-matched examples that for this volume, the SF begins to fold over itself in a process known as *operculization* (shown with white arrows), which is early for a 183 days old fetus [14]. Although it is challenging to manually compare the degree of operculization across the shown samples in a 2D slice, the large prediction error could also suggest advanced development of the SF for this fetus. As across the healthy population there will be inter-subject anatomical variations, it is expected that some cases naturally have advanced or delayed gyrification. For the outlier shown for the POF, it can be observed that only the fissure on one side of the hemisphere is well visible, which might contribute to the larger error as not all anatomical information is present. Furthermore, for the CLC, the outlier (a) seems to resemble (b) more than the age-matched (c), therefore it is

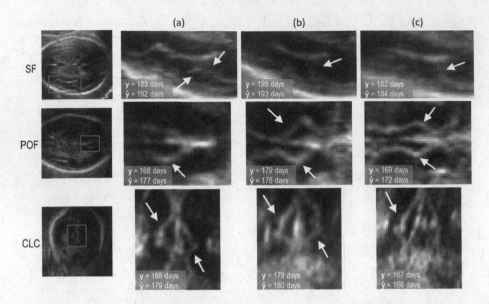

Fig. 3. Predictions for all fissures; (a) outlier predictions having a large error, (b/c) predictions with a small error, approximately matched to (b) the predicted GA of a, and (c) the true GA of a. The white arrows indicate key age-dependant areas.

unsurprising that for (a) and (b) the same GA is predicted. This suggests that the fetus in (a) might have delayed CLC development.

3.3 Network Attention Maps

To investigate which image areas the trained CNNs were focusing on, we generated attention maps for our input volumes during inference. For all three fissures, attention maps of varying GA are shown in Fig. 4. Interestingly, for the SF, it can be clearly seen that the network's attention changes with increasing GA. At earlier ages the network traces the cortical plate, focusing around the insula, whereas for older ages the attention is on the SF's operculization. These observations mimic what neurosonographers look for and are recorded in SF grading charts [16]. Although less distinct for the POF and CLC, also for these fissures it can be observed that part of the network focus follows the fissure boundaries.

From the attention maps it can also be observed that the network focuses on some areas outside of the fissure itself (orange arrows). A clear focus area in the fetal skull can be observed in the SF patch, indicating that the skull contributes to the output prediction. For the CLC, a clear spot is consistently present in the left bottom corner of the image, suggesting that the presence or absence of a certain structure at that location, and thus the exact location of the bounding box might influence the final predictions. Based on these observations, we implemented several control experiments, presented in the next section.

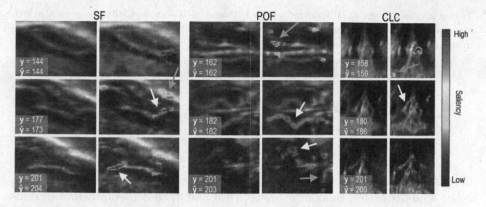

Fig. 4. Attention maps for each of the three fissures across different GAs. The arrows indicate regions of high saliency for clinically meaningful parts of the fissure (white) and areas outside the fissure of interest (orange), i.e. the skull. (Color figure online)

Next to providing valuable information regarding network performance, the attention maps might also provide improved interpretability of the age prediction. However, as saliency methods have shown to produce unreliable predictions under certain conditions [1], more thorough investigation is necessary to validate the attention maps, which is planned for further work.

3.4 Control Experiments

Whole Brain Masking. The SF bounding boxes contained part of the fetal skull, which the network was found to focus on (see Fig. 4). The fetal skull can potentially provide GA information in two ways: Firstly, due to the high intensities of the skull, interpolation patterns introduced when up-sampling all volumes to the same size are expected to be especially pronounced in this area. Although we compensate for the interpolation by applying random sampling operations during training, some interpolation information might still be picked up on by the network. Secondly, with increasing GA, the skull becomes thicker and undergoes calcification, changing its appearance on US volumes. Although the skull appearance is anatomical information, we wish to decouple it from the SF's appearance in our GA predictions. For these reasons, we retrained our ResNet-10 for the SF using brain volumes with the fetal skull masked out. Masking was performed by applying whole brain masks, eroded with 10 pixels to ensure that no skull was included (Fig. 1). The results of this are depicted in Table 1 (+*brain mask*), showing that brain masking significantly improved the prediction performance by 0.7 days ($p < 0.005$). Although the shape of the brain mask itself could also contain GA information, this experiments does show that the age estimation does not rely heavily on skull appearance.

Table 2. Mean absolute difference in days between each volume without augmentation, and the same volume subject to a range of augmentations. Lower values indicate better robustness. Values between brackets are the standard deviations.

	Rotation	Scaling	Translation	Combination
SF	0.9 (0.6)	0.4 (0.3)	0.7 (0.4)	2.3 (0.8)
CLC	0.6 (0.5)	0.3 (0.3)	0.6 (0.5)	2.2 (1.3)
POF	0.7 (0.6)	0.4 (0.3)	0.6 (0.4)	2.2 (1.2)

Gaussian Mask. For the POF and CLC, we applied Gaussian masks to the input volumes with a random σ between 1.3 and 3, chosen empirically to minimise the intensities at the border areas surrounding the fissures whilst highlighting the centre of the bounding box (Fig. 1). Despite the smoothing, the CLC performance remained the same, whilst the POF reduced by 0.3 days, showing the border information is not required for age prediction (Table 1 +*Gaussian mask*).

Inference Augmentation. To investigate the robustness of our networks with respect to bounding box position during inference, we applied test-time augmentation. For each volume, a range of random augmentations, within the same range as during training, were applied, and the mean absolute difference was computed between the GA predicted from each augmented volume and the volume without augmentation as shown in Table 2.

Across the three fissures, the variation in age prediction was less than a day after applying either scaling, translation or rotation to the bounding boxes, with rotation performing slightly worse. This suggests that the network is more robust to the location of the fissures than the angle, which notably changes across gestation. However, a larger mean difference of >2 days was found when applying all three augmentations, which is likely caused by key areas of the fissures being cropped from the bounding box area. This experiment demonstrates that our method is robust to bounding box location to a certain degree, however, in further work, more investigation is necessary to confirm this.

4   Conclusion

We have shown that it is possible to accurately predict the fetal GA based on the morphology of individual fissures, with the goal of assessing regional fetal development. Studying each fissure individually provides crucial insights into regional brain development, which ensures local changes do not cancel each other out. Our automated method has the potential to replace manual scoring systems that exhibit significant intra- and inter-observer variability, allowing fast, accurate and highly reproducible age predictions. This could be used clinically to ensure the brain is developing at the expected rate. In future work, better validated attention maps could also improve clinicians' trust and understanding of the method, by visualizing what the network's prediction is based on.

Acknowledgements. MW is supported by the Engineering and Physical Sciences Research Council (EPSRC) and Medical Research Council (MRC). LH acknowledges the support of the UK EPSRC Doctoral Training Award.MJ is supported by the National Institute for Health Research (NIHR) Oxford Biomedical Research Centre (BRC), and this research was funded by the Well- come Trust [215573/Z/19/Z]. The Wellcome Centre for Integrative Neuroimaging is supported by core funding from the Wellcome Trust [203139/Z/16/Z]. AN is grateful for support from the UK Royal Academy of Engineering under the Engineering for Development Research Fellowships scheme, and to St Hilda's College, Oxford.

References

1. Adebayo, J., Gilmer, J., Muelly, M., Goodfellow, I., Hardt, M., Kim, B.: Sanity checks for saliency maps. arXiv preprint arXiv:1810.03292 (2018)
2. Clouchoux, C., et al.: Delayed cortical development in fetuses with complex congenital heart disease. Cereb. Cortex **23**(12), 2932–2943 (2013)
3. Coelho Neto, M., Roncato, P., Nastri, C.O., Martins, W.D.P.: True reproducibility of ultrasound techniques (trust): systematic review of reliability studies in obstetrics and gynecology. Ultrasound Obstet. Gynecol. **46**(1), 14–20 (2015)
4. Dinsdale, N.K., et al.: Learning patterns of the ageing brain in MRI using deep convolutional networks. Neuroimage **224**, 117401 (2021)
5. Gholipour, A., et al.: A normative spatiotemporal MRI atlas of the fetal brain for automatic segmentation and analysis of early brain growth. Sci. Rep. **7**(1), 1–13 (2017)
6. Gotkowski, K., Gonzalez, C., Bucher, A., Mukhopadhyay, A.: M3D-CAM: a pytorch library to generate 3D data attention maps for medical deep learning (2020)
7. Hara, K., Kataoka, H., Satoh, Y.: Learning spatio-temporal features with 3D residual networks for action recognition. In: Proceedings of the IEEE International Conference on Computer Vision Workshops, pp. 3154–3160 (2017)
8. He, K., Zhang, X., Ren, S., Sun, J.: Deep residual learning for image recognition. In: Proceedings of the IEEE Conference on Computer Vision and Pattern Recognition, pp. 770–778 (2016)
9. Liao, L., et al.: Multi-branch deformable convolutional neural network with label distribution learning for fetal brain age prediction. In: 2020 IEEE 17th International Symposium on Biomedical Imaging (ISBI), pp. 424–427. IEEE (2020)
10. Namburete, A.I., Stebbing, R.V., Kemp, B., Yaqub, M., Papageorghiou, A.T., Noble, J.A.: Learning-based prediction of gestational age from ultrasound images of the fetal brain. Med. Image Anal. **21**(1), 72–86 (2015)
11. Namburete, A.I., Xie, W., Yaqub, M., Zisserman, A., Noble, J.A.: Fully-automated alignment of 3D fetal brain ultrasound to a canonical reference space using multitask learning. Med. Image Anal. **46**, 1–14 (2018)
12. Paladini, D., et al.: ISUOG practice guidelines (updated): sonographic examination of the fetal central nervous system. Part 2: Performance of targeted neurosonography. Ultrasound Obstet. Gynecol. **57**(4), 661–671 (2021)

13. Pérez-García, F., Sparks, R., Ourselin, S.: TorchIO: a python library for efficient loading, preprocessing, augmentation and patch-based sampling of medical images in deep learning. Comput. Methods Programs Biomed. 106236 (2021)

14. Pistorius, L., et al.: Grade and symmetry of normal fetal cortical development: a longitudinal two-and three-dimensional ultrasound study. Ultrasound Obstet. Gynecol. 36(6), 700–708 (2010)

15. Poon, L.C., et al.: Transvaginal three-dimensional ultrasound assessment of sylvian fissures at 18–30 weeks' gestation. Ultrasound Obstet. Gynecol. 54(2), 190–198 (2019)

16. Quarello, E., Stirnemann, J., Ville, Y., Guibaud, L.: Assessment of fetal sylvian fissure operculization between 22 and 32 weeks: a subjective approach. Ultrasound Obstet. Gynecol. Official J. Int. Soc. Ultrasound Obstet. Gynecol. 32(1), 44–49 (2008)

17. Rajagopalan, V., et al.: Local tissue growth patterns underlying normal fetal human brain gyrification quantified in utero. J. Neurosci. 31(8), 2878–2887 (2011)

18. Simonyan, K., Zisserman, A.: Very deep convolutional networks for large-scale image recognition. arXiv preprint arXiv:1409.1556 (2014)

19. Springenberg, J.T., Dosovitskiy, A., Brox, T., Riedmiller, M.: Striving for simplicity: the all convolutional net. arXiv preprint arXiv:1412.6806 (2014)

Texture-Based Analysis of Fetal Organs in Fetal Growth Restriction

Aya Mutaz Zeidan[1](✉), Paula Ramirez Gilliland[1](✉), Ashay Patel[1],
Zhanchong Ou[1], Dimitra Flouri[1], Nada Mufti[1,2], Kasia Maksym[2],
Rosalind Aughwane[2], Sébastien Ourselin[1], Anna L. David[2],
and Andrew Melbourne[1]

[1] School of Biomedical Engineering and Imaging Sciences, King's College London,
London, UK
{aya.zeidan,paula.ramirez_gilliland}@kcl.ac.uk
[2] Institute for Women's Health, University College London, London, UK

Abstract. Fetal growth restriction (FGR) is common, affecting around 10% of all pregnancies. Growth restricted fetuses fail to achieve their genetically predetermined size and often weigh <10th centile for gestation. However, even appropriately grown fetuses can be affected, with the diagnosis of FGR missed before birth. Babies with FGR have a higher rate of stillbirth, neonatal morbidity such as breathing problems, and neurodevelopmental delay. FGR is usually due to placental insufficiency leading to poor placental perfusion and fetal hypoxia. MRI is increasingly used to image the fetus and placenta. Here we explore the use of novel multi-compartment Intravoxel Incoherent Motion Model (IVIM)-based models for MRI fetal and placental analysis, to improve understanding of FGR and quantify abnormalities and biomarkers in fetal organs. In 12 normally grown and 12 FGR gestational-age matched pregnancies (Median 28^{+4}wks$\pm3^{+3}$wks) we acquired T_2 relaxometry and diffusion MRI datasets. Decreased perfusion, pseudo-diffusion coefficient, and fetal blood T_2 values in the placenta and fetal liver were significant features distinguishing between FGR and normal controls (p-value <0.05). This may be related to the preferential shunting of fetal blood away from the fetal liver to the fetal brain that occurs in placental insufficiency. These features were used to predict FGR diagnosis and gestational age at delivery using simple machine learning models. Texture analysis was explored to compare Haralick features between control and FGR fetuses, with the placenta and liver yielding the most significant differences between the groups. This project provides insights into the effect of FGR on fetal organs emphasizing the significant impact on the fetal liver and placenta, and the potential of an automated approach to diagnosis by leveraging simple machine learning models.

Keywords: FGR severity assessment · Multi-compartment models

The first two authors contributed equally.

© Springer Nature Switzerland AG 2021
C. H. Sudre et al. (Eds.): UNSURE 2021/PIPPI 2021, LNCS 12959, pp. 253–262, 2021.
https://doi.org/10.1007/978-3-030-87735-4_24

1 Introduction

The term Fetal Growth Restriction (FGR) is used to describe a fetus that has not reached their genetic growth potential, due to placental insufficiency causing inadequate supply of oxygen and nutrients [1]. FGR is a clinical diagnosis, defined by the Delphi consensus standardised definitions [2], and is divided into two different phenotypes, with onset either early (less than 32 weeks gestational age) or late in gestation. It is associated with high rates of stillbirth [3] and neonatal morbidity including increased rates of cerebral palsy, bronchopulmonary dysplasia, and cardiovascular disease long term [4]. There is currently no treatment for FGR, therefore clinicians must weigh the risks of prematurity against the risk of hypoxia and death in utero to determine the optimal delivery time. There are limited clinical tools to do this, so at present, clinicians follow national guidelines to make this decision [5].

MRI is increasingly used to image the placental circulation. The DECIDE multi-compartment model separates fetal and maternal flow characteristics of the placenta allowing measurement of the relative proportions of vascular spaces [6,7]. When applied in early-onset FGR, it identified reduced feto-placental blood oxygen saturation, where the degree of abnormality correlated with disease severity defined by ultrasound fetal and maternal arterial Doppler findings [8].

The motivation for this research was to compare MR derived parameters relating to perfusion and oxygenation within the placenta and three fetal organs (the brain, liver and lungs) between normally grown pregnancies and those complicated by early onset FGR, through multi-compartment models and texture analysis. Distinguishing features were then used to predict FGR diagnosis and gestational age (GA) at delivery via simple machine learning models.

2 Methods

2.1 Data

Patient MRI scans of voxel resolution $1.9 \times 1.9 \times 6$ mm were acquired using the acquisition parameters from [6] (enabling both T_2 relaxometry and diffusion MRI fitting), using a 1.5 T Siemens Avanto and performed under free-breathing. The dataset consisted of 12 early-onset FGR [2] ranging between $[24^{+2}, 33^{+6}]$ gestation weeks^{+days}, and 12 control pregnancies with MR data ranged between $[25^{+1}, 34^{+0}]$ GA interval, (Median 28^{+4}wks$\pm3^{+3}$wks) respectively. Specific details on subject inclusion criteria are available in [6]. The study was approved by the UK National Research Ethics Service and all participants gave written informed consent (REC reference 15/LO/1488).

There are biological mechanisms that may cause differences in the distribution of blood perfusion throughout the fetus in FGR. To investigate this, manual segmentation of the placenta, liver, lungs and brain was accomplished using itk-SNAP software. The resultant 3D mask files were used within the *NiftyFit* package for multi-parametric model-fitting [6], and to perform texture analysis.

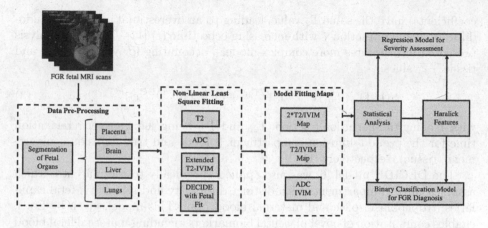

Fig. 1. Succeeding pre-processing of the data, model fitting techniques were applied to yield parameters describing various signals extracted from the placenta and fetal organs of interest. These parameters were then employed to perform texture analysis from multi-contrast MRI modelling. Results from the model fitting were used as inputs to the classifier and regressor to predict a diagnosis of FGR and the gestational age at delivery.

2.2 Model Fitting

Model fitting techniques were applied to each organ segmentation over the averaged ROI signal and on a voxelwise scale, yielding quantitative metrics for both approaches. Non-linear least squares were used to perform the fitting, with voxelwise fitting being initialised with the ROI parameter estimates - enhancing SNR. A range of models were explored, including simple T_2 and Apparent Diffusion Coefficient (ADC) estimation, as well as more complex models based on Intravoxel Incoherent Motion Model (IVIM) [9] and DECIDE [6] (Fig. 1).

The IVIM model describes perfusion as a pseudodiffusion process (represented by a pseudodiffusion coefficient, D^*), by characterising the collective motion of blood water molecules within the vessel network as a random walk. The IVIM model also incorporates "true" diffusion of water molecules (ADC), modelling the signal as

$$S = S_0[fe^{-bD^*} + (1-f)e^{-b\text{ADC}}], \tag{1}$$

where f is the perfusion fraction (volume occupied by incoherently flowing blood in a given voxel), b is the b-value, S is the measured signal and S_0 the baseline signal, [10]. This can be extended to incorporate T_2 relaxometry as

$$S = S_0e^{-t/T_2}[fe^{-bD^*} + (1-f)e^{-b\text{ADC}}]. \tag{2}$$

However, this model presents inherent limitations, as it assumes both vascular and tissue compartments (parametrised by pseudo-diffusion and true diffusion coefficients) have the same T_2 value, leading to an overestimation of the pseudo-diffusion volume fraction f with increasing echo time (t) [11]. Thus, the analysis presented incorporates more complex models, accounting for varying blood and tissue T_2 values:

$$S(\mathbf{b},\mathbf{t}) = S_0[fe^{-bD^*}e^{-t/T_{2p}} + (1-f)e^{-bADC}e^{-t/T_{2t}}], \tag{3}$$

with f being the perfusion fraction, T_{2p} and T_{2t} being the transverse relaxation time for the pseudo-diffusion compartment (blood) and true diffusion compartment (tissue), respectively [11].

The DECIDE model [6] was also applied specifically to the placenta, which assumes three compartments with distinct diffusivity and relaxivity: fetal capillaries, trophoblast space and maternal blood pool. This model, given by Eq. 4, enables computation of novel placental biomarkers including maternal fetal blood volume ratio and fetal blood saturation.

$$S(\mathbf{b},\mathbf{t}) = S_0 \left[fe^{-bD^*-tR_2^{fb}} + (1-f)e^{-bADC} \left(\nu e^{-tR_2^{mb}} + (1-\nu)e^{-tR_2^{ts}} \right) \right]. \tag{4}$$

Here, R_2^{fb}, R_2^{mb} and R_2^{ts} represent the inverse of relaxation transverse relaxation times for fetal blood, maternal blood and trophoblast space, respectively; and ν is the maternal blood volume fraction.

2.3   Texture Analysis

The aim of texture analysis was to examine the spatial arrangement of intensities in the segmented organs. To achieve this, Haralick features were extracted from the grey level co-occurrence matrix to describe the overall image texture using measures encompassing energy, entropy, correlation, contrast, variance, and homogeneity [12].

These features were computed for each subject on all model fitting maps, as well as the original IVIM T_2-weighted MRI scan, to yield interpretable texture descriptors [12,13]. The images were quantised into grey level bins of fixed equal width for between-subject texture feature value comparisons. Single-factor analysis of each feature was conducted between the FGR and control patients.

2.4   Statistical Methods and Feature Selection

Statistical analysis was performed on the model fitting maps to identify the most significant parameters in differentiating between the control and FGR cohorts. A Shapiro-Wilk test was used to confirm normality of the results. T-tests were then carried out between the cohorts for all the model fitted parameters, Haralick features, and organ ratio parameters. Results with p-value less than 0.05 indicated statistically significant differences between the control and FGR group means.

2.5   Machine Learning for FGR Diagnosis and Severity Assessment

Binary Classification Model. A linear logistic regression model was trained to predict healthy or FGR using an 80/20% train-test split. The best regularisation parameters were derived by performing a grid search and 3-fold cross-validation, yielding a final model with regularisation strength of 0.001 and an L1 ratio of 0, i.e. L2 regularisation.

Regression Model. A linear regression model was trained to predict the scan-to-delivery interval, using clinical records data including the date of birth and the date on which the MRI scan was taken. The trained model had an elastic-net regularisation with an L1 ratio of 0.22 (i.e. a combination of L1 and L2 regularisation at a ratio of 0.22:0.78 respectively) and a regularisation strength of 0.0061.

3   Results

3.1   Model Fitting

Figure 2 depicts examples of the parameter maps obtained from the model fitting techniques. The lower parameter map intensities in FGR compared to that in the controls is indicative of hypoperfusion and low oxygen saturation levels in these fetal organs. The T_2 maps display pronounced differences in the signal intensities of both cohorts. The most significant ROI and voxelwise parameters in identifying differences between controls and FGR fetuses were the perfusion fraction, S_0, pseudo-diffusion coefficient (D^*), and T_2 (as given in Tables 1 and 2). The placenta and liver were determined to be the most influential organs in diagnosing FGR.

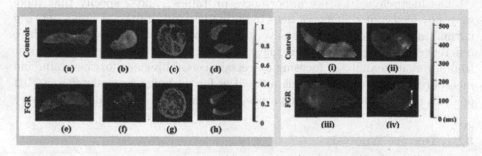

Fig. 2. Green shaded area plots: Perfusion fraction layer in the model fitting maps each taken from a single slice in the MRI scan. These correspond to ((a), (e)) placenta, ((b), (f)) liver, ((c), (g)) brain and ((d), (h)) lungs. **Grey shaded area plots:** T_2 maps for placenta ((i), (iii)), and **liver** ((ii), (iv)) from a single slice. In all cases top and bottom rows correspond to controls and FGR, respectively. (Color figure online)

Table 1. Hierarchy of parameter feature importances of the ROI measurements.

Model Fitting Technique	Parameter	Average Metric	Pairwise Group Comparison	Organ	T Statistic	P-Value
Standard IVIM	Perfusion Fraction	Mean	Control vs FGR	liver	4.114582746	0.000494053
T2 Dependent IVIM	Perfusion Fraction	Mean	Control vs FGR	placenta	2.885534725	0.008849743
DECIDE	Perfusion Fraction	Mean	Control vs FGR	placenta	2.776531433	0.011308604
Standard IVIM	S0	Min	Control vs FGR	placenta	2.469368562	0.022196615
Standard IVIM	Perfusion Fraction	Mode	Control vs FGR	placenta	2.440225199	0.023629376
ADC	Average S0	Median	Control vs FGR	placenta	1.985288036	0.060326406
Average T2 over ROI	Average T2	Min	Control vs FGR	placenta	1.928174375	0.067464017
T2 Dependent IVIM	T2	Median	Control vs FGR	placenta	1.926088314	0.067738281
Extended 2xT2 IVIM	Perfusion Fraction	Min	Control vs FGR	brain	1.866272068	0.078382926
T2 Dependent IVIM	S0	Mode	Control vs FGR	placenta	1.815709446	0.083721797

Table 2. Hierarchy of parameter feature importances of the voxelwise measurements.

Model Fitting Technique	Parameter	Average Metric	Pairwise Group Comparison	Organ	T Statistic	P-Value
Dependent IVIM	D*	Mean	Control vs FGR	placenta	-4.597300242	0.00015589
Extended 2xT2 Dependent IVIM	D*	Mean	Control vs FGR	placenta	-4.560436097	0.000170214
DECIDE Model (Voxelwise Measurements)	D*	Mean	Control vs FGR	placenta	-4.205788361	0.00039723
Extended 2xT2 Dependent IVIM	Perfusion Fraction	Min	Control vs FGR	placenta	3.725183003	0.001250966
Extended 2xT2 Dependent IVIM	Perfusion Fraction	Mode	Control vs FGR	placenta	3.725183003	0.001250966
Standard IVIM	Perfusion Fraction	Median	Control vs FGR	liver	3.624757118	0.001587669
T2 Dependent IVIM	T2	Min	Control vs FGR	placenta	3.463092031	0.002326109
Extended 2xT2 Dependent IVIM	Perfusion Fraction	Median	Control vs FGR	placenta	3.27041186	0.003653498
T2 Dependent IVIM	Perfusion Fraction	Min	Control vs FGR	placenta	3.249455242	0.003836258
T2 Dependent IVIM	Perfusion Fraction	Mode	Control vs FGR	placenta	3.249455242	0.003836258

The hierarchy of feature importances in Tables 1 and 2 specify that there were no significant differences detectable in the fetal brain and lungs between normal and FGR fetuses, especially compared to the placenta and liver, where differences were significant. This suggests that the brain and lungs may benefit from alternative analysis, focusing on certain cortical regions for the brain, and incorporating alternative imaging modalities for the lungs, as model fitting MRI analysis may not be the most appropriate technique for this fluid-filled organ.

3.2   Texture Analysis

Texture analysis was conducted on the most significant parameter maps for each organ, as determined by the t-tests. Results from the texture analysis were then concatenated by considering the mean and max of each Haralick feature.

Evaluation of the resulting Haralick features corroborated the degree of effect on the placenta in FGR, particularly using the Extended T_2 IVIM map and its mean variance. The brain was the least significantly different organ in this anal-

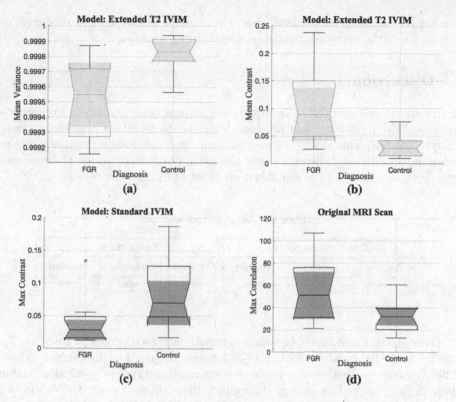

Fig. 3. Notched box plots of the most significant placental (pink) and liver (blue) Haralick features: **(a, b)** mean variance and contrast of the perfusion fraction in the Extended T_2 IVIM model, **(c)** max contrast of the Standard IVIM model, and **(d)** max correlation computed from the original scan. (Color figure online)

ysis. The notches in the box plots delineate the extent of significant difference in the medians of the investigated features by representing the confidence interval of the metric. Greater mean variance in the signal from the Extended T_2 IVIM model of the healthy cohort (refer to Fig. 3(a)), is indicative of increased heterogeneity in FGR placentas. Max correlation of the liver perfusion fraction in the controls in Fig. 3(d) reflects larger intensity differences compared to FGR. This is a significant feature to consider in the Standard IVIM model when studying the liver in FGR, especially given that the notches do not overlap between the cohorts.

3.3 Machine Learning

The classifier achieved a prediction accuracy of 100% in testing (refer to Table 3). Regarding the regressor, an RMSE of 0.02 weeks was achieved with the training data (N = 18), in contrast with the RMSE of 3.06 weeks obtained on the test set (N = 5). A generalisation error was evident, indicating a great degree of overfitting of the regression model on the training samples.

4   Discussion

In this study, we combined model fitting techniques, texture analysis from multi-contrast MRI modelling, and machine learning, to facilitate multi-fetal organ analysis of FGR. This provided a more holistic approach to imaging this common pregnancy condition. Differences were observed, particularly in the placenta and fetal liver, emphasising the significant effect of FGR on these organs.

Table 3. Classification results.

Cross Validation (N = 18)			Testing (N = 5)		
Accuracy	Sensitivity	Specificity	Accuracy	Sensitivity	Specificity
89%	78%	100%	100%	100%	100%

Overall, the fitted model parameters reveal decreased perfusion fraction, T_2, and D^* in the liver and placenta in FGR fetuses compared to the controls. These differences are indicative of a reduced oxygen saturation and perfusion within these organs, as well as abnormal capillary blood flow motion [8]. We did not observe significant differences in the properties of fetal brains and lungs between the FGR and control groups.

The machine learning analysis on these results supports the potential use of these parametric biomarkers in measuring FGR and providing an estimate of severity, including an indication of the likely GA at delivery. The classifier achieved 100% accuracy on testing data, indicating the model features are powerful indicators for FGR detection. But these results require prospective validation in a larger study population due to the small test group size in this proof-of-concept study. Moreover, a larger dataset would permit the transition into more complex prediction models in future research.

The RMSE of 3.06 weeks for the regressor's predictive performance encodes a large window in terms of fetal development. This method must therefore be refined before translation to a clinical environment. However, it may serve as a guide on condition severity. In practice though, this tool would also be used in conjunction with a wide range of information, including ultrasound data on fetal size, and maternal and fetal Doppler analysis of vascular resistance, which we have not included so far in this work.

The most influential Haralick features were extracted from the perfusion fraction measurements, particularly computed from the Extended T_2 IVIM and Standard IVIM models. Another important parameter determined by the Haralick features was T_2, attributed to its correlation with oxygen saturation (lower T_2 reflects a lower oxygen saturation [14]).

The placenta was established as the organ with most significant textural differences between the FGR and control groups. Variance, contrast, entropy and energy in placental perfusion fraction maps were the most significant textural differences between FGR and controls. This may be related to differences in the presence of maternal and fetal vascular malformation [15,16].

The second organ with greatest textural differences between both cohorts was the liver, particularly the pseudo-diffusion coefficient (D^*) maps (contrast, correlation, and energy), indicating spatial differences in the incoherent fetal capillary blood motion in this organ. This may indicate an abnormal blood motion in the liver compared to a healthy developing organ, affecting nutrient supply to this organ and may be related to the role of the ductus venosus in redistributing blood to the heart under the influence of increasing hypoxia [17]. Energy was heavily influenced by the number of grey levels and was, therefore, a significant feature for the placenta, lungs and brain, due to the presence of similar intensity voxels within local regions. Correlation was affected by the noise present in the image, which explains the notable correlation differences found in the liver, being the organ with the lowest SNR.

Analysis on parameter correlations indicated that as the perfusion fraction in the liver and placenta decreased, the more severely growth-restricted the FGR fetuses were. This corroborated our initial hypotheses for selecting the fetal liver and placenta as severely-affected organs in FGR, with SNR perhaps too low and variability too high to observe differences in the fetal brain and lung. Despite this, further work is needed to refine the analysis of the signals from these organs to better study the impact of FGR.

5   Conclusion

This study demonstrated the potential of MRI to improve holistic assessment of the fetus in FGR by assessing the vascular properties of highly-perfused fetal organs, via a multi-compartmental model fitting approach and texture analysis. The placenta and liver were prominent organs in identifying FGR fetuses, with key parametric features indicating a reduced perfusion, oxygenation and fetal capillary blood motion in these organs. Future work into multi-organ fetal analysis will extend these techniques to other placental complications in a larger-scale study.

Acknowledgements. This research was supported by the Wellcome Trust (210182/Z/18/Z and Wellcome Trust/EPSRC NS/A000027/1) and the Radiological Research Trust. The funders had no direction in the study design, data collection, data analysis, manuscript preparation or publication decision.

References

1. Lyall, F., Robson, S.C., Bulmer, J.N.: Spiral artery remodeling and trophoblast invasion in preeclampsia and fetal growth restriction: relationship to clinical outcome. Hypertension **62**(6), 1046–1054 (2013)

2. Gordijn, S.J., et al.: Consensus definition of fetal growth restriction: a Delphi procedure. Ultrasound Obstet. Gynecol. Official J. Int. Soc. Ultrasound Obstet. Gynecol. (2016)

3. Gardosi, J., Madurasinghe, V., Williams, M., Malik, A., Francis, A.: Maternal and fetal risk factors for stillbirth: population based study. BMJ (Online) **346**(7893) (2013)

4. Colella, M., Frérot, A., Novais, A.R.B., Baud, O.: Neonatal and long-term consequences of fetal growth restriction. Curr. Pediatr. Rev. **14**(4), 212–218 (2018)

5. Green-top Guideline No. The investigation and management of the small-for-gestational-age fetus (2002)

6. Melbourne, A., et al.: Separating fetal and maternal placenta circulations using multiparametric MRI. Magn. Reson. Med. **81**(1), 350–361 (2019)

7. Couper, S., et al.: The effects of maternal position, in late gestation pregnancy, on placental blood flow and oxygenation: an MRI study. J. Physiol. (2020)

8. Aughwane, R., et al.: MRI measurement of placental perfusion and oxygen saturation in early onset fetal growth restriction. BJOG: Int. J. Obstet. Gynaecol. 1471–0528.16387 (2020)

9. Le Bihan, D., Breton, E., Lallemand, D., Grenier, P., Cabanis, E., Laval-Jeantet, M.: MR imaging of intravoxel incoherent motions: application to diffusion and perfusion in neurologic disorders. Radiology **161**(2), 401–407 (1986)

10. Le Bihan, D.: What can we see with IVIM MRI? Neuroimage **187**, 56–67 (2019)

11. Jerome, N.P., et al.: Extended T2-IVIM model for correction of TE dependence of pseudo-diffusion volume fraction in clinical diffusion-weighted magnetic resonance imaging. Phys. Med. Biol. **61**(24), N667 (2016)

12. Haralick, R.M., Shanmugam, K., Dinstein, H.: Textural features for image classification. IEEE Trans. Syst. Man Cybern. **1**(6), 610–621 (1973)

13. Bharati, M.H., Liu, J.J., MacGregor, J.F.: Image texture analysis: methods and comparisons. Chemometr. Intell. Lab. Syst. **72**(1), 57–71 (2004)

14. Portnoy, S., Osmond, M., Zhu, M.Y., Seed, M., Sled, J.G., Macgowan, C.K.: Relaxation properties of human umbilical cord blood at 1.5 tesla. Magn. Reson. Med. **77**(4), 1678–1690 (2017)

15. Mifsud, W., Sebire, N.J.: Placental pathology in early-onset and late-onset fetal growth restriction. Fetal Diagn. Ther. **36**, 117–128 (2014)

16. Burton, G.J., Woods, A.W., Jauniaux, E., Kingdom, J.C.P.: Rheological and physiological consequences of conversion of the maternal spiral arteries for uteroplacental blood flow during human pregnancy. Placenta (2009)

17. Mifsud, W., Sebire, N.J.: Placental pathology in early-onset and late-onset fetal growth restriction. Fetal Diagn. Ther. **36**(2), 117–128 (2014)

Distributionally Robust Segmentation of Abnormal Fetal Brain 3D MRI

Lucas Fidon[1(✉)], Michael Aertsen[2], Nada Mufti[1,3,4], Thomas Deprest[2],
Doaa Emam[4,6], Frédéric Guffens[2], Ernst Schwartz[5], Michael Ebner[1],
Daniela Prayer[5], Gregor Kasprian[5], Anna L. David[3,4], Andrew Melbourne[1],
Sébastien Ourselin[1], Jan Deprest[2,3,4], Georg Langs[5], and Tom Vercauteren[1]

[1] School of Biomedical Engineering and Imaging Sciences, King's College London,
London, UK
lucas.fidon@kcl.ac.uk
[2] Department of Radiology, University Hospitals Leuven, Leuven, Belgium
[3] Institute for Women's Health, University College London, London, UK
[4] Department of Obstetrics and Gynaecology, University Hospitals Leuven,
Leuven, Belgium
[5] Department of Biomedical Imaging and Image-guided Therapy Medical University
of Vienna, Vienna, Austria
[6] Department of Gynecology and Obstetrics, University Hospitals Tanta,
Tanta, Egypt

Abstract. The performance of deep neural networks typically increases
with the number of training images. However, not all images have the
same importance towards improved performance and robustness. In fetal
brain MRI, abnormalities exacerbate the variability of the developing
brain anatomy compared to non-pathological cases. A small number of
abnormal cases, as is typically available in clinical datasets used for training, are unlikely to fairly represent the rich variability of abnormal developing brains. This leads machine learning systems trained by maximizing
the average performance to be biased toward non-pathological cases. This
problem was recently referred to as hidden stratification. To be suited
for clinical use, automatic segmentation methods need to reliably achieve
high-quality segmentation outcomes also for pathological cases. In this
paper, we show that the state-of-the-art deep learning pipeline nnU-Net
has difficulties to generalize to unseen abnormal cases. To mitigate this
problem, we propose to train a deep neural network to minimize a percentile of the distribution of per-volume loss over the dataset. We show
that this can be achieved by using Distributionally Robust Optimization
(DRO). DRO automatically reweights the training samples with lower
performance, encouraging nnU-Net to perform more consistently on all
cases. We validated our approach using a dataset of 368 fetal brain T2w
MRIs, including 124 MRIs of open spina bifida cases and 51 MRIs of
cases with other severe abnormalities of brain development.

Electronic supplementary material The online version of this chapter (https://
doi.org/10.1007/978-3-030-87735-4_25) contains supplementary material, which is
available to authorized users.

C. H. Sudre et al. (Eds.): UNSURE 2021/PIPPI 2021, LNCS 12959, pp. 263–273, 2021.
https://doi.org/10.1007/978-3-030-87735-4_25

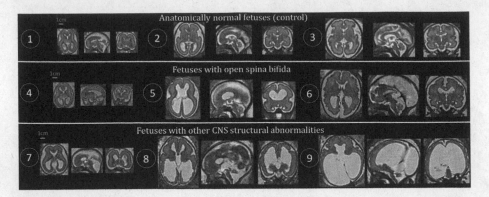

Fig. 1. Illustration of the anatomical variability in fetal brain across gestational ages and diagnostics. 1: Control (22 weeks); 2: Control (26 weeks); 3: Control (29 weeks); 4: Spina bifida (19 weeks); 5: Spina bifida (26 weeks); 6: Spina bifida (32 weeks); 7: Dandy-walker malformation with corpus callosum abnormality (23 weeks); 8: Dandy-walker malformation with ventriculomegaly and periventricular nodular heterotopia (27 weeks); 9: Aqueductal stenosis (34 weeks).

1   Introduction

The segmentation of fetal brain tissues in MRI is essential for the study of abnormal fetal brain developments [2]. Fetal brain structures segmentation could also support the evaluation and prediction of surgery outcome for open spina bifida [1,4,16,21,22]. Accurate and automatic methods for fetal brain segmentation are necessary as manual segmentation is very time-consuming and suffers from high inter- and intra-rater variability. Recently, deep neural network-based methods for fetal brain T2w MRI segmentation have been proposed [7,8,15,18,19]. On average, deep learning currently achieves state-of-the-art segmentation performance. However, those studies do not evaluate specifically the generalization and robustness properties when applied to fetuses with a pathological central nervous system.

Datasets used to train deep neural networks typically contain some under-represented subsets of cases. These cases are not specifically dealt with by the training algorithms currently used for deep neural networks. This problem has been referred to as hidden stratification [17]. Hidden stratification has been shown to lead to deep learning models with good average performance but poor performance on some clinically relevant subsets of the population [17]. While uncovering the issue, the study of [17], which is limited to classification, does not study the cause or propose a method to mitigate this problem. Cases with abnormal fetal brain development are likely to suffer from hidden stratification effects for two reasons: 1) The presence of abnormalities exacerbates the anatomical variability of the fetal brain between 18 weeks and 38 weeks of gestation, as illustrated in Fig. 1; and 2) The prevalence of those diseases is typically below 1/1000 births [1].

In this work, we study the problem of hidden stratification in fetal brain MRI segmentation using deep learning. We claim that the methodology currently used to train deep neural networks, that is maximizing the average performance across the training volumes, is at the root of the hidden stratification problem. Instead of the average empirical risk, training safe and robust deep learning models requires an asymmetric measure of risk that gives higher weights to the cases for which the algorithm fails (hard examples). Percentiles, also known as value-at-risk, is such a measure of risk that has even been adopted in industry regulations [13]. Given a per-volume fetal brain MRI segmentation metric such as the Dice score and an algorithm, the percentile at 5% is the value of the score below which 5% of the cases fall, i.e. perform worse than the percentile. The percentile relates to hidden stratification effects as it informs us of how badly worst-case examples are performing. Our contributions are four-fold. 1) We empirically show that the state-of-the-art deep learning pipeline nnU-Net [14] trained by maximizing the average segmentation performance leads to clinically significant failures for fetal brain MRI segmentation. 2) We propose to use percentiles of the Dice score on clinically relevant subpopulations as a measure of hidden stratification effects. 3) We propose to train a deep learning network to minimize a percentile of the per-volume loss function. 4) We propose a relaxation of this optimization problem based on distributionally robust optimization that can be solved efficiently in practice. We evaluate the proposed methodology for the automatic segmentation of white matter, ventricles, and cerebellum based on fetal brain 3D T2w MRI. We used a total of 368 fetal brain 3D MRIs including anatomically normal fetuses, fetuses with open spina bifida, and fetuses with other central nervous system pathologies for gestational ages ranging from 19 weeks to 39 weeks. Our empirical results suggests that the proposed training method based on distributionally robust optimization leads to better percentiles values for abnormal fetuses. In addition, qualitative results shows that distributionally robust optimization allows to reduce the number of clinically relevant failures of nnU-Net.

2 Minimization of a Percentile Loss Using Distributionally Robust Optimization

In this section, we study how a deep neural network can be trained to minimize percentiles of the loss function using a distributionally robust optimization (DRO) approach [10].

Standard deep learning training consists in optimizing the parameters θ of a deep neural network $f(\cdot; \theta)$ by minimizing the average per-example loss $\mathcal{L}$

$$\min_{\theta} \frac{1}{n} \sum_{i=1}^{n} \mathcal{L}\left(f(\boldsymbol{x}_i; \theta), \boldsymbol{y}_i\right) \tag{1}$$

Within this empirical risk minimization framework, $f(\cdot; \theta)$ is typically a Convolutional Neural Network (CNN), $\mathcal{L}$ is a smooth per-volume loss function, and $\{(\boldsymbol{x}_i, \boldsymbol{y}_i)\}_{i=1}^{n}$ is the training dataset.

In our case, $\boldsymbol{x}_i$ are the input 3D fetal brain T2w MRI volumes and $\boldsymbol{y}_i$ are the ground-truth manual segmentations. This approach is the one used to train state-of-the-art deep learning methods for segmentation using stochastic gradient descent [14]. Due to the scarcity and the higher anatomical variability of abnormal cases illustrated in Fig. 1, we cannot assume that the set of all possible fetal brain anatomies is sampled uniformly in the training dataset. However, in (1), all brain volumes are given the same weight equal to $\frac{1}{n}$.

Instead of the average per-volume loss, for robust and safe segmentation, we argue that it might be more interesting to minimize the percentile l_α at α (e.g. 5%) of the per-volume loss function. Formally, this corresponds to the minimization problem

$$\min_{\theta, l_\alpha} \quad l_\alpha \quad \text{such that} \quad \mathbb{P}\left(\mathcal{L}\left(f(\mathrm{x}; \boldsymbol{\theta}), \mathrm{y}\right) \geq l_\alpha\right) \leq \alpha \tag{2}$$

where $\mathbb{P}$ is the empirical distribution defined by the training dataset. In other words, if $\alpha = 0.05$, the optimal $l_\alpha^*(\boldsymbol{\theta})$ of (2) for a given value set of parameters $\boldsymbol{\theta}$ is the value of the loss such that the per-volume loss function is worse than $l_\alpha^*(\boldsymbol{\theta})$ 5% of the time. As a result, training the deep neural network using (2) corresponds to minimizing the percentile of the per-volume loss function $l_\alpha^*(\boldsymbol{\theta})$.

Unfortunately, the minimization problem (2) cannot be solved directly using stochastic gradient descent to train a deep neural network. We now propose a tractable upper bound for $l_\alpha^*(\boldsymbol{\theta})$ and show that it can be solved in practice using distributionally robust optimization [10].

The Chernoff bound [3] applied to the per-volume loss function and the empirical training data distribution states that for all l_α and $\beta > 0$

$$\mathbb{P}\left(\mathcal{L}\left(f(\mathrm{x}; \boldsymbol{\theta}), \mathrm{y}\right) \geq l_\alpha\right) \leq \frac{\exp\left(-\beta l_\alpha\right)}{n} \sum_{i=1}^{n} \exp\left(\beta \mathcal{L}\left(f(\boldsymbol{x}_i; \boldsymbol{\theta}), \boldsymbol{y}_i\right)\right) \tag{3}$$

To link this inequality to the minimization problem (2), we set β such that

$$\alpha = \frac{\exp\left(-\beta \hat{l}_\alpha(\boldsymbol{\theta})\right)}{n} \sum_{i=1}^{n} \exp\left(\beta \mathcal{L}\left(f(\boldsymbol{x}_i; \boldsymbol{\theta}), \boldsymbol{y}_i\right)\right) \tag{4}$$

$$\iff \hat{l}_\alpha(\boldsymbol{\theta}) = \frac{1}{\beta} \log\left(\frac{1}{\alpha n} \sum_{i=1}^{n} \exp\left(\beta \mathcal{L}\left(f(\boldsymbol{x}_i; \boldsymbol{\theta}), \boldsymbol{y}_i\right)\right)\right) \tag{5}$$

$\hat{l}_\alpha(\boldsymbol{\theta})$ is therefore an upper bound for $l_\alpha^*(\boldsymbol{\theta})$, independently to the value of $\boldsymbol{\theta}$. We propose to relax the minimization problem (2) by

$$\min_{\theta} \frac{1}{\beta} \log\left(\sum_{i=1}^{n} \exp\left(\beta \mathcal{L}\left(f(\boldsymbol{x}_i; \boldsymbol{\theta}), \boldsymbol{y}_i\right)\right)\right) \tag{6}$$

where $\beta > 0$ is a hyperparameter, and where the term $\frac{1}{\beta} \log\left(\frac{1}{\alpha n}\right)$ was dropped as being independent of $\boldsymbol{\theta}$. While in (6), α does not appear in the optimization

Table 1. Training and testing dataset details. Other Abn: other brain structural abnormalities. There is no overlap of subjects between training and testing.

Train/Test	Origin	Condition	Volumes	Gestational age (in weeks)
Training	Atlas [12]	Control	18	[21, 38]
Training	FeTA [18]	Control	5	[22, 28]
Training	UHL and MUV	Control	116	[20, 35]
Training	UHL and MUV	Spina Bifida	28	[22, 34]
Training	UHL and MUV	Other Abn	10	[23, 35]
Testing	FeTA [18]	Control	28	[20, 34]
Testing	FeTA [18]	Spina Bifida	31	[22, 31]
Testing	FeTA [18]	Other Abn	16	[20, 34]
Testing	UHL and MUV	Control	26	[26, 37]
Testing	UHL and MUV	Spina Bifida	65	[19, 33]
Testing	UHL and MUV	Other Abn	25	[21, 40]

problem directly anymore, β essentially acts as a substitute for α. The higher the value of β, the higher weights the per-volume losses with a high value will have in (6).

We give a proof in the supplementary material[1] that (6) is equivalent to solving the distributionally robust optimization problem

$$\min_{\theta} \max_{q \in \Delta_n} \left(\sum_{i=1}^{n} q_i \, \mathcal{L}\left(f(x_i; \theta), y_i \right) - \frac{1}{\beta} D_{KL}\left(q \,\middle\|\, \frac{1}{n}\mathbf{1} \right) \right) \qquad (7)$$

where a new unknown probabilities vector parameter q is introduced, $\frac{1}{n}\mathbf{1}$ denotes the uniform probability vector $\left(\frac{1}{n}, \ldots, \frac{1}{n} \right)$, D_{KL} is the Kullback-Leibler divergence, Δ_n is the unit n-simplex, and $\beta > 0$ is a hyperparameter. D_{KL} measures the dissimilarity between q and the uniform probability vector $\frac{1}{n}\mathbf{1}$ that corresponds to assign the same weight $\frac{1}{n}$ to each sample. Therefore, β controls how much the samples with a relatively high loss value (hard examples) are weighted.

Recently, hardness weighted sampling [10] was introduced as a principled hard example mining method to solve (7). Here, we proved that it can be used to minimize the proposed relaxed minimization (6) of the percentile loss problem.

3   Anatomically Abnormal Fetal Brain T2w MRI Dataset

In this section, we give details about the fetal brain 3D MRI data, the labelling protocol, and the pre-processing used in our experiments.

---

[1] Please see the arxiv version for the supplementary material http://arxiv.org/abs/2108.04175.

Public Fetal Brain Datasets. We used the 18 control fetal brain 3D MRI volumes of the spatio-temporal fetal brain atlas[2] [12] for gestational ages ranging from 21 weeks to 38 weeks. We also used 80 volumes from the publicly available FeTA MICCAI challenge dataset[3] [18]. For the 40 MIAL 3D MRIs, corrections of the segmentations were performed by authors MA, LF, and PD to reduce the variability against the published segmentation guidelines that was released with the FeTA dataset [18]. Those corrections were performed as part of our previous work [8] and are publicly available[4]. Brain masks for the FeTA data were obtained via affine registration using two fetal brain atlases[5] [11,12].

Image Acquisition and Preprocessing for the Private Dataset. All images in the private dataset were part of routine clinical care and were acquired at UHL and MUV due to congenital malformations seen on ultrasound.

In total, 93 cases with open spina bifida, 35 cases with other central nervous system pathologies, and 142 cases with other malformations, though with normal brain, and referred as controls, were included. The gestational age at MRI ranged from 19 weeks to 40 weeks. We have started to make fetal brain T2w 3D MRIs publicly available[6]. For each study, at least three orthogonal T2-weighted HASTE series of the fetal brain were collected on a 1.5T scanner using an echo time of 133 ms, a repetition time of 1000 ms, with no slice overlap nor gap, pixel size 0.39 mm to 1.48 mm, and slice thickness 2.50 mm to 4.40 mm. A radiologist attended all the acquisitions for quality control.

The reconstructed fetal brain 3D MRIs were obtained using `NiftyMIC` [6] a state-of-the-art super resolution and reconstruction algorithm. The volumes were all reconstructed to a resolution of 0.8 mm isotropic and registered to a fetal brain atlas [12]. Our pre-processing improves the resolution, and removes motion between neighboring slices and motion artefacts present in the original 2D slices [6]. We used volumetric brain masks to mask the tissues outside the fetal brain. Those brain masks were obtained using the automatic segmentation method described in [6,20].

Labelling Protocol. The labelling protocol used for white matter, ventricles and cerebellum is the same as in [18]. The three tissue types were segmented for our private dataset by a trained obstetrician and medical students under the supervision of a paediatric radiologist specialized in fetal brain anatomy, who quality controlled and corrected all manual segmentations.

Separation of the Data into Training and Testing. A summary of the number of fetal brain 3D MRIs used at training and testing for each central

---

[2] http://crl.med.harvard.edu/research/fetal_brain_atlas/.

[3] DOI: 10.7303/syn25649159.

[4] DOI: 10.5281/zenodo.5148611.

[5] DOI: 10.7303/syn25887675.

[6] https://www.cir.meduniwien.ac.at/research/fetal/.

Ground-truth segmentation   nnU-Net   nnU-Net-DRO (ours)

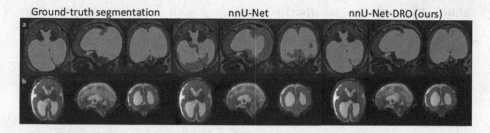

Fig. 2. Qualitative results. a) Fetus with aqueductal stenosis (34 weeks). b) Fetus with open spina bifida (27 weeks). For those two cases, nnU-Net [14] misses completly the cerebellum and achieves poor segmentation for the white matter and the ventricles. Our nnU-Net-DRO achieves satisfactory segmentation for the cerebellum for the two cases, and for all tissue types for the aqueductal stenosis case.

nervous system condition can be found in Table 1. The training dataset contains a total of 177 cases with a majority of 139 controls and only 38 abnormal cases which is typical in clinical datasets. Five controls from the FeTA dataset were added in the training dataset because we found in preliminary experiments that nnU-Net [14] fails on most of the FeTA data at testing when it is trained using only data from UHL and MUV and the fetal brain atlas [12]. The testing dataset contains 193 volumes with a majority of abnormal cases which is necessary to cover the anatomical variability of abnormal cases in our evaluation.

4 Experiments

Common Deep Learning Pipeline. We used nnU-Net [14], a generic deep learning pipeline for medical image segmentation, that has been shown to outperform other deep learning pipelines on 23 public datasets without the need to tune the loss function or the deep neural network architecture. Specifically, we used nnU-Net version 2 in 3D-full-resolution mode which is the recommended mode for isotropic 3D MRI data. nnU-Net automatically splits the training data into 5 folds 80% training/20% validation used to train 5 networks for each method. The predicted class probability maps of the 5 models are averaged at inference to improve robustness [14]. We used NVIDIA Tesla V100 GPUs with 16 GB of memory. Training each network took from 4 to 6 days.

Specificities of Each Method. The baseline consists in using nnU-Net [14] without any modification. Our method, nnU-Net-DRO, also uses nnU-Net. The only difference is that we changed the sampling strategy to use the hardness weighted sampler for DRO [10]. We used the default hyper-parameter values for the hardness weighted sampler, i.e. $\beta = 100$ with importance sampling and clipping values $w_{min} = 0.1$ and $w_{max} = 10$ as described in [10]. No other values were tested. Our implementation of the nnU-Net-DRO training procedure is publicly available at https://github.com/LucasFidon/HardnessWeightedSampler. It provides an implementation of the hardness weighted sampler described in [10].

Table 2. Evaluation of distribution robustness with respect to the pathology (193 3D MRIs). WM: White matter, Vent: Ventricles, Cer: Cerebellum. p_X: X^{th} percentile of the Dice score distribution in percentage. Best values are in bold.

Method	CNS	ROI	Dice Score (%)					
			Mean	Std	p_{50}	p_{25}	p_{10}	p_5
(baseline) nnU-Net	Controls (54 cases)	WM	**93.9**	2.9	94.1	**91.5**	**90.6**	**89.3**
		Vent	87.8	6.8	89.7	82.1	78.1	**76.8**
		Cer	**94.5**	3.2	94.6	92.4	**90.7**	89.8
	Spina Bifida (98 cases)	WM	89.9	7.9	92.5	89.1	79.9	73.4
		Vent	90.6	10.6	93.0	88.6	84.8	80.7
		Cer	78.2	28.7	**89.8**	**84.2**	13.9	0.0
	Other Abn. (41 cases)	WM	90.3	9.8	**92.7**	**89.7**	**82.7**	70.1
		Vent	87.1	7.3	87.1	82.5	77.7	75.2
		Cer	89.7	14.7	**92.8**	89.4	85.1	81.6
(ours) nnU-Net-DRO	Controls (54 cases)	WM	93.8	3.0	93.9	91.2	90.1	89.2
		Vent	**87.9**	**6.7**	**89.9**	**82.6**	**78.3**	76.7
		Cer	94.4	**3.1**	94.6	**92.6**	**90.7**	89.5
	Spina Bifida (98 cases)	WM	**90.3**	**7.5**	**92.9**	**89.2**	**81.5**	**73.7**
		Vent	**90.9**	**10.3**	**93.2**	**89.2**	**85.1**	**81.7**
		Cer	**79.7**	**27.6**	89.7	84.1	**40.4**	0.0
	Other Abn. (41 cases)	WM	**90.3**	**9.5**	92.5	89.6	82.5	**72.0**
		Vent	**87.5**	**7.1**	87.5	**82.7**	**80.4**	**76.7**
		Cer	**90.6**	**10.5**	**92.8**	**89.8**	**85.5**	**82.9**

Evaluation Method. We evaluate the quality of the automatic fetal brain MRI segmentations using the Dice score [5,9]. We are particularly interested in measuring the statistical risk of the results as a way to evaluate the robustness of the different methods. To this end, in addition to the mean and standard deviation, we also report the percentiles of the Dice score at 50%, 25%, 10%, and 5%. In Table 2, we report those quantities for the Dice scores of the three tissue types white matter, ventricular system, and cerebellum.

For each method, nnU-Net is trained 5 times using different train/validation splits and different random initializations. The 5 same splits, computed randomly, are used for the two methods. The results in Table 2 are for the ensemble of the 5 3D U-Nets. Ensembling is known to increase the robustness of deep learning methods for segmentation [14]. It also makes the evaluation less sensitive to the random initialization and to the stochastic optimization.

Evaluation of nnU-Net and nnU-Net-DRO. Quantitative evaluation of nnU-Net and nnU-Net-DRO for the three different central nervous system conditions control, spina bifida, and other abnormalities can be found in Table 2.

For spina bifida and other brain abnormalities, the proposed nnU-Net-DRO achieves same or higher mean Dice scores and lower standard deviations than nnU-Net [14] for the three tissue types. For controls, the mean Dice scores and standard deviation of nnU-Net-DRO and nnU-Net differ by less than 0.1 percentage points (pp) for the three tissue types.

The comparison of the percentiles of the Dice score allows us to compare methods at the tail of the Dice scores distribution where segmentation methods reach their worst-case performance. For spina bifida, nnU-Net-DRO achieves higher values of percentiles than nnU-Net for the white matter ($+0.6$pp for $\mathbf{p}_{10}$), for the ventricular system ($+1.0$pp for $\mathbf{p}_5$), and for the cerebellum ($+26.5$pp for $\mathbf{p}_{10}$). And for other brain abnormalities, nnU-Net-DRO achieves higher values of percentiles than nnU-Net for the white matter ($+1.9$pp for $\mathbf{p}_5$), for the ventricular system ($+1.5$pp for $\mathbf{p}_5$ and $+2.7$pp for $\mathbf{p}_{10}$), and for the cerebellum ($+1.3$pp for $\mathbf{p}_5$). All the other percentile values differ by less than 0.5pp of Dice score between the two methods. This suggests that nnU-Net-DRO achieves better worst case performance than nnU-Net for abnormal cases.

It is worth noting that the Dice scores decrease for the white matter and the cerebellum between controls and spina bifida and abnormal cases. It was expected due to the higher anatomical variability in pathological cases. However, the Dice scores for the ventricular system tend to be higher for abnormal cases than for controls. This can be attributed to the large proportion of pathological cases with enlarged ventricles because the Dice score values tend to be higher for larger region of interests.

As can be seen in the qualitative results of Table 2, there are cases for which nnU-Net predicts an empty cerebellum segmentation while nnU-Net-DRO achieves satisfactory cerebellum segmentation. There were no cases for which the converse was true. Robust segmentation of the cerebellum for spina bifida is particularly relevant for the evaluation of fetal brain surgery for open spina bifida [1,4,21]. Additional qualitative results in the supplementary material[7] illustrates 5 other cases for which nnU-Net-DRO outperforms nnU-Net.

5   Conclusion

The high anatomical variability of the developing fetal brain across gestational ages and pathologies hampers the robustness of deep neural networks trained by maximizing the average per-volume performance. Specifically, it limits the generalization of deep neural networks to abnormal cases for which few cases are available during training. In this paper, we propose to mitigate this problem by training deep neural networks to minimize a percentile of the per-volume performance rather than the average. To allow to do this in practice, we propose to train deep neural networks with Distributionally Robust Optimization (DRO) and we show that the DRO objective is a relaxation of the per-volume loss percentile. We have validated the proposed training method on a multicentric dataset of 368 fetal brain T2w 3D MRIs with various diagnostics. nnU-Net trained with DRO achieved improved segmentation results for pathological

---

[7] Please see the arxiv version for the supplementary material http://arxiv.org/abs/2108.04175.

cases as compared to the unmodified nnU-Net, while achieving similar segmentation performance for the neurotypical cases. Our results suggest that nnU-Net trained with DRO is more robust to anatomical variabilities than the original nnU-Net.

Acknowledgments. This project has received funding from the European Union's Horizon 2020 research and innovation program under the Marie Skłodowska-Curie grant agreement TRABIT No 765148. This work was supported by core and project funding from the Wellcome [203148/Z/16/Z; 203145Z/16/Z; WT101957], and EPSRC [NS/A000049/1; NS/A000050/1; NS/A000027/1]. TV is supported by a Medtronic/RAEng Research Chair [RCSRF1819\7\34].

References

1. Aertsen, M., et al.: Reliability of MR imaging-based posterior fossa and brain stem measurements in open spinal dysraphism in the era of fetal surgery. Am. J. Neuroradiol. **40**(1), 191–198 (2019)
2. Benkarim, O.M., et al.: Toward the automatic quantification of in utero brain development in 3D structural MRI: a review. Hum. Brain Mapp. **38**(5), 2772–2787 (2017)
3. Chernoff, H., et al.: A measure of asymptotic efficiency for tests of a hypothesis based on the sum of observations. Ann. Math. Stat. **23**(4), 493–507 (1952)
4. Danzer, E., Joyeux, L., Flake, A.W., Deprest, J.: Fetal surgical intervention for myelomeningocele: lessons learned, outcomes, and future implications. Dev. Medi. Child Neurol. **62**(4), 417–425 (2020)
5. Dice, L.R.: Measures of the amount of ecologic association between species. Ecology **26**(3), 297–302 (1945)
6. Ebner, M., et al.: An automated framework for localization, segmentation and super-resolution reconstruction of fetal brain MRI. Neuroimage **206**, 116324 (2020)
7. Fetit, A.E., et al.: A deep learning approach to segmentation of the developing cortex in fetal brain MRI with minimal manual labeling. In: Medical Imaging with Deep Learning, pp. 241–261. PMLR (2020)
8. Fidon, L., et al.: Label-set loss functions for partial supervision: application to fetal brain 3D MRI parcellation. arXiv preprint arXiv:2107.03846 (2021)
9. Fidon, L., et al.: Generalised Wasserstein dice score for imbalanced multi-class segmentation using holistic convolutional networks. In: Crimi, A., Bakas, S., Kuijf, H., Menze, B., Reyes, M. (eds.) BrainLes 2017. LNCS, vol. 10670, pp. 64–76. Springer, Cham (2018). https://doi.org/10.1007/978-3-319-75238-9_6
10. Fidon, L., Ourselin, S., Vercauteren, T.: Distributionally robust deep learning using hardness weighted sampling. arXiv preprint arXiv:2001.02658 (2020)
11. Fidon, L., et al.: A spatio-temporal atlas of the developing fetal brain with spina bifida aperta. Open Res. Europe (2021)
12. Gholipour, A., et al.: A normative spatiotemporal MRI atlas of the fetal brain for automatic segmentation and analysis of early brain growth. Sci. Rep. **7**(1), 1–13 (2017)
13. Holton, G.: Value at Risk: Theory and Practice. Academic Press (2003)
14. Isensee, F., Jaeger, P.F., Kohl, S.A., Petersen, J., Maier-Hein, K.H.: nnU-Net: a self-configuring method for deep learning-based biomedical image segmentation. Nat. Methods **18**(2), 203–211 (2021)

15. Khalili, N., et al.: Automatic brain tissue segmentation in fetal MRI using convolutional neural networks. Magn. Reson. Imaging **64**, 77–89 (2019)
16. Mufti, N., et al.: Cortical spectral matching and shape and volume analysis of the fetal brain pre-and post-fetal surgery for spina bifida: a retrospective study. Neuroradiology 1–14 (2021)
17. Oakden-Rayner, L., Dunnmon, J., Carneiro, G., Ré, C.: Hidden stratification causes clinically meaningful failures in machine learning for medical imaging. In: Proceedings of the ACM Conference on Health, Inference, and Learning, pp. 151–159 (2020)
18. Payette, K., et al.: An automatic multi-tissue human fetal brain segmentation benchmark using the fetal tissue annotation dataset. Sci. Data **8**(1), 1–14 (2021)
19. Payette, K., et al.: Longitudinal analysis of fetal MRI in patients with prenatal spina bifida repair. In: Wang, Q., et al. (eds.) PIPPI/SUSI -2019. LNCS, vol. 11798, pp. 161–170. Springer, Cham (2019). https://doi.org/10.1007/978-3-030-32875-7_18
20. Ranzini, M., Fidon, L., Ourselin, S., Modat, M., Vercauteren, T.: MONAIfbs: MONAI-based fetal brain MRI deep learning segmentation. arXiv preprint arXiv:2103.13314 (2021)
21. Sacco, A., et al.: Fetal surgery for open spina bifida. Obstetrician Gynaecol. **21**(4), 271 (2019)
22. Zarutskie, A., et al.: Prenatal brain imaging for predicting need for postnatal hydrocephalus treatment in fetuses that had neural tube defect repair in utero. Ultrasound Obstet. Gynecol. **53**(3), 324–334 (2019)

Analysis of the Anatomical Variability of Fetal Brains with Corpus Callosum Agenesis

Fleur Gaudfernau[1]([⊠]), Eléonore Blondiaux[2], and Stéphanie Allassonnière[1]

[1] CRC, Université de Paris, INRIA EPI HeKa, INSERM UMR 1138, Sorbonne Université, Paris, France
`fleur.gaudfernau@etu.u-paris.fr`
[2] Service de Radiologie, Hôpital Armand-Trousseau, APHP, Paris, France

Abstract. Corpus Callosum Agenesis (CCA), one of the most common congenital anomalies, has uncertain neurodevelopmental outcome, especially when the disease is isolated. To provide parents with informed counselling, it is crucial to identify anatomical markers linked to a predicted outcome early in pregnancy. Quantitative exploration of fetal brains with CCA is rare and has been mostly limited to the study of specific brain structures. Here, we propose to characterize the anatomical variability of fetal brains with CCA using a shape analysis pipeline based on diffeomorphic transformation. 38 MRIs from healthy fetuses and 73 from fetuses with CCA are retrospectively selected and volume reconstructed. A healthy template is registered to each fetal brain to quantify deviations from normal development at a *global scale*. Deformations are parallel transported to the same space to smooth age differences between fetuses. Deformation modes specific to CCA are identified using Principal Component Analysis (PCA) and classification. In accordance with more local analyses, the most relevant deformation mode for classification combines well-known alterations of brains with CCA. This preliminary work is promising for the quantitative exploration of abnormal fetal brains and will be used in the future to identify anatomical features correlated to poor clinical outcome.

Keywords: Corpus callosum agenesis · Fetal magnetic resonance imaging · Diffeomorphic registration

1 Introduction

Corpus callosum agenesis (CCA) is one of the most common congenital brain anomalies, with a prevalence at birth of 0.02% [13]. It is characterized by the total or partial absence of the largest commissure of the brain, responsible for the transmission of sensory, motor and cognitive information between hemispheres [13]. Diagnosis is usually suspected during the second-trimester routine ultrasound, and confirmed by fetal Magnetic Resonance Imaging (MRI) [13]. In

© Springer Nature Switzerland AG 2021
C. H. Sudre et al. (Eds.): UNSURE 2021/PIPPI 2021, LNCS 12959, pp. 274–283, 2021.
https://doi.org/10.1007/978-3-030-87735-4_26

complement with genetic screening, MRI is valuable to provide clinicians with additional information, as the presence of other anomalies is the only consensual prognosis factor for neurodevelopmental delays [17]. In the presence of associated defects, accounting for 45% cases [17], the outcome is usually poor, with impairments affecting motor control, coordination and language [5]. Predicting the outcome is challenging in isolated CCA, where 20–30% children demonstrate a broad spectrum of cognitive deficits [5,17], resulting in heterogenous medical counselling across hospitals and countries [7]. To provide parents with informed counselling, it is crucial to identify anatomical markers linked to neurodevelopmental outcome as early as possible during pregnancy.

Quantitative analysis of fetal brains has long been limited by the scarcity of fetal MRI and its restriction to 2D slices [4]. Most studies focused on characterizing healthy brain growth and cortical folding. Only few studies have investigated quantitatively anatomical alterations in fetuses with CCA [12,14,18,19], and their focus was on specific brain structures rather than global trends. Another limitation is the difficulty to compare fetal brains of different gestational ages (GA), since they undergo rapid and drastic changes across pregnancy [10].

Whole brain shape analysis can provide information about which structures are impaired along with corpus callosum. To perform such global analysis, one can think of image registration, which maps a population average brain template onto individual images in order to measure a distance from normality. In a clinical setting, functions called diffeomorphisms are an appropriate choice for computing shape changes, as they are high dimensional, topology-preserving, and sensitive to small anatomical variations. The Large Deformation Diffeomorphic Metric Mapping (LDDMM) setting [3,20] is a powerful method for computing such functions, which are seen as geodesics on a Riemanian manifold. Diffeomorphisms can be efficiently computed through a discrete parametrization [8]. The LDDMM framework also provides geometrical tools such as parallel transport, which enables comparing subjects of different developmental stages. Diffeomorphisms have proven useful in the quantification and classification of disorders such as Alzheimer's disease [6,15]. To our knowledge, deformation models have never been applied to abnormal fetal brains.

Here, we propose to explore the anatomical variability of fetal brains diagnosed with CCA using diffeomorphic brain mapping. After registration to a template brain, age-related differences between fetuses will be erased by transporting deformations to a common space. CCA specific deformations will be identified using Principal Component Analysis (PCA) and classification.

2   Materials and Methods

2.1   Image Acquisition and Preprocessing

Data. Data consists of retrospectively selected fetal MRIs from hospital [anonymous], performed between 2006 and 2019. Abnormality of the corpus callosum was identified at second or third trimester screening ultrasound examinations, followed by expert ultrasound assessment to investigate other associated fetal anomalies before the fetal MRI. Corpus callosum anomalies were defined as: 1. complete CCA defined as the complete absence of the corpus callosum and 2. partial CCA defined as the absence of one or more of the five segments of the corpus callosum resulting in an abnormally shaped corpus callosum. Inclusion criteria were: fetuses affected by isolated or associated partial or complete corpus callosum abnormalities and fetuses with normal central nervous system findings at MRI examination. The database contains 38 healthy fetuses scanned at GA between 26 and 37 weeks (mean $= 32.4 \pm 1.69$) and 73 fetuses diagnosed with CCA scanned at GA between 25 and 37 weeks (mean $= 31.63 \pm 2.09$). In the latter group, 51 fetuses have partial CCA and 22 complete CCA.

Image Acquisition. Fetal brain MRI was performed using repeated T2 half-Fourier Single Shot Fast Spin Echo (SSFSE), or Single-Shot half-Fourier Turbo Spin Echo (SshTSE). Fetal MRIs were performed on a 1.5 T MRI system Achieva Philips (Best, the Netherlands) before 2016 and Optima MR450w General Electric (Waukesha, WI, USA), since 2016. Maternal sedation was systematically offered to reduce fetal motion artefacts. Scan acquisitions were performed in the three orthogonal planes. Scanning parameters were as follows: field of view: 256×256 or 512×512 mm; echo time: 150–200 ms; repetition time: 3500–4000 ms; slice thickness: 4 mm; flip angle: 90°; acquisition matrix: 320×320.

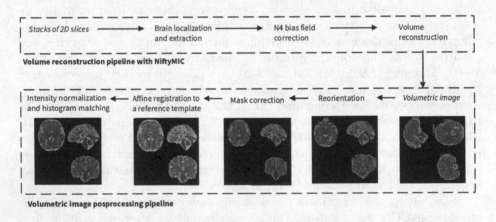

Fig. 1. Fetal MRI preprocessing steps

Image Processing. Isotropic high resolution 3D volume reconstruction of fetal brains is performed using a state-of-the-art algorithm [9] and followed by additional processing steps as described in Fig. 2. The main reconstruction steps comprise localization and segmentation of the fetal brain, bias field correction and outlier-robust super-resolution reconstruction. We process further the resulting image in order to enable inter-subjects comparisons (see Fig. 1). As the fetus orientation is unknown during image acquisition, the coronal, sagittal and axial planes are automatically identified based on length and symmetry measurements, and flipped in the right direction by minimizing the sum of squared differences between all possible orientations and a reference fetal brain at 31 weeks of GA from [10] atlas. The subject brains are aligned and cropped to a size of $105 \times 100 \times 120$ voxels. The algorithm used for brain segmentation [9] often misclassifies voxels belonging to the skull or the placenta as brain (see Fig. 1). A correct brain mask is extracted from the reference brain, rigidly registered to each erroneous fetal brain mask, and used to re-mask the fetal brain. In cases where brain over-detection is too important, manual refinement of the masks is performed using ITK-SNAP, Version 3.6 [22]. To enable inter-subjects comparisons and eliminate position and size differences, fetal brains are transported to a common anatomical space by performing affine registration to the previously used reference template. Finally, intensity normalization and histogram matching to the template brain are performed.

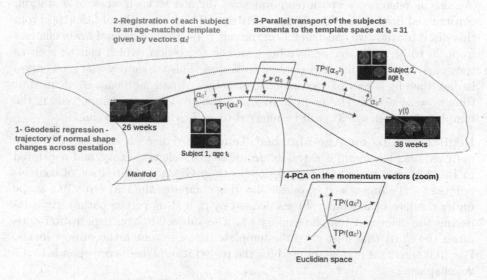

Fig. 2. Shape analysis pipeline (Color figure online)

2.2   Shape Analysis

Shape Analysis Pipeline. Registering a reference average brain, called template, to healthy or pathological brains, yields transformations that encode subject-specific anatomical deviations from normality. As brains undergo important structural changes during gestation, we compare each fetal brain to a healthy template brain of the same age using registration. To enable inter-subjects comparisons, deformations are transported to a common space using parallel transport. PCA is applied to the transported subject deformations to reduce dimension and extract relevant features. Finally, these features are fed to a Support Vector Machine (SVM) to perform patient classification. The steps of our shape analysis pipeline are summarized in Fig. 2 and detailed below.

We perform shape analysis computations in the LDDMM framework [3,20], where shapes are seen as objects on a Riemanian Manifold and transformations belong to groups of diffeomorphims. Diffeomorphisms can be entirely parametrized by a finite set of initial momentum vectors α_0 attached to control points c_0 [8]. Evolution of $\alpha(t)$ and $c(t)$, described by Hamiltonian equations, defines a time-varying velocity field, whose integration yields a flow of diffeomorphisms [8]. This framework provides tools for computational anatomy, namely geodesic regression, geodesic shooting and parallel transport. Optimization is performed with a gradient-descent algorithm. Shape comparisons are computed using the open-source software Deformetrica [2].

Geodesic Regression of Template Brains. To adjust for anatomical changes during gestation, each fetal brain is compared to an age-matched healthy brain. We use as reference a spatiotemporal atlas defined at each week of gestation, constructed from 81 healthy fetuses between 19 and 39 weeks of GA [10]. From this discrete atlas, we construct a continuous trajectory of normal brain changes from 26 to 38 weeks by performing geodesic regression, which can be seen as the generalisation of linear regression to shapes. This trajectory $\gamma(t)$ (red curve) is described by a pair of vectors: control points c_0 and momenta α_0 defined at time $t_0 = 31$ weeks. The point from which the geodesic was computed, i.e. the template brain at age 31, will be referred to as T_{ref} in the following.

Registration to an Age-Matched Template. For each subject i, the age-matched template brain is extracted from the geodesic trajectory, and registered to the subject's brain using geodesic shooting. Given an initial set of controls points c_0^i and momenta α_0^i, geodesic shooting computes the trajectory of a shape under the flow of diffeomorphisms defined by c_0^i and α_0^i (green paths). By comparing the deformed template image and the subject image, registration optimizes the c_0^i, α_0^i that best warp the template image to match the subject image. P = 10,000 control points are used for the registration, which corresponds to a 5 voxel spacing.

Parallel Transport. The diffeomorphism encodes, for each subject, the differences between its anatomy and that of an age-matched healthy brain. To enable comparisons between subjects, transformations need to exist in the same space. The momenta parametrizing each deformation are parallel transported to the tangent space of T_{ref}. In brief, parallel transport translates the deformation towards subject i, defined by c_0^i and α_0^i, at any time point along the trajectory $\gamma(t)$ (blue arrows). It adjusts for anatomical differences related to GA while preserving components of the transformation non-related to age.

PCA. Given the high dimension of the transformations (P = 10,000 momenta) and the low sample size (N = 111), the momenta cannot be used as features to perform prediction. To reduce feature space and extract interpretable deformation modes, PCA is applied to the data. We denote by β_i the 3P transported momentum vector for subject i. Let X be the N by 3P data observation matrix: $X = (\beta_1, \ldots, \beta_N)^T$. X is mean-centered, and the empirical covariance matrix is given by $\Sigma = X^T X$. Eigendecomposition of Σ is performed in the form of

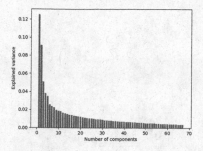
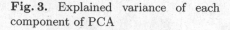

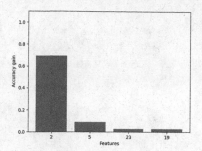

Fig. 3. Explained variance of each component of PCA

Fig. 4. Accuracy gain for each feature added to the model

$\Sigma = U\Lambda U^{-1}$ in which U is a matrix of size 3PxN, whose columns $(U_1, \ldots, U_n)$ are the eigenvectors of Σ, and Λ a diagonal matrix of size NxN, whose diagonal elements $(\lambda_1, \ldots, \lambda_n)$ are the eigenvalues of Σ. We extract the first 67 components that characterize 90 % of the sample shape variability (see Fig. 3).

Being a linear combination of momentum vectors, each eigenvector can generate a diffeomorphism, called deformation mode, which represents how the template brain anatomy varies within the population. The ith mode is given by: $m_i = \bar{X} + c\sigma_i U_i$, with $c \in [-4, -2, 0, 2, 4]$, $\sigma_i = \sqrt{\lambda_i}$, and U_i the ith eigenvector. Geodesic shooting applies the generated diffeomorphism to T_{ref} in order to visualize the deformation mode. In T2 fetal MRI, thinness and hypointensity of the corpus callosum make it difficult to discern. Geodesic shooting is performed on the template segmentation image as provided by [10] to make corpus callosum deformations discernible. Projection of the momenta of subject j on deformation i is computed as follows: $P_{\beta_j} = \beta_j^T U_i$. P_{β_j} can be seen as a score quantifying how much β_j is represented by the ith deformation mode.

Classification. To assess whether or not the deformation modes can discriminate between controls and fetuses with CCA, we perform classification with a SVM equipped with a RBF kernel. SVM parameters (width of the gaussian kernel and penalty) are tuned using grid-search. The dataset is randomly split into a training (70% of the data) and a test set (30% of the data) to perform 5-fold cross validation. While modes with the highest eigenvalues are those that explain best the variability of the data, they don't only encode shape variations related to CCA, but also components of rigid registration correction and inter-subjects variability. To extract deformation modes specific to CCA, we perform forward feature selection: starting from an initial model with no input features, we train the model with each of the 67 principal deformations independently and keep the one that best enhances the model accuracy. This process is repeated iteratively until the addition of a new deformation does not augment the accuracy. This leads to the selection of 4 deformation modes as indicated in Fig. 4.

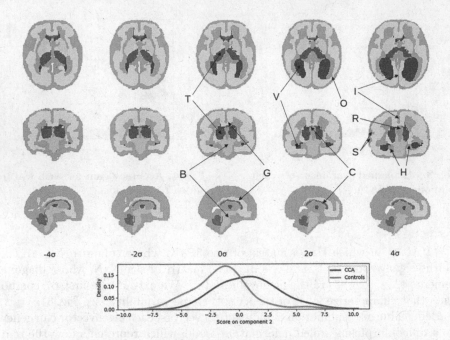

Fig. 5. Second mode of deformation applied to the segmentation of the template brain at age 31. Top three rows: axial, coronal and sagittal views. Bottom row: distribution of the subjects scores on component 2. B: brainstem. C: corpus callosum. G: cingulate gyrus. H: hippocampi. I: interhemispheric fissure. O: occipital cortex. R: roof of the third ventricle. S: superior and inferior temporal sulci. T: thalami. V: lateral ventricles.

3   Results

The final classification model reaches a 90% (±7%) accuracy. Interestingly, feature selection did not retain the first component of PCA. Visual inspection of the related deformation mode (not shown here) indicates it corrects for brain misalignment and characterizes subjects with large ventricles.

We present here the second component, which drives to most of the model accuracy (see Fig. 4). Subjects with CCA score generally higher on this component (see Fig. 5). The corresponding deformation mode reveals a thinning and a shortening of corpus callosum (C) on sagittal view. It is folded into a V-like shape, with a stronger distortion towards its posterior part. Volume of the cingulate gyrus (G) is also reduced. Lateral ventricles (V) are widely spaced and parallel, with prominent occipital horns and atrium, corresponding to colpocephaly. Dilation is slightly stronger in the right ventricle. The interhemispheric fissure (I) is wide. Shape of the third ventricle is abnormal, with an expanded roof (R). Volume of the occipital cortical and subcortical region (O) is reduced, especially in the right hemisphere. Hippocampi (H) appear thinner and verticalized. The superior and inferior temporal sulci (S) seem less pronounced. On coronal view, thalami (T) are parallelized and displaced away from the interhemispheric fissure. Shape of the brainstem (B) is abnormal on sagittal view, with prominent pons and midbrain.

4 Discussion and Conclusion

In this work, we addressed the challenge of exploring *quantitatively* alterations in abnormal fetal brains. We developed a shape analysis pipeline adapted to the specificities of fetal MRI and extracted anatomical deformations correlated to CCA. Geometrical models based on diffeomorphisms, that were originally designed for postnatal imaging, enabled us to compare fetuses of different ages and investigate brain alterations *globally*, without requiring any prior assumption. Such models are adapted to the scarcity of medical data and to the need for interpretable results. This preliminary work opens new perspectives for the quantitative analysis of fetal brains with malformations.

In contrast to previous studies, which had fewer data and often targeted specific brain areas or structures [1,12,14,19,21], our method extracted global deformations that correlate together, as they belong to the same deformation mode. These alterations revealed well-known defects of brains with CCA. As expected, the corpus callosum had abnormal shape and size. It was especially distorted in its posterior segment, the splenium, which is usually the missing part in partial CCA [16]. The cingulate gyrus, commonly absent in CCA [1], was also reduced. As our dataset comprised fetuses with complete and partial CCA, it cannot be known whether these patterns reflect a reduction or an absence of both structures. CCA is often accompanied by the development of a pair of aberrant callosal fibers, called Probst bundles, that run parallel to the midline, and a rearrangement of the midline cerebral structures [13]. The most common alterations include colpocephaly, elevation of the third ventricle, and widening of the interhemispheric fissure [1,13], all of which were clearly visible in the second deformation mode. Ventricles dilation and volume reduction of the occipital cortical and subcortical brain matter were uneven across hemispheres, which may reflect a tendency for abnormal brain asymmetry, frequently encountered in fetuses with CCA [11,18]. The observed volume reduction of the occipital region coincides with findings of decreased thickness of the cerebral wall in the lateral occipital region [18]. Consistent with findings of abnormal shape and rotation of the hippocampi in fetuses with CCA [11,12], we observed verticalized hippocampi, probably because of the extension of the temporal ventricular horns into the parahippocampal gyri. Both observations might be related to reduced volume of the ventral cingulum bundle, the fibers of which normally have an initial course below the body of the corpus callosum and then course within the parahippocampal gyrus in the inferior and medial temporal lobe [14]. We also observed underdeveloped temporal sulci, which might be related to delayed sulcation [21] or altered cortical folding [19], commonly observed in fetuses with CCA. Verticalization and displacement of the thalami, which are not reported in the literature, probably result from the widening of the interhemispheric fissure. It has been suggested that in CCA other interhemispheric connections such as indirect thalamic nuclei connections supply the absence of callosal fibers [1]. Understanding whether the displacement of the thalami is a marker of the absence or presence of such indirect connections and related to neurodevelopmental outcome could help understand the differences in outcome of patients with apparently isolated CCA. Surprinsingly, we observed a strong deformation of the brainstem, which is not a typical feature of CCA. This result likely originates from inaccurate segmentation of the brainstem during image processing, which tended to exclude the medulla.

Together, our findings draw a typical profile of brains with CCA, which is in agreement with the results of more local methods, validating our approach. Our method could help understand the mechanisms of the rearrangements linked to CCA, and, above all, identify the anatomical defects related to poor clinical outcome in isolated CCA.

This work has several limitations. Fetal brains undergo important and rapid changes across gestation. The majority of fetuses in our dataset had a GA between 30 and 34 weeks and alterations on late developing structures may have been missed. Speed of growth across gestation and structures is not constant, contrary to the assumption made by parallel transport. To strenghen the methodology, spatiotemporal models [6] could be adapted to take into account regional and temporal differences in growth rate. As the registration was computed in the space of the healthy template brain using topology-preserving deformations, structures specific to CCA brains such as Probst's bundles could not be studied. Furthermore, the small spacing between control points yielded unregular deformations, that can be anatomically inaccurate.

In the future, we will study the correlation between anatomical alterations in CCA brains and clinical outcome, with a focus on isolated cases. As partial and complete CCA can affect brains differently [18], deformations specific to each subgroup will be extracted. CCA-specific anomalies such as Probst's bundles will be further studied using atlas estimation. Efforts will be made towards increasing the sample size in order to extract more robust features, and registration will be adapted to the matching of complex biological structures.

Acknowledgement. This work was partly funded by the last author's chair in the PRAIRIE institute funded by the French national agency ANR as part of the "Investissements d'avenir" programme under the reference ANR-19- P3IA-0001.

References

1. Benezit, A., et al.: Organising white matter in a brain without corpus callosum fibres. Cortex **63**, 155–171 (2014)
2. Bône, A., Louis, M., Martin, B., Durrleman, S.: Deformetrica 4: an open-source software for statistical shape analysis. In: Reuter, M., Wachinger, C., Lombaert, H., Paniagua, B., Lüthi, M., Egger, B. (eds.) ShapeMI 2018. LNCS, vol. 11167, pp. 3–13. Springer, Cham (2018). https://doi.org/10.1007/978-3-030-04747-4_1
3. Christensen, G., Rabbitt, R., Miller, M.: Deformable template using large deformation kinematics. IEEE Trans. Image Process. **5**, 1435–1447 (1996)

4. Clouchoux, C., et al.: Quantitative in vivo MRI measurement of cortical development in the fetus. Brain Struct. Function **217**, 127–139 (2011)
5. D'Antonio, F., et al.: Outcomes associated with isolated agenesis of the corpus callosum: a meta-analysis. Pediatrics **138**, e20160445 (2016)
6. Debavelaere, V., Durrleman, S., Allassonnière, S.: Learning the clustering of longitudinal shape data sets into a mixture of independent or branching trajectories. Int. J. Comput. Vis. **128**, 2794–2809 (2020)
7. des Portes, V., et al.: Outcome of isolated agenesis of the corpus callosum: a population-based prospective study. Eur. J. Paediatr. Neurol. **22**, 82–92 (2017)
8. Durrleman, S., Allassonnière, S., Joshi, S.: Sparse adaptive parameterization of variability in image ensembles. Int. J. Comput. Vis. **101**, 1–23 (2012)
9. Ebner, M., et al.: An automated framework for localization, segmentation and super-resolution reconstruction of fetal brain MRI. Neuroimage **206**, 116324 (2020)
10. Gholipour, A., et al.: A normative spatiotemporal MRI atlas of the fetal brain for automatic segmentation and analysis of early brain growth. Sci. Rep. **7**, 1–13 (2017)
11. Glatter, S., et al.: Beyond isolated and associated: a novel fetal MR imaging-based scoring system helps in the prenatal prognostication of callosal agenesis. Am. J. Neuroradiol. **42**, 782–786 (2021)
12. Knezović, V., et al.: Underdevelopment of the human hippocampus in callosal agenesis: an in vivo fetal MRI study. Am. J. Neuroradiol. **40**, 576–581 (2019)
13. Leombroni, M., et al.: Fetal midline anomalies: Diagnosis and counselling part 1: corpus callosum anomalies. Eur. J. Paediatr. Neurol. **22**(6), 951–962 (2018)
14. Nakata, Y., et al.: Diffusion abnormalities and reduced volume of the ventral cingulum bundle in agenesis of the corpus callosum: a 3T imaging study. AJNR Am. J. Neuroradiol. **30**, 1142–1148 (2009)
15. Qiu, A., et al.: Parallel transport in diffeomorphisms distinguishes the time-dependent pattern of hippocampal surface deformation due to healthy aging and the dementia of the Alzheimer's type. Neuroimage **40**, 68–76 (2008)
16. Raybaud, C.: The corpus callosum, the other great forebrain commissures, and the septum pellucidum: anatomy, development, and malformation. Neuroradiology **52**, 447–477 (2010)
17. Santo, S., et al.: Counseling in fetal medicine: agenesis of the corpus callosum. Ultrasound Obstet. Gynecol. **40**(5), 513–521 (2012)
18. Schwartz, E., et al.: The prenatal morphomechanic impact of agenesis of the corpus callosum on human brain structure and asymmetry. Cerebral Cortex (New York, N.Y.: 1991) (2021)
19. Tarui, T., et al.: Disorganized patterns of sulcal position in fetal brains with agenesis of corpus callosum. Cerebral Cortex (New York, N.Y.: 1991) **28**, 3192–3203 (2018)
20. Trouve, A.: Diffeomorphisms groups and pattern matching in image analysis. Int. J. Comput. Vis. **28**, 213–221 (1998)
21. Warren, D., Connolly, D., Griffiths, P.: Assessment of sulcation of the fetal brain in cases of isolated agenesis of the corpus callosum using in utero MR imaging. AJNR Am. J. Neuroradiol. **31**, 1085–1090 (2010)
22. Yushkevich, P., Gao, Y., Gerig, G.: ITK-SNAP: an interactive tool for semi-automatic segmentation of multi-modality biomedical images, vol. 2016, pp. 3342–3345 (2016)

Predicting Preterm Birth Using Multimodal Fetal Imaging

Riine Heinsalu[1], Logan Williams[1,2], Aditi Ranjan[1,2], Carla Avena Zampieri[1,2], Alena Uus[1,2], Emma Claire Robinson[1,2], Mary Ann Rutherford[1,2], Lisa Story[1,3], and Jana Hutter[1,2(✉)]

[1] Centre for the Developing Brain, School of Biomedical Engineering and Imaging Sciences, King's College London, London, UK
jana.hutter@kcl.ac.uk
[2] Department of Biomedical Engineering, School of Biomedical Engineering and Imaging Sciences, King's College London, London, UK
[3] Department of Women and Children's Health, King's College London, London, UK

Abstract. Preterm birth (PTB) (<37 weeks' gestational age (GA)) is associated with increased risk of short- and long-term sequelae. Accurate predictive tools allow to improve the outcomes of those born preterm by offering early obstetric interventions to mothers at high-risk of PTB.

Methods: This study combines a wide range of structural and functional MRI parameters, from the fetal head, lung, placenta with clinically available Ultrasound and outcome data. A preprocessing pipeline adapted to the special requirements of the often incomplete and highly GA dependant data and a supervised machine learning model based on these derived markers derived is proposed. Data from 58 preterm and 217 term-born neonates were analysed.

Results: The best SVR model achieved an R^2 value of 0.67 and correctly predicted 92% of true preterm cases using a combination of two maternal and four fetal features.

Conclusion: The significance of this study is uncovering the potential of markers derived from multi-modal imaging data in the prediction of PTB using large-scale fetal studies. This study paves the way for future studies focusing on at-risk women to further enhance the data set and thus predictive power.

Keywords: Preterm · MRI · Prediction

1 Introduction

Preterm birth (PTB), affecting 8% of all deliveries in the UK, poses a significant challenge to healthcare services due to the complex and multifaceted nature of the condition. The burden is prevalent not only in the perinatal period but throughout life, with those born preterm having higher risk of neurodevelopmental delay and motor impairment compared to their term-born counterparts

---

Lisa Story and Jana Hutter are joint senior authors.

© Springer Nature Switzerland AG 2021
C. H. Sudre et al. (Eds.): UNSURE 2021/PIPPI 2021, LNCS 12959, pp. 284–293, 2021.
https://doi.org/10.1007/978-3-030-87735-4_27

(Luu et al. 2017). Developing better diagnostic and predictive tools can help patients receive early, targeted support leading to improved outcomes (WHO 2020). However, current predictive capabilities are limited (Suff et al. 2019).

Most commonly, a history of previous PTB and cervical length (McIntosh et al. 2016) are used in a clinical setting. Recently, (Watson et al. 2019b) combined risk factors such as previous preterm births and multiple pregnancy ($\geq$2 fetuses), with clinical investigations such as fetal fibronectin values and cervical length measurements were employed to predict whether a woman is high risk for preterm birth (Watson et al. 2019b). For women with symptoms of threatened preterm labour, the model combining risk factors and fetal fibronectin predicted 77%–96% of the cases correctly depending on the GA.

However, most screening tools for preterm birth are limited to ultrasound (US) derived cervical length and biochemical markers and fail to match the complex etiology of PTB by not including placental or other fetal parameters. While US and Doppler US (DUS) are the mainstream screening techniques during pregnancy, they are operator-dependent methods that have limited utility in some clinical populations e.g. mother's with increased body mass index (BMI). Fetal magnetic resonance imaging (MRI) is increasingly used both for research and clinical use especially in high risk populations (mother's with increased BMI). It also provides both structural and functional information in an operator-dependent manner, covering the entire uterus even in late gestation. Studies using fetal MRI to investigate preterm birth have found decreased thymus volumes (Story ct al. 2020b), smaller lungs (Story et al. 2020a) and a reduction in cortical and extra-axial cerebrospinal fluid volumes (Story et al. 2021) in fetuses who subsequently deliver preterm compared to those who deliver at term.

Previous *in utero* functional MRI studies have employed both diffusion MRI (Slator et al. 2021), which provides information about tissue microstructure and T2* relaxometry, which provides an indirect measure of tissue oxygenation via the blood-oxygen-level-dependent (BOLD) effect (Sorensen et al. 2020). Decreased placental T2* has been correlated with low birth weight (Sorensen et al. 2020), pre-eclampsia (Ho et al. 2020) and fetal growth restriction. However, there is a paucity of literature using *in utero* functional MRI to investigate preterm birth. To our knowledge no previous MRI studies have combined multiple functional and structural measures to predict preterm birth.

Data-driven methods are therefore ideally suited for the data set obtained. Identification of the features which hold the highest predictive power can provide valuable clinical insight and lead to improved targeting, monitoring and outcomes for high-risk women and their babies. This study aims to leverage the data available from large scale fetal MRI studies, together with available clinical background and US information, to build supervised machine learning models capable of predicting whether a fetus will be born preterm.

2   Methods

The steps in Fig. 1, from data collection to model evaluation will be detailed in the following.

2.1 Data

The data sets analysed here are combined from multiple ongoing large-scale fetal research studies with similar protocols. These studies are: the Cardiac and Placental Imaging Project (CARP), the Placental Imaging Project (PIP) and the infection study for patients with prolonged preterm rupture of membranes (PPROM). Data was collected from a combined total of 275 patients, and can be divided into the following five categories:

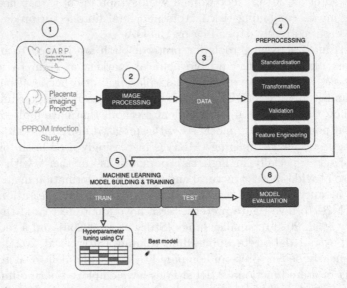

Fig. 1. Illustration of the workflow for the study. All six stages from data collection to model evaluation are graphically depicted.

1. **Structural MRI data:** automatic and manual segmentation of MRI scans to obtain imaging features e.g. volumes of different brain regions or bi-parietal diameter of the fetal head. (in red in Fig. 2B)
2. **Functional MRI data:** functional imaging features derived from the MRI data e.g. mean placental T2* (in red, italic and bold in Fig. 2B).
3. **Ultrasonographic data:** measurements such as the expected fetal weight (in blue in Fig. 2B)
4. **Medical history and demographic data:** e.g. maternal age, previous preterm deliveries and smoking status from patient records.
5. **Pregnancy outcome data:** gestational age at birth, birth weight, placental histopathology.

Structural and Functional MRI Data. After informed consent, all women where scanned in supine position on either a 3T Philips Achieva scanner or a 1.5 T Philips Ingenia scanner (Hughes et al. 2021) under constant monitoring of vital signs including blood pressure, oxygen saturation and heart rate, with frequent verbal interaction. After survey and calibration scans, T2-weighted Turbo Spin Echo images ($1.25 \times 1.25 \times 2\,mm^3$ resolution) were acquired in 3–5 orientations, covering the uterus and fetal head in sagittal and coronal planes. A 30 s coronal Multi-Echo Gradient Echo scan (T2ME), covering the entire uterus ($3 \times 3 \times 3\,mm^3$ resolution), was acquired. Furthermore, diffusion, perfusion, angiographic and other sequences were acquired, however, the present work here focuses on the T2 weighted and T2* scans.

The T2 weighted scans were employed to obtain 3D reconstructions of the brain and lung using slice-to-volume techniques (Uus et al. 2020). The T2ME data was fitted to the mono-exponential decay model, resulting in quantitative T2* maps. These were either manually segmented (placenta, brain) or further processed by a 3D model (lungs) and then segmented.

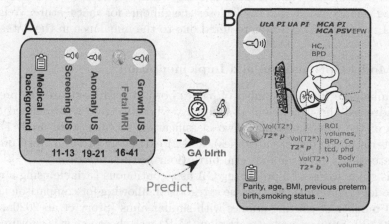

Fig. 2. Overview over the (A) time course considered and (B) attributes considered for this study.

Ultrasonographic Data and Medical History. A growth ultrasound was taken within one week of the MRI. In addition, the data from the screening and anomaly scan were available for this study. The following measurements were obtained: pulsatility indices of the uterine, umbilical and middle cerebral arteries and morphological measurements including abdominal circumference, femur length, expected fetal weight, head circumference and bi-parietal diameter.

Maternal age, body-mass index, parity, previous preterm birth, smoking status, medication status and diagnosis with gestational diabetes mellitus, fetal growth restriction or pre-eclampsia were recorded. At the time of delivery the birth weight, birth weight centile, head circumference and APGAR score at one and five minutes were included. Where available, histopathological information was recorded, most notably the placental weight, the presence of chorioamnionitis and maternal and fetal villi malperfusion.

2.2 Preprocessing

The main concerns for this specific dataset are the following: 1) a large proportion of missing values; 2) age-dependent features 3) imbalance in the dataset between preterm and term babies; 4) the relatively small size of the dataset. As the size of the dataset cannot be changed, it is important to preserve all present data points. The imbalance of the data will be dealt with during model training. The following describes the preprocessing performed for 1) and 2).

Z-scores were calculated (DeVore 2017) for all time-dependent variables using the control group as basis for the transformation. Z-score transformation was performed before imputation to ensure that only measured and no imputed values are included when finding the mean and standard deviation regression lines. This aims to limit any systematic error that could be introduced through Z-score transformation. k-nearest neighbour was then performed on all numerical features with missing values. Each missing value was thereby replaced with a weighted average value from the k-closest neighbours for that feature. Weighting by the Euclidean distance was required due to the imbalance in the dataset.

2.3 Model Optimisation and Implementation

Class imbalance (greater number of babies born at term compared to preterm) was addressed through weighted sampling, where weights were defined as the inverse of the class frequency. Two classification schemes were used: 1) term vs. preterm birth (binary categories), and 2) extremely preterm, very preterm, moderate-to-late preterm and term birth (four birth categories).

This study focuses on predicting GA as a continuous variable using a regression model. The results can then be categorised, allowing for comparison against a small number of existing studies with similar aims (Story et al. 2020a, Story et al. 2020b). Support vector regression (SVR) was chosen as it is captures non-linearity, is capable of dealing with many features and the flexibility to define error margins, which is essential when dealing with low signal-to-noise data such as fetal MRI. A split of 80/20 was used for stratified train/test. Feature selection and exploration was performed by computing the correlation between each feature and GA at birth and then converting into an F statistic. Features with the 19 highest F statistics, and parity (categorical), were selected. Next, Feature x feature interactions were explored between the top 20 features (19 continuous + 1 categorical). Features with >60% missing were excluded from further processing.

3 Results

3.1 Preprocessing

Results from before (Fig. 3) and after (Fig. 4) Z-score transformation, demonstrated exemplarily for placental mean $T2^*$, illustrate the change from a negative linear relationship to close to constant evolution over GA.

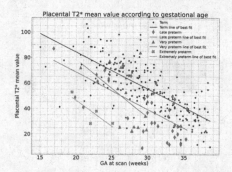

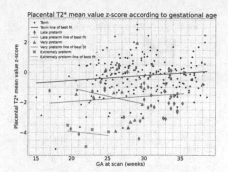

Fig. 3. Placental T2* mean values and the line of best fit over GA (term (blue), extremely preterm (orange), very preterm (green) and late preterm birth (pink)). (Color figure online)

Fig. 4. Placental T2* mean value after z-score transformation (term (blue), extremely preterm (orange), very preterm (green) and late preterm birth (pink)). (Color figure online)

Feature selection was performed in three steps, with the mean placental T2* performing best. The predicted GA at birth for all cases in the test dataset using the best model was further divided into four birth categories ($<28^{+0}, 28^{+0} - 33^{+6}, 34^{+0} - 36^{+6}$ and $\geq 37^{+0}$ weeks' GA) and binary birth categories (preterm vs. term) to show the confusion matrices in Fig. 5 and Fig. 6 respectively. Figure 5 indicates that the only fetus in the test set born extremely preterm was correctly predicted by the model. For the very preterm group, one fetus was correctly predicted by the model while two instances were predicted to be born late preterm rather than very preterm. For the late preterm group, seven out of eight children were correctly predicted by the model. 65.5% of the instances were correctly predicted by the model to be term babies while 12.7% who were also term-born were incorrectly predicted to be late preterm. Similarly, when the prediction results were divided into term and preterm, the number of correctly diagnosed term instances was 36 or 65.5% while there were 11 or 20% of correctly diagnosed preterm instances. The number of false positives or instances which were predicted to be preterm but were actually born at term was seven or 12.7%. Only one instance or 1.8% was predicted to be term while they were actually born preterm.

Figure 7 illustrates the R^2 values using the best model for all possible combinations of two features among the best 19 continuous features and the categorical parity feature. The mean placental T2* score ($R^2 \in [0.34, 0.6]$) followed by the pulsatility index of the uterine artery and the body volume with R^2 values $\in$ [0.1–0.5]. A number of features, which did not individually result in high R^2 scores display high R^2 scores when combined with other features. An example of this would be the mean brain T2* value, which alone yields a R^2 value of 0.018 but paired with the placental mean T2* score the R^2 increases to 0.4. The

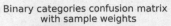

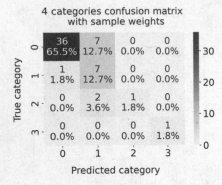

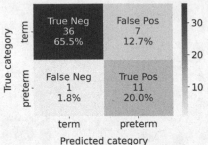

Fig. 5. Confusion matrix for the best SVR model, obtained using sampling weights with four birth categories. All fields add up to 100%.

Fig. 6. Confusion matrix for the best SVR model with sampling weights with binary categories. All fields add up to 100%.

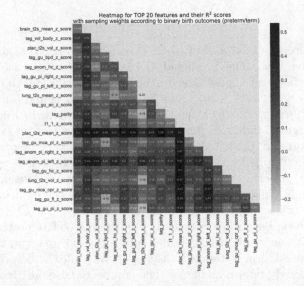

Fig. 7. Heatmap matrix for 20 features with each element corresponding to the R^2 value for the test set with the best model for any given pair of features. The diagonal of the matrix gives the R^2 for the best model based on the individual features.

most extreme case is the parity which raises its R^2 value from -0.031 to 0.54 if combined with the mean placental T2* score.

From the top 19 continuous features with the highest R^2 scores and the parity feature, any feature with a missing value percentage of more than 60% was removed. The top 10 features were then the following: Mean placental T2* score, CPR score, head circumference, abdominal circumference, femur length, pulsatility index uterine artery, bi-parietal diameter from the growth ultrasound

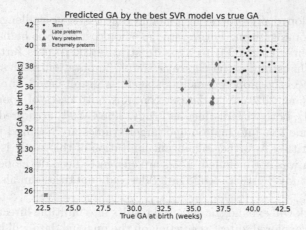

Fig. 8. Results of the SVR model with the highest R^2 score ($C = 100$, $\epsilon = 0.1$, $\gamma = 0.1$, kernel = sigmoid). Predicted GA from the best SVR model is plotted vs true GA at birth for the test dataset. The features used were the placenta T2* mean, the head and abdominal circumference, the femur length and the pulsatility index of the right uterine artery from the growth ultrasound and the number of previous preterm births.

and the parity. For the top 10 features, models were trained and tested with all combinations of features, with the number of features ranging from 1 to 10. The best five models all used the same following parameters: $C = 100$, degree = 2, $\gamma = 0.1$ and kernel = sigmoid. The best model uses six features and results in a R^2 of 0.665 and a mean absolute error of 1.6 weeks. Figure 8 shows the predicted GA at birth for the test set using the best model compared to the true GA. The R^2 value for the best model was 0.665, the mean absolute error was 1.6 weeks and the root mean squared error was 2.0 weeks.

4   Discussion and Conclusion

The present study exploits a comprehensive dataset containing clinical, US and multimodal fetal MRI data to predict the GA, and thus ultimately preterm birth. The results reflect that, in order to accurately predict preterm birth, acquiring datasets that capture the multifactorial nature of preterm birth are essential. As preterm birth is still poorly understood, acquiring detailed datasets provides an opportunity to better investigate the aetiology and pathophysiology of preterm birth. This study is however merely a first attempt to combine such large and diverse derived parameters.

There are a number of important limitations. These include the number of available datasets. While the collection is big for obstetric comprehensive datasets, it is small for ML standards. It is not well balanced between PTB and term-born cases and includes data from a range of different pregnancy complications, all with their own disease aetiology and progression. The required and here developed pre-processing pipeline reflects these challenges and works towards overcoming them. Future studies should include a higher number of women with threatened PTB to allow to stratify these cases further. Another limitation of this study is the choice of simple imputation method. Next steps can include recently proposed methods such as graph-based imputation techniques (You et al. 2020). The dataset contains both cases of spontaneous and iatrogenic PTB, both with distinct aetiology. The GA at birth prediction results thus also include this information and larger studies are required to treat these as different entities. A further significant limitation of this study is the fact that cervical length was not included. Tools are currently been developed to add this into a future study. Further second order derived quantities can also be included in a next step.

Future work will expand the achieved results into multiple directions. Further models will be explored, direct prediction on the imaging data will be explored to include whether further characteristics such as the heterogeneity of the placenta further increases the ability to predict PTB and further cohorts will be recruited, such as these with previous cervical surgeries or overt signs of inflammation.

References

DeVore, G.: Computing the Z score and centiles for cross-sectional analysis: a practical approach. J. Ultrasound Med. **36**, 459–473 (2017)

McIntosh, J., Feltovich, H., Berghella, V., Manuck, T., Society for Maternal-Fetal Medicine (SMFM): The role of routine cervical length screening in selected high- and low-risk women for preterm birth prevention. Am. J. Obstet. Gynecol. **215**, B2–B7 (2016)

Hughes, E.J., Price, A.N., McCabe, L., et al.: The effect of maternal position on venous return for pregnant women during MRI. NMR Biomed. **34**, e4475 (2021)

Story, L., et al.: Brain volumetry in fetuses that deliver very preterm. NeuroImage Clin. **30**, 102650 (2021)

Story, L., et al.: Foetal lung volumes in pregnant women who deliver very pretermy. Pediatr. Res. **87**, 1066–1071 (2020a)

Uus, A., et al.: Deformable slice-to-volume registration for reconstruction of quantitative T2* placental and fetal MRI. In: Hu, Y., et al. (eds.) ASMUS/PIPPI -2020. LNCS, vol. 12437, pp. 222–232. Springer, Cham (2020). https://doi.org/10.1007/978-3-030-60334-2_22

Story, L., Zhang, T., Uus, A., et al.: Antenatal thymus volumes in fetuses that delivered <32 weeks' gestation: an MRI pilot study. Acta Obstet. Gynecol. Scand. **100**, 1040–1050 (2020b). https://doi.org/10.1111/aogs.13983

Suff, N., Story, L., Shennan, A.: The prediction of preterm delivery: what is new? Semin. Fetal Neonatal Med. **24**(1), 27–32 (2019)

Luu, T.M., Rehman Mian, M.O., Nuyt, A.M.: Long-term impact of preterm birth: neurodevelopmental and physical health outcomes. Clin. Perinatol. **44**(2), 305–314 (2017). Delivery in the Periviable Period

You, J., Ma, X., Yi, D., Ding, Y., Kochenderfer, M., Leskovec, J.: Handling missing data with graph representation learning. In: NeurpIPS 2020 Proceedings (2020)

Slator, P.J., et al.: Data-driven multi-contrast spectral microstructure imaging with InSpect. Med. Image Anal. **71**, 102045 (2021)

Sørensen, A., Hutter, J., Seed, M., Grant, P.E., Gowland, P.: T2*-weighted placental MRI: basic research tool or emerging clinical test for placental dysfunction? Ultrasound Obstet. Gynecol. **55**, 293–302 (2020)

Ho, A.E.P., et al.: T2* placental magnetic resonance imaging in preterm preeclampsia an observational cohort study. Hypertension **75**, 1523–1531 (2020)

Watson, H., et al.: Development and validation of predictive models for QUiPP App vol 2: tool for predicting preterm birth in asymptomatic high?risk women. USOG **55**, 348–356 (2019b)

WHO: Preterm birth (2018). https://www.who.int/en/news-room/fact-sheets/detail/preterm-birth. Accessed 24 Jan 2021

WHO: The top 10 causes of death (2020). https://www.who.int/news-room/fact-sheets/detail/the-top-10-causes-of-death. Accessed 28 Mar 2021

Author Index

Printed in the United States
by Baker & Taylor Publisher Services